Machines as the Measure of Men

Cornell Studies in Comparative History

George Fredrickson and Theda Skocpol, *editors*

Empires, by Michael W. Doyle

Machines as the Measure of Men: Science, Technology, and Ideologies of Western Dominance, by Michael Adas

MICHAEL ADAS

MACHINES AS THE MEASURE OF MEN

Science, Technology, and Ideologies

of Western Dominance

CORNELL UNIVERSITY PRESS

ITHACA AND LONDON

First published 1989 by Cornell University Press.

International Standard Book Number 0-8014-2303-1
Library of Congress Catalog Card Number 89-845

Printed in the United States of America

Librarians: Library of Congress cataloging information
appears on the last page of the book.
The paper in this book is acid-free and meets the guidelines for permanence
and durability of the Committee on Production Guidelines for Book Longevity
of the Council on Library Resources.

For John Smail
As teacher and scholar,
a model for us all

CONTENTS

MAPS AND ILLUSTRATIONS

ACKNOWLEDGMENTS

THE PATTERNS I explore in this book began to coalesce decades ago when as a graduate student at the University of Wisconsin I was intrigued by the similarities between the arguments of modernization theorists, whose works were obligatory reading for non-Western specialists at the time, and of nineteenth-century advocates of European colonization, the focus of much of my own research and writing. The comparative methodology and cross-cultural orientation I gained in those years from participating in the Comparative Tropical History Program, which Philip Curtin had so skillfully put together in Madison, are as fundamental to this book as they have been to my previous work. In the ten years since I began sustained research for *Machines as the Measure of Men*, I have received generous support, useful critiques, and timely encouragement from a variety of sources. The John Simon Guggenheim Memorial Foundation and the Rutgers Research Council provided funding for two consecutive years of essential research and writing. Philip Curtin, Theda Skocpol, Joseph Miller, and Peter Stearns gave me useful advice and support in the formative stages of the project. I also benefited greatly at that point from responses to portions of the work by members of the history departments at Duke University, the University of Pennsylvania, and the School of Oriental and African Studies at the University of London. The collections and able staffs of the British Library, the India Office Records and Library, and the School of Oriental and African Studies greatly facilitated the often daunting task of research on themes covering several civilizations over five centuries. I owe a special debt to colleagues at Rutgers and other universities who have read and carefully criticized portions of the manuscript. Philip Curtin, Robert Gottfried, Reese Jenkins, David Ludden, Philip Pauley, Kevin Reilly, Herb Rowen, and Traian Stoianovich provided detailed responses to the early sections. Scott Cook, Frederick Cooper, Victoria de Grazia, Michael Geyer, Allen Howard, David Leverenz, David Levering Lewis, Arthur Mitzman, Richard Rathbone, James C. Scott, Michael Seidman, Laura Tabili, Mark Wasserman, and Lewis Wurgaft made useful suggestions for revision on the sections dealing with the nineteenth century. I am especially indebted to Ramchandra Guha, who read the entire manuscript and supplied invaluable leads and criticisms from an ecological and developmental perspective. John Ackerman of Cornell University Press provided timely encouragement and useful suggestions over the last half decade, and Patricia Sterling's rigorous and thoughtful editing should make things a good deal easier for the reader. From my friend and co-teacher, Lloyd Gardner, I learned much about American variations on the themes considered, and I benefited greatly from the stylistic suggestions of two other friends: Karen Wunsch, and my wife, Jane. Jane and my children, Joel and Claire, provided encouragement and the companionship I needed to see me through a project that at times seemed boundless and impossible. Claire also directed me to some literary sources that wonderfully illustrated the themes I was exploring.

My dependence on the research and scholarship of specialists in the areas compared in the book and of those who have preceded me in the exploration of cross-cultural themes will be obvious to the reader. I have tried to indicate in the footnotes particularly influential authors and works. As the notes suggest, the literature relating to the patterns considered is vast. I am well aware that a single book cannot possibly cover all the relevant sources or treat comprehensively all the issues related to the complex processes and broad themes addressed here. I cannot pretend that I have found definitive answers to the many questions my research has raised. I can only hope that I have got some of the questions right and have begun the search for answers that will help us to better understand the central process of modern history, the expansion of the West and Euroamerican interaction with the other cultures and civilizations of the globe.

MICHAEL ADAS

Rutgers University

Machines as the Measure of Men

They are landing with rulers, squares, compasses,
Sextants
White skin fair eyes, naked word and thin lips
Thunder on their ships.

Léopold Sédar Senghor, "Ethiopiques"

INTRODUCTION

IN THE 1740s while the ship on which he was traveling was at anchor off the mouth of the Gambia River, William Smith went ashore to instruct one of the ship's mates in the use of surveying instruments. On a stretch of beach near a small town, Smith had begun to demonstrate how one could measure distances with his theodolite and hodometer when he noticed a sizable band of armed Africans gathering nearby. Troubled by their hostile gestures, Smith questioned the ship's slave, who had come along to help him operate the instruments, as to why they were so vexed by activities that Smith regarded as entirely peaceful and nonthreatening. The slave explained that the "foolish natives" were alarmed by Smith's strange devices, which they believed he would use to bewitch them. The Africans had driven off their cattle, sent the women and children from the town to hide in a nearby forest, and marched out to confront the dangerous strangers who had appeared so suddenly in their midst. Concluding that attempts to reason with "ignorant savages" would be futile, and observing that the Africans were afraid to approach the surveying party, Smith continued with his instruction—though he cautioned the mate to keep his blunderbuss ready. As the slave measured the distance along the beach with the hodometer, the frightened warriors tried to stop him by running in his path—but they were careful not to touch the wheel. The slave amused himself by trying to run into them. Smith and his companions found the Africans' fear of the wheel and their scrambling to avoid it a source of merriment which they wished the "other gentlemen" on board had been present to enjoy. But when Smith stopped to rest in the shade and sent the mate and slave off to make punch, the townsmen grew more and

I

more threatening. Alone and surrounded by the angry and well-armed warriors, Smith began to fear that he might be attacked. Just as it occurred to him that he could be "cruelly torn asunder," the mate returned, and together they chased away the band of warriors by making threatening gestures and discharging the weapons that the terrified Africans had left behind on the beach. The mate pleaded to be allowed to pursue them, but Smith insisted on returning to the safety of the ship.[1]

Though in itself a relatively insignificant encounter, this incident reveals much about European attitudes toward and interaction with non-Western peoples in the centuries since the fifteenth-century voyages of exploration. It was the Europeans who went out to the peoples of Africa, Asia, and the Americas, and never the reverse—though at times Africans and Amerindians were carried back to Europe to be exhibited and studied. For the Africans the ship from which Smith and his small party disembarked was a marvel of design and workmanship. It represented an area of technology in which the Europeans had few rivals by the fifteenth century and in which they reigned supreme by the seventeenth. Sailing ships with superior maneuverability and armament permitted the Europeans to explore, trade, and conquer all around the world. Smith's instruments and his reason for taking them ashore convey both a sense of the curiosity that provided a major motivation for the Europeans' overseas expansion and their compulsion to measure and catalogue the worlds they were "discovering." With little thought for the reaction of the people who lived there, Smith set out to measure a tiny portion of the vast continent he was exploring. Much more than his physical appearance and dress, it was Smith's unknown instruments and what the Africans perceived as strange behavior in employing these devices that became the focus of their concern. He delighted in dazzling and terrifying the townspeople with his strange machines and continued his activities despite their obvious hostility to his intrusion. When threatened, he relied on the Africans' fear of his technology to drive them off. As he informs the reader, the guns that the warriors dropped and he and the ship's mate fired into the air were of European manufacture.

Our only source of information about this encounter is what we are told by Smith. It is the European observer who describes the Africans' reactions, interprets their motives, and even speculates about their underlying belief systems—in this case, with the assistance of a black slave whose views are filtered through Smith's own perceptions and memo-

---

[1]William Smith, *A New Voyage to Guinea* (London, 1744), pp. 15–20.

ry. The explorer stresses that the awe and fear evoked by his innocuous surveying equipment is the main source of the power he is able to exert over the African townspeople. He suggests that they are too backward to have any comprehension of the use of these devices and too ignorant for him to attempt an explanation. He concurs with the slave's assumption that because the Africans have no natural frame of reference from which to comprehend these wondrous devices and cannot imagine humans creating them, they resort to superstitious notions rooted in witchcraft to explain them. The black slave, who is depicted as a loyal servant, has nothing but contempt for the Africans. The surveying instruments both tie him to his white masters and set him off from the African townspeople. Though he is supervised by Smith, his ability to use the hodometer places him above the "ignorant Savages," and he readily joins in the merriment caused by what he regards as their unfounded fear of the surveying equipment. It is implicit in Smith's account that the slave himself could not conduct sophisticated surveying operations, much less devise the instruments that make them possible. But he has been instructed in their use and understands that they are human fabrications, not the products of magic or witchcraft.

As Smith relates this encounter, it is superior technology—the surveying instruments and firearms—that set the European traveler and his companions off from the Africans and give them the upper hand in the confrontation that ensues. Taken in isolation, the incident overstates the importance of technology in an age when religion was still the chief source of western Europeans' sense of superiority. But it indicates how influential achievements in material culture had become, especially those relating to technology and science, in shaping European perceptions of non-Western peoples even before the Industrial Revolution. From the very first decades of overseas expansion in the fifteenth century, European explorers and missionaries displayed a great interest in the ships, tools, weapons, and engineering techniques of the societies they encountered. They often compared these with their own and increasingly regarded technological and scientific accomplishments as significant measures of the overall level of development attained by non-Western cultures. By the mid-eighteenth century, scientific and technological gauges were playing a major and at times dominant role in European thinking about such civilizations as those of India and China and had begun to shape European policies on issues as critical as the fate of the African slave trade. In the industrial era, scientific and technological measures of human worth and potential dominated European thinking on issues ranging from racism to colonial education. They also provided key components of the civilizing-mission ideology that both justified Europe's global hegemony and vitally influenced the ways in which European power was exercised.

In view of their importance, it is remarkable that scientific and technological accomplishments as measures of European superiority and as gauges of the abilities of non-Western peoples have been so little studied. Most authors who have dealt with European attitudes toward African and Asian peoples in the industrial era acknowledge that Europe's transformation and the power differential that it created had much to do with the hardening of European assumptions of racial supremacy in the late nineteenth century.[2] But few writers have examined these complex connections in any detail, and in all cases consideration of them has been subordinated to discussion of racist issues. The rare works that deal in any depth with the pervasive effects of the scientific and industrial revolutions on European perceptions of non-Western peoples are focused on Africa, the geographical area that elicited the most extreme responses.[3] Because these studies cover a wide range of topics beyond the impact of European advances in science and technology, even for Africa we have only a partial view of one of the most critical dimensions of European interaction with non-Western peoples in the modern era. For China, India, the Islamic empires, and the Amerindian civilizations of the New World, we have little more than chance comments on the superiority of European weapons, tools, and mathematical techniques. The accounts that deal with these observations often give little sense of the material conditions and the cultural and ideological milieus that shaped them or their place in the broader, ongoing process of European exchange with non-Western peoples which has spanned the last half-millennium.

This book examines the ways in which Europeans' perceptions of the material superiority of their own cultures, particularly as manifested in scientific thought and technological innovation, shaped their attitudes toward and interaction with peoples they encountered overseas. It is not a work in the history of science or technology as those fields are usually defined. The processes of invention and of scientific investigation which have traditionally occupied scholars in these fields and the patterns of

[2]See, e.g., Ronald Hyam, *Britain's Imperial Century, 1815–1914* (New York, 1976), pp. 47–50; Christine Bolt, *Victorian Attitudes towards Race* (London, 1971), pp. 27–28, 111–12, 211; Gérard Leclerc, *Anthropologie et colonialisme: Essai sur l'histoire de l'africanisme* (Paris, 1972), pp. 26–28; and Francis G. Hutchins, *The Illusion of Permanence: British Imperialism in India* (Princeton, N.J., 1967), pp. 121–24.

[3]The best of these include Philip Curtin, *The Image of Africa* (Madison, Wis., 1964); H. A. C. Cairns, *Prelude to Imperialism: British Reactions to Central African Society, 1840–1890* (London, 1965); and William B. Cohen, *The French Encounter with Africans: White Response to Blacks, 1530–1880* (Bloomington, Ind., 1980).

institutional and disciplinary development which have more recently come into favor are crucial to the themes I explore. But for my purposes, these discoveries and developments are important only insofar as they influenced the ways in which Europeans viewed non-Western peoples and cultures and as these perceptions affected European policies toward the African and Asian societies they came to dominate in the industrial era. Though varying forms of interaction—including trade, proselytization, and colonial conquest—often resulted in the diffusion of European tools and scientific learning to overseas areas, my main concerns are the attitudes and ideologies that either promoted or impeded these transfers rather than the actual processes of diffusion. Because the spread of European science and technology has been central to the global transformations that Western expansion set in motion, the assumptions and policies that determined which and how many discoveries, machines, and techniques would be shared with which non-Western peoples have been critical determinants of the contemporary world order.

As I stress in the early chapters on the growing impact of material achievement on European perceptions of non-Western peoples and societies, the meanings of "science" and "technology" changed considerably over the centuries covered in this book. It is therefore necessary to indicate at the outset how I define these terms for the purposes of the study as a whole. Though contemporary scholars continue to debate how they ought to be understood and struggle to delineate the boundaries between them,[4] I have adopted broad definitions combining elements suggested by A. R. Hall and Edwin Layton. Hall terms scientific those endeavors that are aimed at gaining a knowledge of the natural environment, while he views technology as efforts to exercise a "working control" over that environment. Layton elaborates upon similar definitions: he sees the search for the understanding of fundamental entities as the essence of science, whereas techology seeks to solve more practical and immediate problems. Science may be theoretical or applied, but it is oriented toward systematic experimentation and the discovery of underlying principles. The primary objective of technology,

---

[4]As well as those between "pure" and "applied" science. For an introduction to many of the issues contested, see Robert Multhauf, "The Scientist as an 'Improver' of Technology," *Technology and Culture* 1/1 (1959), 38–47; the essays by Peter Drucker and James Feibleman in ibid. 2/4 (1961); the contributions by A. R. Hall and Peter Mathias in Mathias, ed., *Science and Society, 1600–1900* (Cambridge, Eng., 1972); A. R. Hall, *The Historical Relations of Science and Technology* (London, 1963); and Charles C. Gillispie, "The Natural History of Industry," in A. E. Musson, ed., *Science, Technology, and Economic Growth in the Eighteenth Century* (London, 1972).

though it may often involve theory and experimentation, is design, the application of rules to human artifice.[5]

I had originally intended to focus this study exclusively on the industrial age, when scientific and technological measures of human capacity peaked in importance. It soon became clear, however, that the impact of these standards in the industrial period could not be understood without some sense of their influence in the preceding centuries of European overseas expansion. European observers did not suddenly begin in the industrial era to distinguish their cultures from all others on the basis of material achievement; they had stressed the uniqueness of the extent and quality of their scientific knowledge and mechanical contrivances for centuries. In the early phase of overseas expansion, European travelers and missionaries took pride in the superiority of their technology and their understanding of the natural world. Their evaluations of the tools and scientific learning of the peoples they encountered shaped their general estimates of the relative abilities of these peoples.

Still, throughout most of the preindustrial period, scientific and technological accomplishments remained subordinate among the standards by which Europeans judged and compared non-Western cultures. Religion, physical appearance, and social patterns dominate accounts of the areas explored and colonized. When discussed, science and technology are generally treated as part of a larger configuration of material culture. Within this configuration, monumental architecture, sailing vessels, and even housing were often more critical than tools or astronomical concepts in determining European attitudes toward different non-Western peoples.

Throughout the centuries covered by this study, European judgments about the level of development attained by non-Western peoples were grounded in the presuppositions that there are transcendent truths and an underlying physical reality which exist independent of humans, and that both are equally valid for all peoples. Further, most of the travelers, social theorists, and colonial officials who wrote about non-Western societies assumed that Europeans better understood these truths or had probed more deeply into the patterns of the natural world which manifested the underlying reality. In the early centuries of overseas expansion, considered in Chapter 1, the Europeans' sense of superiority was anchored in the conviction that because they were Christian, they best understood the transcendent truths. Thus, right thinking on religious

[5]A. R. Hall, "Science, Technology, and Utopia in the Seventeenth Century," in Mathias, *Science and Society*, pp. 33–53; and Edwin Layton, "Mirror-Image Twins: The Communities of Science and Technology in 19th-Century America," *Technology and Culture* 12/4 (1971), 562–80.

questions took precedence over mastery of the mundane world in setting the standards by which human cultures were viewed and compared. The Scientific Revolution did not end the reliance on Christian standards. In fact, to the present day they remain paramount for certain groups and individuals, most obviously Christian missionaries. But as I suggest in Chapter 2, religious measures of the attainments of overseas peoples diminished in importance for many European observers beginning in the eighteenth century, while scientific and technological criteria became increasingly decisive.

The rise to predominance of scientific and technological measures of human capacity during the industrial era, which is discussed in Chapter 3, owed much to the fact that they could be empirically demonstrated. In the late eighteenth and nineteenth centuries, most European thinkers concluded that the unprecedented control over nature made possible by Western science and technology proved that European modes of thought and social organization corresponded much more closely to the underlying realities of the universe than did those of any other people or society, past or present. In Chapter 4 I examine the ways in which this assumption shaped ideologies of European imperialism. Chapter 5 focuses on two closely related themes: the impact of European scientific and technological superiority on arguments of white racial supremacy, and the ways in which European advantages in these fields influenced the educational policies by which European colonizers proposed to refashion non-Western societies.

In Chapter 6 I explore the reasons why the trench slaughter on the Western Front in World War I caused many European thinkers to challenge the assumption that better machines and equations demonstrated privileged access to physical as well as transcendent truths. In the Epilogue, I suggest some of the reasons why doubts about scientific and technological measures of human accomplishment were less pronounced in American intellectual circles after the war. I then consider the longstanding and increasing American addiction to technological innovation and the ways in which it contributed to the rise of modernization theory. This paradigm, resting on an assumed dichotomy between traditional and modern societies, represented a reassertion of scientific and technological standards. Its popularity in the post–World War II era reflected a restored confidence in the premise that there was close correspondence between Western thinking and external reality.

As these patterns suggest, European responses to non-Western peoples and cultures over the past five centuries have been strongly influenced by advances in Western understanding of and control over the material world. But the links between material advance and shifts in perception or judgment were not always clear or direct. Both the conflicting views held by different thinkers or groups during the same "phase" of development, and the lag between changes in European material conditions and shifts in European ideas about non-Western peoples caution against attempts to periodize rigidly or to treat prevailing views as the consensus of a given age.[6] Though I have identified general phases, I have tried to show that the boundaries between them are blurred. Ideas that were dominant in one era persisted but played lesser roles in the next, and various authors writing in the same period could draw widely varying conclusions from the same evidence. Thus, for example, reports of African material backwardness were cited in the eighteenth century both by writers who sought to prove the racial inferiority of Africans—and thereby justify their enslavement—and by abolitionists who argued that the Africans' vulnerability, reflected in their low level of development, made it morally imperative for Europeans to protect rather than exploit them. Some centuries later, European intellectuals proposed a range of often contradictory solutions to the crisis of Western civilization brought on by World War I. These included, on the one hand, assaults on science and industry, which were blamed for the horrific magnitude of the war, and, on the other, visions of Americanized technocracies of the future.

In each of the phases considered, I have attempted to examine these conflicting responses and to weigh their impact on European views of and interaction with non-Western peoples. I have also sought to avoid reducing the factors that shaped European attitudes to those involving material accomplishment by comparing the influence of these gauges in each period with the major alternatives to them, including physical appearance, religious beliefs, and social customs. Finally, I have had to take into account the fact that the impact of European scientific and technological breakthroughs on shifts in European responses to non-Western peoples was often not felt until decades later. The failure of sixteenth-century European explorers and missionaries to appreciate fully the advantages that the mechanical innovations of medieval artisans had bequeathed to them provides a major example of this lag. Another is illustrated by the fact that the eighteenth-century rage for *chinoiseries* peaked in the very decades when a number of French and British authors, who were attuned to the latest European advances in the sciences and familiar with the writings of the Jesuit missionaries on

[6]My thinking on these issues has been strongly influenced by John Greene's superb essays on approaches to the history of science; see esp. "Objectives and Methods in Intellectual History" and "The Kuhnian Paradigm," both reprinted in *Science, Ideology, and World View* (Berkeley, Calif., 1981).

China, had begun to dismiss the "Middle Kingdom" as despotic, superstition-ridden, stagnant, and hopelessly behind Europe in civilized attainments. For all these reasons, the phases and patterns I identify arise not from the delusion that the "messy realities" of history can be reduced to a rigid hierarchy of factors and precise categories but from an effort to give analytical coherence to the large and complex questions I address.

My central concerns are the attitudes toward non-Western peoples and cultures which were held by literate members of the upper and middle classes of western European societies, and the ways in which these attitudes shaped ideologies of Western dominance and informed colonial policy-making. Though these ideas often influenced the actual social interaction of all classes of Europeans with Africans and Asians, I deal only indirectly and peripherally with what George Frederickson has termed the "societal" dimensions of contacts between European and non-Western peoples.[7] This approach reflects my agreement with Theda Skocpol that ideologies ought to be distinguished from cultural idioms. Arguments for or against the abolition of the slave trade, appeals to the "civilizing mission," and competing approaches to modernization theory were (or are) all "idea systems deployed as self-conscious political arguments by identifiable political actors."[8] These ideologies tended to be less temporally specific and at times more oriented to intellectual and moral disputes than Skocpol's exclusively political definition would allow, but I strongly concur with her contention that they must be distinguished from the less consciously fashioned and more anonymous ideas and values that are constants in all cultural systems. Therefore, when I write of "European" views and responses, I am (unless I indicate otherwise) referring collectively to the ideas and arguments of those members of the "articulate classes"[9] of western Europe who concerned themselves with issues relating to European involvement overseas. Most of the authors who dealt with these issues can at best be characterized as middle-level intellectuals, and some were little more than polemicists or popularizers in the worst sense of the term; only a handful—including Voltaire, John Stuart Mill, and René Guenon—were major thinkers.

As I seek to demonstrate in the book's early chapters, both the class

[7]George Frederickson, "Toward a Social Interpretation of the Development of American Racism," in Nathan I. Huggins, Martin Kilson, and Daniel M. Fox, *Key Issues in the Afro-American Experience* (New York, 1971), pp. 240–54.

[8]See the stimulating exchange on these issues between Theda Skocpol and William H. Sewell, Jr., in *Journal of Modern History* 57/1 (1985), 57–96 (quoted portion, p. 91).

[9]As G. M. Young has so aptly labeled them in *Victorian England: Portrait of an Age* (Oxford, 1964), p. 6.

and occupational background of those who wrote about overseas areas shifted considerably during the centuries covered. The bourgeoisie steadily increased in numbers and influence in the mix of aristocratic and middle-class observers. Explorers, traders, missionaries, and writers of fiction dominated the discourse on African and Asian lands in the early decades of expansion. Though they remained important, from the late eighteenth century on, natural scientists, colonial administrators, social theorists, and anthropologists became the leading experts on matters relating to the non-Western world. As the occupational backgrounds of these writers suggest, both thinkers within Europe itself and Europeans engaged in diverse enterprises overseas played critical roles in shaping responses to non-European peoples and cultures. From the first decades of expansion the two were constantly interacting. Medieval accounts of the fabled Orient and the African empire of Prester John aroused the expectations of early explorers, missionaries, and conquistadores. Their accounts of the worlds they had "discovered" provided the basis for the works of authors in Europe, from the philosophical tracts of Montesquieu and Voltaire to the disquisitions of naturalists such as Julien Virey and Johann Blumenbach. These works, and those by such later authors as James Mill and John Barrow, in turn shaped the attitudes of Westerners who went out to colonize or Christianize African and Asian lands in the nineteenth century, and who described them in unprecedented detail for the rapidly growing readership back home.

Because the British and French were prominent among the European nations involved in overseas expansion in each of the phases I consider, and because they were the foremost imperialist powers of the nineteenth century, the travelers and administrators and social theorists of these two nations have been by far the most important sources of information and opinion about the non-Western world. Both countries were also leading centers of scientific investigation and technological innovation throughout the centuries in question. Even though France was slower to industrialize than Great Britain, the French were as sensitive as the English to the profound differences, created by the scientific and industrial revolutions, between western Europe and the rest of the world. For these reasons, I concentrate on British and French writings in all but the earliest period and the latter half of the twentieth century. In dealing with the sixteenth and seventeenth centuries, I also make use of Iberian and Dutch and to a lesser extent Italian and German descriptions of overseas lands and cultures because accounts by explorers and travelers from these areas were among the most influential in this era. In the Epilogue I compare nineteenth-century European ideas with those of twentieth-century American social scientists, who have dominated post–World War II thinking on the relevance of Western science and technology for the Third World.

Comparison of shared and divergent British and French responses throughout all the different phases discussed serves to identify both the assumptions that writers from the more advanced nations of western Europe held in common, and areas where perceptions and policies differed by nationality. A comparative approach applied also to the areas to which European observers were responding makes possible the identification of generalized patterns of European perception and policy as well as variations in European responses to specific cultures and the sources of those differences.

Among the many culture areas with which the Europeans interacted, I have concentrated on three: sub-Saharan Africa, India, and China. Not only have my teaching and previous research given me some familiarity with these areas, but each has proved ideal for testing the themes I am examining. Though their interaction with the agents of an expansive European civilization differed considerably, they were all major targets of early European exploration and remained primary centers of European overseas trade, proselytization, and conquest or informal domination. European observers saw in these three culture areas major examples of the differing levels of social development that eighteenth- and nineteenth-century writers sought to locate on a variety of evaluative scales. European thinkers also judged that the peoples of each had reached a different level of scientific understanding and technological mastery. Careful examination of these areas soon impresses one with the great diversity within each one, but in European thinking they were often treated as single civilizations or their achievements regarded as those of a single "race" or people. Thus, although I have noted important variations in cases where these differences were vital to the issues under consideration, I have generally followed my sources in comparing each with the others as a single and discrete entity.

Other culture areas, particularly Japan and various centers of Islamic and Amerindian civilization, have great potential for comparison and might well warrant examination in subsequent studies, but none proved as suitable for the present work as the three I have chosen. Some, such as Japan and Polynesia, were not at all or only marginally in contact with the Europeans during key phases of the centuries considered and thus were not consistently major objects of European intellectual inquiry. Others—the Middle Eastern centers of Islamic civilization, for example—not only shared the Mediterranean heritage of western Europe but had long been rivals of the Europeans and had maintained significant contacts for centuries through trade, war, and cultural exchange. As a result, the Muslims were never "discovered" like the Indians and Chinese, and the Semitic origins of the Arabs tended to muddy discussions of "racial" characteristics. This and the fact that they had bequeathed to the Europeans, whether as originators or as go-betweens, some of the technology and a good deal of the basic mathematical and scientific learning vital to the West's transformation from backward outlyer to global hegemon make it difficult to distinguish clearly between the achievements of the two civilizations. The early phases of European interaction with the peoples of the New World produced patterns of response comparable to those discussed in Chapter 1, and I have been strongly tempted to include one of the Amerindian civlizations as a fourth case study. But the early conquest of the New World societies and the demographic catastrophes that followed, coupled with the early and relatively large migration of Europeans to the Americas, gave the patterns of thought and domination that I examine very different meanings in New World contexts. A consideration of these contrasts would have greatly extended the scope and length of the present work.

In view of the issues that have preoccupied writers on related subjects, it is vital that I indicate a number of things that this book does not attempt to do. It is not a study of racism or racial prejudice per se, even though the patterns I explore converge with racist ideologies in each phase. But the impact of racism in the only sense in which it has been a meaningful concept at the level of intellectual discourse—the belief that there are innate, biologically based differences in abilities between rather arbitrarily delineated human groups—varied greatly from one time period to another. Terms such as ethnocentrism, cultural chauvinism, and physical narcissism more aptly characterize European responses in the early centuries of overseas contact, and they remain more important than racism in much of the literature on two of the three culture areas considered. Though scientific and technological measures of human potential were used to support racist ideologies, particularly in the nineteenth century, these gauges were widely applied long before racist ideas were first systematically expounded by such writers as Edward Long and S. T. Soemmering in the late 1700s. Even in the nineteenth century, when racist theories relating to non-Western peoples won their widest acceptance among the articulate classes of Europe, many thinkers gave credence to scientific and technological proofs of Western superiority while rejecting those based on racist arguments. These patterns underscore one of the major findings of my research: racism should be viewed as a subordinate rather than the dominant theme in European intellectual discourse on non-Western peoples.

In this work I do not attempt to determine the accuracy of either individual or collective European assessments of African and Asian technology and scientific thought at different points in time. Rather, my aims are to trace the history of these assessments, to give some sense of the conditions in Europe and overseas that influenced the choice of items selected for comment and how these were regarded, to explore how both objects of interest and evaluations changed over time, and to examine the impact of these changes on broader European attitudes toward non-Western peoples and on the formulation of ideologies of Western dominance. A determination of the validity of European commentary in different periods on the quality of African tools or the accuracy of Chinese astronomical calculations would entail a very different sort of inquiry.[10] It would require extensive comparisons of European accounts with whatever contemporary writings are available from each culture area, and with the findings of research carried out in the past three or four decades by scholars working on the history of science and technology in China, Africa, and India.[11] Therefore, unless correctives were provided by contemporaries, I have refrained from specific commentary on the accuracy of European assessments of differing non-Western peoples' conceptions of the natural world and their level of material culture. However, in my more general discussions of European interaction with African and Asian peoples at different points in time, I have tried to indicate where statements and impressions unduly distort the actual relationship between the Europeans and the culture area in question.

As "the measure of men" in the title is intended to suggest, scientific and technological standards have been, with rare exceptions, applied by males to activities presumed to be dominated by males. The Marquise

[10]Some sense of the size of such a task can be gained from the detailed notes that J. L. Cranmer-Byng has appended as editor to Lord Macartney's journal of his visit to China in the 1790s; see *An Embassy to China* (London, 1962), pp. 355–98.

[11]The most important work to appear thus far on non-Western science and technology is the monumental, multivolume study by Joseph Needham (assisted by Wang Ling), *Science and Civilization in China* (Cambridge, Eng., 1954–).

On China, see also the useful essays in Nathan Sivin, ed., *Technology in East Asia* (New York, 1977); and Sivin and Shigeru Nakayama, eds., *Chinese Science: Explorations of an Ancient Tradition* (Cambridge, Mass., 1973). For an overview of scientific investigation in India, see the contributions in D. M. Bose, S. N. Sen, and B. V. Subarayappa, *A Concise History of Science in India* (New Delhi, 1971). David Pingree's *Census of the Exact Sciences in Sanskrit* (Philadelphia, 1970–81) conveys a sense of the depth and range of Indian scientific learning, while Shiv Visvanathan's monograph *Organizing for Science* (New Delhi, 1985) provides numerous insights into the nature and organization of industrial research in modern India. Robin Horton's essay "African Traditional Thought and Western Science," *Africa* 37/1–2 (1967), 51–71, 155–87, is a good place to begin an inquiry into African approaches to the natural world. Jack Goody's *Technology, Tradition, and the State in Africa* (London, 1971) provides a provocative interpretation of the role of technology in African history. See also Ralph A. Austin and Daniel Headrick, "The Role of Technology in the African Past," *African Studies Review* 26/3–4 (1983), 163–84.

du Chatelet and Marie Curie nothwithstanding, European and North American thinkers have assumed that the unprecedented achievements in experiment and invention which they invoked to demonstrate Western superiority, as well as the African and Asian scientific learning and tools with which these accomplishments were compared, were the products of male ingenuity and male artifice. Colonial proposals to train physicians and railway engineers were drawn up with male students in mind, just as colonial development schemes and post-independence modernization proposals (both capitalist- and socialist-inspired) have been for the most part male-oriented.[12] Throughout the five centuries surveyed here, male attainments and male potential were being measured; better machines and equations were being invoked to demonstrate that men of one type were superior to those of another.

The phrase "ideologies of dominance" in the subtitle indicates that assessments of African tools, Chinese timepieces, and the Indians' capacity to run steam locomotives were not simply academic exercises. They were expressions of power relationships. Especially in the industrial era, science and technology were sources of both Western dominance over African and Asian peoples, male and female, and of males over females in European and American societies. As I note in Chapter 5, at times the parallels between European women and non-Western "races" in this regard were explicitly stated. But usually it was simply assumed that women knew and cared to know little about mathematics and engineering and that the power derived from superiority in these fields should be monopolized by white males.

Machines as the Measure of Men is not intended to be an exercise in antiscientific or antiindustrial polemic. In fact, it has occurred to me as I work at my personal computer—surely one of the more remarkable products of Western (and increasingly Japanese) inquiry and innovation—that it would be hypocritical to engage in such an exercise. I have no utopian system to propose as a replacement for the scientific-industrial order, nor do I believe that the non-Western rivals it has come to domi-

---

[12]Ester Boserup's *Woman's Role in Economic Development* (New York, 1970) pioneered the study of the impact of colonial development and postcolonial "modernization" schemes on the women of Africa and Asia. For a recent appraisal of Boserup's work which takes into account the considerable research conducted since *Woman's Role* first appeared, see Lourdes Beneria and Gita Sen, "Accumulation, Reproduction, and Women's Role in Economic Development: Boserup Revisited," *Signs* 7/2 (1981), 279–98. For additional studies on these issues, see esp. Barbara Rogers, *The Domestication of Women: Discrimination in Developing Societies* (London, 1980); Maxine Molyneux, "Women in Socialist Societies: Problems of Theory and Practice," in Kate Young, Carol Wolkowitz, and Roslyn McCullagh, *Of Marriage and the Market* (London, 1981), pp. 167–202; the essays in the symposium published in *Signs* 7/2 (1981); and the earlier collection, "Women and National Development," in *Signs* 3/1 (1977).

nate were intrinsically better. For all the problems associated with scientific and technological innovations, they remain the only way we have yet discovered to provide a decent standard of living for a high proportion of the populations of human societies. That all societies or all groups within industrialized societies have not equally enjoyed these benefits is a matter for continued reform efforts but not in itself cause to conclude that science and technology have led humankind down the wrong path.

Nevertheless, as I seek to demonstate, evidence of scientific and technological superiority has often been put to questionable use by Europeans and North Americans interested in non-Western peoples and cultures. It has prompted disdain for African and Asian accomplishments, buttressed critiques of non-Western value systems and modes of organization, and legitimized efforts to demonstrate the innate superiority of the white "race" over the black, red, brown, and yellow. The application of technological and scientific gauges of human potential has also vitally affected Western policies regarding education and technological diffusion which go far to explain the varying levels of underdevelopment in the Third World today.

The misuse of these standards has not only impeded and selectively channeled the spread of Western knowledge, skills, and machines; it has also undermined techniques of production and ways of thinking about the natural world indigenous to African and Asian societies. Concern for the decline of these alternatives is not simply a matter of relativistic affirmation of the need to preserve difference and heterogeneity. Their demise means the neglect or loss of values, understandings, and methods that might have enriched and modified the course of development dominated by Western science and technology. The possibilities of alternative systems are suggested, for example, by the recent Western recognition of the efficacy of Chinese acupuncture, as well as Indian, African, and Amerindian healing techniques. As we better understand the attitudes toward the environment and material acquisition that were fostered by non-Western philosophical and religious systems, we also begin to appreciate how they might have tempered the Western obsession with material mastery and its consequences: pollution, the squandering of finite resources, and the potential for global destruction. It is, I think, significant that a passage from the *Bhagavad-Gita* "floated through the mind" of the "father" of the atomic bomb, Robert Oppenheimer, as he witnessed the detonation of the first of these weapons: "I am become death, the shatterer of worlds."[13]

[13]Quoted in Peter Goodchild, *J. Robert Oppenheimer: The Shatterer of Worlds* (Boston, 1981), p. 162.

Less arrogance and greater sensitivity to African and Asian thought systems, techniques of production, and patterns of social organization would also have enhanced the possibility of working out alternative approaches to development in non-Western areas, approaches that might have proved better suited to Third World societies than the scientific-industrial model in either its Western or its Soviet guise. At the very least, the first generations of Western-educated leaders in the newly independent states of Africa and Asia would have been more aware of the possibilities offered by their own cultures and less committed to full-scale industrialization, which most of them viewed as essential for social and economic reconstruction. The reappraisal in recent decades of Gandhian social and economic philosophy, which was long a favorite target for the sarcastic barbs of development specialists, reflects a growing recognition that the paths followed by western Europeans, North Americans, and the Soviets are not the only possible routes to national solvency and material well-being.[14]

[14]For an early defense of Gandhi's economic thinking, see Shiva Nand Jha, *A Critical Study of Gandhian Economic Thought* (Agra, 1955), esp. chap. 4. For later reappraisals, see A. K. N. Reddy, "Alternative Technology: A View from India," *Social Studies of Science* 5/3 (1975), 331–42; and Abdul Aziz, "Gandhian Economic System: Its Relevance to Contemporary India," in J. T. Patel, ed., *Studies on Gandhi* (New Delhi, 1983).

PART ONE

BEFORE THE
INDUSTRIAL
REVOLUTION

If any man should make a collection of all the inventions and all the productions that every nation, which now is, or ever has been; upon the face of the globe, the whole would fall short, either as to number or quality, of what is to be met with in China.

Isaac Vossius (1618–89), quoted in
John Barrow, *Travels in China* (1804)

If the renowned sciences of the ancient Bragmanes of the Indies consisted of all of the extravagant follies which I have detailed, mankind have indeed been deceived in the exalted opinion they have long entertained of their [the Indians'] wisdom.

François Bernier, *Travels in
the Mogol Empire* (1656–68)

I conversed with great numbers of the northern and western nations of Europe; the nations which are now in possession of all power and all knowledge; whose armies are irresistible, and whose fleets command the remotest parts of the globe. When I compare these men with the natives of our own kingdom, and those that surround us, they appear almost another order of beings. In their countries it is difficult to wish for anything that may not be ordained: a thousand arts, of which we have never heard, are continually labouring for their convenience and pleasure; and whatever their own climate has denied them is supplied by their commerce.

Samuel Johnson, *The History of Rasselas* (1759)

Sketch from Pieter de Marees's *Beschryvinghe ende historische verhael van het gout koninckrijck van Gunea* (1602) illustrating African tools, weapons, modes of transportation, and scant clothing. The drawing focuses on aspects of material culture that were of great interest to early explorers and merchants. (Reproduced by courtesy of the Trustees of the British Museum)

Late seventeenth-century engraving of Adam Schall, one of the most prominent members of the Jesuit mission to China. He is depicted in full Chinese scholar-gentry regalia among the European globes, maps, and astronomical instruments that had proved so critical in the Jesuits' efforts to win access to the Ming and Qing courts. (Reproduced by courtesy of the Trustees of the British Museum)

An engraving of the Hindu observatory at Banaras from Robert Barker's *Account of the Brahmins' Observatory at Benares* (1777). The careful illustration of the astronomical instruments, whose functions and dimensions Barker describes in detail, was prompted by a growing European interest in Indian scientific learning during this period. (Reproduced by courtesy of the Trustees of the British Museum)

First Encounters: Impressions of Material Culture in an Age of Exploration

ACCORDING TO estimates made in recent decades, by the fifteenth century the peoples of western Europe possessed an advantage of three or four to one over the Chinese in per capita capacity to tap animal and inanimate sources of power.[1] Though the poor quality of the data for both civilizations renders these estimates rough approximations, the comparison suggests just how far the Europeans had advanced in technological mastery during the medieval period. Among all preindustrial civilizations only western Europe could rival China, which had excelled in invention for millennia, in the application of technology to everything from farming and transportation to scholarship, bureaucracy, and war.[2] Without the agricultural and mechanical innovations of the Middle Ages and the development of new instruments in the Renaissance, the Europeans would not have had the means to undertake the explorations that culminated in the voyages of Columbus and Vasco da Gama. Advances in weaponry, shipbuilding, and manufacturing were equally vital to the efforts of Europeans to project their influence overseas through trade and warfare from the sixteenth century onward.

These patterns suggest that evidence of material achievement ought to have had a major impact on European attitudes toward the peoples and cultures they encountered in the first phase of overseas expansion. In

---

[1] Estimates by historians of the *Annales* school. See Pierre Chaunu, *L'expansion européenne du XIIIe au XVe siècle* (Paris, 1969), pp. 336–39, who interpolates from Fernand Braudel's rather impressionistic eighteenth-century statistics in *The Structures of Everyday Life: Civilization and Capitalism*, vol. 1 (New York, 1981), esp. chap. 5.

[2] See Joseph Needham, *Science and Civilization in China* (Cambridge, Eng., 1954–), esp. vol. 4, pt. 2 on mechanical engineering, and pt. 3 on civil engineering.

fact, it provided at best a subordinate standard by which travelers and missionaries assessed the attainments of other cultures and compared them with their own. Tools, modes of transportation, and cropping patterns were mentioned by most sixteenth- and seventeenth-century travelers, but they rarely described African and Asian technology and production techniques in any detail. Even in its most applied forms, scientific knowledge was discussed still less frequently. Most observers treated tools and scientific instruments as individual objects of inquiry. Few viewed them as proof of superior European achievements in science and technology as a whole. In contrast to the practice of the eighteenth and nineteenth centuries, inventiveness and scientific knowledge were rarely stressed as standards by which to judge the level of development attained by African or Asian societies or to evaluate the capacities of non-Western peoples.

A variety of factors account for the Europeans' lack of emphasis, (relative to later centuries) on their technological and scientific accomplishments in the early centuries of expansion. The conditions under which they traveled to Africa and Asia were not conducive to detailed, much less accurate, observation and description. This was particularly true for such aspects of culture as manufacturing techniques and scientific learning, which African and Asian peoples were reluctant to share with outsiders. In addition, most of the Europeans who went overseas had a very limited knowledge of their own societies' achievements in these areas, and few were as interested in the tools and cosmologies of the peoples they encountered as in physical appearance, customs, and ceremonies. Whether they were merchants or missionaries, European travelers in this era viewed their Christian faith, rather than their mastery of the natural world, as the key source of their distinctiveness from and superiority to non-Western peoples. But assessments of the sophistication of African and Asian science and technology as aspects of larger configurations of material culture did affect European attitudes toward different peoples and cultures. This was especially evident in the contrasts they perceived between African and Asian societies and in their tendency to elevate China above all the civilizations they had "discovered." Exploration of both the reasons for the relatively marginal role of scientific and technological measures of human achievement in this era and the situations in which these standards were invoked reveals much about the Europeans' sense of themselves and their own culture. It also tells us a good deal about the nature of their interaction with non-Western peoples in the first phase of overseas expansion.

Between the twelfth century, when the Europeans first employed the sternpost rudder and such navigational instruments as the compass and astrolabe, and the fourteenth and fifteenth centuries, which saw major innovations in hull design and rigging, western Europe's oceangoing ships were transformed from unwieldy tubs that seldom ventured from the sight of land into highly maneuverable vessels capable of transglobal voyages.[3] Despite these improvements, the earliest explorers and merchants went out to Africa and Asia in ships that were shallow-keeled, rather primitively rigged, small (most of them less than thirty meters in length, or the size of a modest modern yacht), and very much at the mercy of the elements. Even more than the crews of the larger and more seaworthy vessels Joseph Conrad immortalized in his sea tales centuries later, the sailors and passengers on the caravels and *naos* that were the mainstay of early exploration efforts were all too aware of the power of stormy seas to "toss and shake" their flimsy craft "like a toy in the hand of a lunatic."[4] Though more sheltered and commodious vessels came into wide use in the early decades of the sixteenth century,[5] crews and travelers were still crammmed for weeks—sometimes months, if the weather was unfavorable—into roach- and rat-infested quarters that stank of the garbage and human waste sloshing about in the bilge water below. In addition to seasickness and dysentery, seamen were vulnerable to contagious diseases that spread quickly through unwashed and closely packed crews. Subsisting on a monotonous diet of salted meat and fish, hardtack, and dried vegetables, many travelers suffered from the painful and potentially lethal bouts of scurvy which ravaged ships' companies that were too long at sea without fresh fruits or vegetables.

Vulnerability to disease and inclement weather was of course shared by the populace of Europe as a whole. Thus, whatever aspirations such European thinkers as Francis Bacon may have had for humans to control their natural environment, until well into the eighteenth century it was not readily apparent that their level of mastery was superior to that of other civilizations, particularly those in Asia.[6] Europeans, even wealthy Europeans, suffered from extremes of heat and cold as much as or a good deal more than most of the peoples they contacted overseas. They had no more potent defenses against disease, as recurrent epidemics of

[3]The best discussions of ship construction and navigational instruments in the medieval period and the early centuries of expansion can be found in J. H. Parry, *The Discovery of the Sea* (Berkeley, Calif., 1974), esp. chaps. 1, 2, and 8; and Chaunu, *L'expansion européenne*, pp. 273–307. For sea weaponry, see Geoffrey Parker, *The Military Revolution* (Cambridge, Eng., 1988), chap. 3.

[4]Joseph Conrad, *The Nigger of the "Narcissus"* (Harmondsworth, Eng., 1968), p. 53. See also the storm sequence in Conrad's *Youth: A Narrative* (1902).

[5]On the drawbacks of the larger sixteenth- and seventeenth-century ship designs, see James Duffy, *Shipwreck and Empire* (Cambridge, Mass., 1955).

[6]The fullest treatment of Bacon's famous aspiration can be found in Carolyn Merchant, *The Death of Nature: Women, Ecology, and the Scientific Revolution* (San Francisco, 1983), chap. 7.

the plague and cholera and the rapid spread of syphilis constantly reminded them. Early modern Europeans, Robert Mandrou has observed, "could neither rationally comprehend nor actively control the world in which [they] lived."[7] European travelers, even educated ones, shared with peoples overseas a sense of the helplessness of humans in the face of nature's awesome power. This attitude contrasts sharply with the Europeans' belief, embraced centuries later when the process of industrialization was under way, that the degree to which a society has mastered its environment reflects the extent to which it has ascended from savagery to civilization.

Especially in the first decades of exploration, European observers went out to lands in Africa and Asia about which they had little prior knowledge. In these regions they were exposed to further diseases, most notably yellow fever and new strains of malaria, against which they had no immunity. They were forced to endure extremes of heat and humidity for which their many layers of close-woven, tight-fitting clothing were particularly ill suited. Travel overland was even slower than by sea, and it was a good deal more arduous and dangerous. Except for missionaries, travelers by both land and sea usually resided only briefly in any one place. This meant that opportunities for careful observation and detailed investigation, much less reflection, were rare. In situations where European visitors were unable to converse in the languages of the peoples they encountered or to read their writings (which if they existed were often forbidden to outsiders), travelers focused their attention and written accounts on social patterns that could be readily observed: marriage customs, modes of warfare, religious ceremonies. Travelers and sea captains on the move usually adopted one of two approaches to those aspects of host societies that required extensive inquiry and a more sophisticated understanding of the larger cultural context. In the case of religious beliefs and philosophical ideas, they resorted to general, often fantastical, descriptions. Matters scientific and technological they often ignored altogether.

In the early centuries of expansion, merchants', missionaries', and explorers' accounts were by far the Europeans' most influential sources of information about overseas lands and peoples. Grand syntheses, such as Peter Heylyn's *Microcosmos: A Little Description of the Great World* (1639), were also written and in some cases widely read. But the authors of such works provided only brief and usually standardized summaries of African and Asian societies, which were heavily dependent on de-

[7]Robert Mandrou, *Introduction to Modern France, 1500–1640* (New York, 1976), pp. 29–32, 239 (portion quoted), 240–42.

scriptions from a limited range of travelers' narratives. Essayists and playwrights such as Montaigne and Dryden made selective use of information gleaned from travel accounts to buttress philosophical positions or to enliven social satire that had much more to do with European conditions and concerns than with the reality of overseas cultures. But in contrast to those of the eighteenth century, earlier European images of Africa and Asia, though invariably ethnocentric, were shaped primarily by the reports of travelers and missionaries, not the needs and whims of European dramatists and philosophers.

So little was known of Africa and Asia when the age of discovery began that the main task for sixteenth- and seventeenth-century observers was often simply to record their impressions of the bewildering variety of strange new worlds that the Europeans had rather abruptly been forced to reconcile with their constricted medieval vision of the earth. The works of such ancient writers as Herodotus and Pliny and a handful of genuine or fabricated medieval travel accounts, which gave scant attention to science or technology, had fixed a variety of images of Africa and Asia in the European imagination. These images helped to filter and give coherence to the bewildering rush of new sights, sounds, and smells (which, if Mandrou is correct, were as important as the new sights)[8] that confronted early European seafarers. Popular legends of oceanic monsters and equatorial waters that boiled like giant cauldrons were gradually discarded as experience proved them false, but myths of men with no heads, eyes on their chests, and bodies covered with fur persisted.[9] Just how long is evidenced by the questions routinely put in the eighteenth century by the eccentric Scottish naturalist Lord Monboddo to all overseas voyagers who appeared in his courtroom as to whether they had seen men with tails, whom Monboddo was convinced would someday be found.[10] In many instances the reality of the peoples and cultures that European travelers encountered appeared to substantiate pre-expansion accounts of the world beyond the Islamic empires that had both threatened medieval Christendom and provided its main source of commercial and cultural exchange. But many of the "discoveries" the Europeans made in the outside world forced them to rethink the place and meaning of their own civilization and provided data that

---

[8]Ibid., pp. 50–52.

[9]Urs Bitterli, *Die Entdeckung des Schwarzen Afrikaners* (Zurich, 1970), pp. 38–39; Léon-François Hoffman, *Le nègre romantique* (Paris, 1973), pp. 16–17; and Eldred Jones, *The Elizabethan Image of Africa* (Charlottesville, Va., 1971), pp. 5–6, 9.

[10]For a wide-ranging discussion of this fascination with the grotesque, see Margaret Hodgen, *Early Anthropology in the Sixteenth and Seventeenth Centuries* (Philadelphia, 1964), pp. 33, 65–69, 115–16, 126, 148, 408–11.

gave added urgency to questions about their religious beliefs as well as their assumptions regarding the workings of the natural order.

The backgrounds of the explorers, merchants, and missionaries who shaped European attitudes toward the world beyond Europe in this era strongly influenced the content of their accounts of African and Asian societies. Some reports amount to little more than sea captains' chronologies of the places they visited, traders' lists of products available for export and goods in demand, or missionaries' estimates of the prospects for Christian conversions.[11] But most accounts cover a wider range of topics. Many include detailed descriptions of African and Asian political systems, warfare, and religious practices. The best also contain extensive descriptions of geography, plant and animal life, social patterns, and material culture. Because few travelers were inventors or engineers, and only a small proportion of missionaries and ship's surgeons had received extensive training in the sciences, overseas observers often have little to say about African and Asian technology or knowledge of the natural world. Still, merchants often took serious interest in the tools and techniques of handicraft producers and the computation methods used by their commercial rivals. Merchants and company officials also attempted, usually unsuccessfully, to visit regions that were reputed to be major mining centers. Only the best educated of the missionaries and such exceptionally well-educated travelers as François Bernier had the background, extended overseas residence, and linguistic skills necessary to explore African and Asian scientific learning seriously. Significantly, they approached astronomy and botany as aspects of what was then called natural philosophy, for well into the eighteenth century there was no science as we understand the concept. The same was true of technology, which in this era meant detailed or systematic examination rather than tools or invention. Our concept of technology did not come into widespread use until the nineteenth century. Early European observers treated what they most commonly called the "useful arts" or simply "the arts" as aspects of material culture more generally.

Whatever their background, Europeans who wrote on overseas lands and peoples, except again the best educated, had a very constricted view of the technological and scientific advances that had occurred in the centuries before the voyages of exploration. Scientific breakthroughs

---

[11]For examples of each type, see Peter Floris, *Voyage to East India in the "Globe," 1611–1615* (London, 1934); François Martin de Vitre, *Description du premier voyage fait aux Indes orientales. . .* (Paris, 1604); Eustache de la Fosse, "Voyage à la côte occidentale d'Afrique en Portugal et en Espagne (1479–1480)," *Revue Hispanique* 4 (1897), 174–201; and R. P. Alexis de Saint-Lô, *Relation du voyage du Cap Verd* (Paris, 1637).

were usually known only to small numbers of Europe's educated elite, while technological innovation was often centered in certain regions. The Iberians, for example, who dominated the first centuries of expansion, frequently commented on the superiority of their ships and nautical instruments to those of the peoples they encountered. But they had little to say about mining or metalworking, despite the steady advances that had been made in both processes since the early Middle Ages. The fact that the Iberians themselves had pioneered many of the improvements in shipping but had contributed little to innovations in mining or metalworking, which were concentrated in regions farther north, goes far to explain why they were so impressed by their advantages in the former but seldom even mentioned the latter.[12]

In the early centuries of expansion most of Europe's technological advantages were less apparent than they would later become, and in such endeavors as cotton textile and porcelain manufacturing the Europeans were actually behind their Asian rivals throughout much of this period. Equally important was the fact that the transformation of medieval technology was the product of many centuries of ceaseless tinkering and small improvements, which periodically coalesced in major breakthroughs in extraction, production, or transportation.[13] Major innovations in agriculture, such as the development of the three-field system of crop rotation and the use of horse collars and horse-shoes, began as early as the ninth and tenth centuries. Their gradual diffusion was complemented by the introduction of numerous machines designed to tap inanimate sources of power. Water mills, which had been used throughout much of the ancient world to grind grain, were put to a wide range of additional tasks, from fulling cloth and tanning leather to driving the triphammers of iron forges. The horizontal-axle or "post" windmill, invented in the North Sea region in the late twelfth century, spread throughout northern Europe in the thirteenth and fourteenth centuries. Mill-driven bellows and forges contributed to advances in toolmaking. Ingenious medieval craftsmen devised cranks, a variety of cams and

[12]For an analysis of medieval and early modern developments in the mining and metallurgical industries, see J. U. Nef, *The Conquest of the Material World* (Chicago, 1964).

[13]E. L. Jones has laid special stress on the gradual nature of technological change in the preexpansion era; see *The European Miracle* (London, 1981), pp. 62–64. The best general discussions of these vital transformations include Lynn White, Jr., *Medieval Technology and Social Change* (London, 1962), and *Medieval Religion and Technology: Collected Essays* (Berkeley, Calif., 1978); Jean Gimpel, *The Medieval Machine* (New York, 1983); and B. H. Slichter van Bath, *The Agrarian History of Western Europe* (London, 1963). Braudel's *Structures of Everyday Life* (pp. 353–61) contains a superb and succinct summary of the spread of mill technology. He estimates that there were from 500,000 to 600,000 water mills in central Europe alone on the eve of the Industrial Revolution.

gears, and laboriously cut screws, which were essential to the development of more complex machines. These and other innovations led to profound transformations in European economic and social life. But these changes, like the innovations themselves, occurred over a time span of six or seven centuries.

As we shall see, the Industrial Revolution was also the product of innovations spread over time. But industrial innovations were more dramatic and more geographically dispersed, and they were compressed into a span of decades rather than centuries. Given the fact that these technological breakthroughs were added to an already impressive European endowment, it is not surprising that Western thinkers in the nineteenth century were a good deal more impressed with Europe's technological superiority over non-Western peoples than their sixteenth- and seventeenth-century counterparts had been. It is also important to keep in mind that the overall pace of technological change in Europe had slowed considerably between 1450 and 1700. There were numerous innovations—including forks, fountain pens, and cut glass—which affected the daily lives of at least the bourgeoisie and nobility. But, as G. N. Clark has observed, "the sum total of these improvements seems . . . small in comparison with the general energy of the age . . ."[14] Except in the mills and mines, the tools of European workers, like those of Africa or Asia, remained predominantly manual—that is, they were implements that acted as extensions of the hand. Whether for peasants or urban artisans, tools changed little in this era from those devised in the Middle Ages or even in ancient times. It was a period of "minor improvements" on devices developed in earlier centuries, an age of overall stagnation.[15] Therefore, even though Europe had moved to an unprecedented level of scientific understanding and technological mastery centuries before the Industrial Revolution, those who compared Europe with other civilizations before the eighteenth century grasped only parts of this advance. They failed to see that these changes, more fundamentally than differences in religion, dress, or facial features, set western Europeans off from all previous or contemporary peoples.

The national and social origins of European observers also influenced the degree to which they were interested in the tools and the organization of production in the societies they visited. Most of the authors of missionary and travel accounts in the early centuries of expansion were from aristocratic or upper bourgeois families. In Iberia and France in

---

[14]G. N. Clark, *The Seventeenth Century* (Oxford, 1957), p. 63 (quoted portion) and chap. 5. See also Braudel, *Structures of Everyday Life*, pp. 27, 371; and Domenico Sella, "European Industries, 1500–1700," in Carlo Cipolla, ed., *The Fontana Economic History of Europe* (Glasgow, 1974), vol. 2, esp. pp. 354–58.

[15]Mandrou, *Modern France*, pp. 141–48.

particular, persons at this level of society harbored a pronounced disdain for artisans or those engaged in the "useful arts." As late as the first decade of the seventeenth century, for example, the jurist and influential social commentator Antoine Loyseau lumped artisans and mechanics with all who labored with their hands as "viles personnes." Citing no less an authority than Cicero, Loyseau averred that a person's honor was diminished in proportion to the extent that he engaged in manual labor.[16] This attitude may help to explain why most French, Spanish, and Portuguese travelers displayed such limited interest in technology. In contrast, Dutch and English observers, in whose societies contempt for manual labor was less pronounced among the upper classes and gentlemen were likely to tinker with mechanical contraptions,[17] often had a good deal more to say about technology than their French or Portuguese counterparts. But except for works by physicians such as Thomas Fryer, English and Dutch accounts usually contain considerably less information on the sciences in Africa and Asia than do those of French savants such as François Bernier and Iberian missionaries such as Martin da Rada.

Only a tiny minority of the Europeans who went overseas in this era knew much about the scientific discoveries that were profoundly transforming Western thinking about the natural world. This is not surprising in view of the fact that the series of breakthroughs which has come to be known as the Scientific Revolution really got under way only a half-century after Vasco da Gama's voyage to India, with the publication in 1543 of Copernicus's *On the Revolutions of the Heavenly Spheres* and Vesalius's *On the Fabric of the Human Body*. The scientific work of the medieval period sometimes anticipated, even though it did not always directly influence, the work of such thinkers as Copernicus and Vesalius. But the new ideas were little known beyond the circles of university scholars who proposed and debated them from the twelfth to

[16]Quoted in Regine Pernoud, *Histoire de la bourgeoisie en France: Les temps modernes* (Paris, 1981), pp. 62–63. For Iberia, see Bartolome Bennassar, *The Spanish Character: Attitudes and Mentalities from the Sixteenth to the Nineteenth Centuries*, trans. B. Keen (Berkeley, Calif., 1979), pp. 17–18, 121, 127; and William Callahan, *Honor, Commerce, and Industry in Eighteenth-Century Spain* (Boston, 1972), pp. 1–7, 45–55. For more general discussions of these patterns, see Mandrou, *Modern France*, pp. 142–43; and Nef, *Material World*, pp. 294–315. This disdain for manual labor did not, of course, prevent the bourgeoisie or nobility from workaholic devotion to administrative or mercantile tasks or from strongly advocating hard work and condemning idleness among the masses, as the career of Jean Baptiste Colbert so amply demonstrates: see Pernoud, *Histoire de la bourgeoisie,*, pp. 127–28.

[17]Nef, *Material World*, pp. 134–35, 294, 313–15, 318–19; A. R. Hall, *The Revolution in Science, 1500–1750* (London, 1983), pp. 243–45; G. N. Clark, *Science and Social Welfare in the Age of Newton* (Oxford, 1937), chaps. 1 and 3; and Marie Boas, *The Scientific Renaissance, 1450–1630* (New York, 1962), pp. 197–201.

the fourteenth century. The generally nonexperimental, nonapplied character of early scientific work limited its impact on western European society beyond this scholarly elite.[18] Even within the universities, scientific exploration was usually a secondary preoccupation of a small number of thinkers who were far more interested in the scholastic and humanistic debates that dominated the scholarship of the period.[19]

In the late Middle Ages and the Renaissance, this educated elite, which increasingly comprised an international community of peripatetic scholars, became deeply committed to the search for and dissemination of classical learning, including scientific and technical works.[20] The appraisal of these works as well as the treatises of the Arabs and early Church fathers still involved only small numbers of thinkers. This pattern persisted into the sixteenth and seventeenth centuries, when major breakthroughs in astronomy, anatomy, and mechanics occurred. Thus, only a fraction of the educated elites in societies engaged in overseas exploration understood, much less contributed to, these advances or the improvements in mathematics and experimentation which had made them possible. Even at the time of Newton's culminating studies in optics, calculus, and mechanics in the last decades of the seventeenth century, most Europeans knew little of the scientific discoveries that had catapulted Europe far ahead of the Islamic world, which had initially contributed so much to European learning, and the civilizations of India and China, which had been major centers of scientific and mathematical inquiry since ancient times.[21] Many who did know about the new ideas were opposed to them. The reasons for their opposition varied, but into the eighteenth century a large segment, perhaps a majority, of the educated elite found itself (says Tillyard) "loth to upset the old order by applying [the new] knowledge."[22]

[18]Hall, *Revolution in Science*, pp. 6–8, 30–31; A. C. Crombie, *Augustine to Galileo* (Harmondsworth, Eng., 1969), vol. 2, pp. 25–26, 37, 125–29; and Nef, *Material World*, p. 35.

[19]For surveys of these patterns, see F. C. Copelston, *Medieval Philosophy* (New York, 1952); and Paul O. Kristeller, *Renaissance Thought* (New York, 1955).

[20]Boas, *Scientific Renaissance*, pp. 22–28; Gimpel, *Medieval Machine*, pp. 175–79; and Robert Mandrou, *From Humanism to Science, 1480–1700* (Harmondsworth, Eng., 1978), pp. 40–65.

[21]Crombie, *Augustine to Galileo*, vol. 1, chap. 2; Joseph Needham, "Poverties and Triumphs of the Chinese Scientific Tradition," in A. C. Crombie, ed., *Scientific Change* (New York, 1963), pp. 117–53; and Needham, *The Great Titration: Science and Society in East and West* (London, 1969).

[22]E. W. M. Tillyard, *The Elizabethan World Picture* (New York, n.d.), p. 8. On resistance to and the slow spread of scientific ideas in the sixteenth century, see Giorgio de Santillana, *The Crime of Galileo* (Chicago, 1955), pp. 3–4; and Edwin Burtt, *The Metaphysical Foundations of Modern Physical Science* (London, 1964), pp. 23–25, 39. Estimates of recognition or acceptance of the new learning at different periods cannot be made with any precision, but A. G. R. Smith offers some broad guesses in his *Science and Society* (London, 1972), esp. pp. 26–27, 169–78.

Virtually all European travelers, including the best educated, viewed the world in ways that fundamentally resembled the outlook of the peoples they encountered overseas. They saw the cosmos as geocentric, fixed, and hierarchic; they believed that supernatural forces could, and regularly did, influence the workings of the natural order. In the early travel accounts one frequently finds, for example, prayers of thanksgiving for safe voyages from one landfall to the next and for escapes from the assaults of wild animals or hostile "natives." Most travelers also believed in the power of ghosts, monsters, witches, and other supernatural creatures to affect their fortunes. Except for the best educated, their understanding of the natural world was very similar to that of the ordinary people of the societies they visited, and it was decidedly inferior on many subjects to that of the scholars and priests of these societies. Perhaps most important, European belief systems were at least as firmly grounded in religion as those of the Africans and Asians. The first phase of expansion coincided after all with the Reformation, the Counter-Reformation, and the centuries of bitter debate, unbounded polemic, and brutal persecution that these movements spawned. European merchants and missionaries also shared with African and Asian peoples a reverence for tradition and ancient authorities, which in the European case included both sacred scripture and Greco-Roman writings. In short, their most valued truths were religious, not scientific. Thus, for almost all European observers, including the well educated, the most decisive distinction between themselves and the peoples they encountered was religious. They were Christians; most Africans and Asians were not. A good deal more space in their accounts was devoted to pointing up differences in religious beliefs and practices than in attempting to explain or compare African and Asian ideas about eclipses or techniques of numerical calculation.[23]

Even the better educated European observers and those best informed about Asian sciences were concerned primarily with religion. The Jesuit missionaries valued their astronomical skills and ability to repair mechanical contrivances like clocks not because they proved how far Europe was ahead of China in mathematics or precision instrumentation but as the means by which they could gain access to rulers, whom they hoped to convert to Christianity. In Europe itself, most early scientists were intensely religious and saw little conflict between their work on

[23]C. S. Lewis, *The Discarded Image* (Cambridge, Eng., 1964), is still useful for a general summary of the medieval viewpoint. For the educated classes generally, see Crombie, *Augustine to Galileo*, vol. 1, pp. 38–39, 90–93; for England, see Tillyard, *Elizabethan World Picture;* for France, see Mandrou, *Modern France* or Jacques Le Goff, *Time, Work, and Culture in the Middle Ages* (Chicago, 1980).

natural phenomena and their belief in the supernatural.[24] In addition, many scientists and scholars—including Paracelsus, Tycho Brahe, and Isaac Newton—continued to take very seriously their work in such fields as alchemy and astrology, which were also considered vital by the pundits of India and the scholar-officials of China.[25] The continued pursuit of astrological knowledge was but one manifestation of the vagueness of the boundaries of science in this period. The men whose discoveries made the Scientific Revolution were not professional scientists in our sense. As physicians, clergymen, magistrates, university lecturers, and gentlemen of independent means, they tackled whatever problems interested them, often in a wide range of what would later be demarcated as distinct disciplines. They believed ethics (or moral philosophy), grammar, and logic to be as scientific as mechanics and astronomy.[26] Thus, the writers who provided the earliest assessments of non-Western learning had neither a rigid sense of what qualified as scientific knowledge as a basis for judgments about other peoples' accomplishments in this realm, nor the conviction that moral philosophy and rhetoric—subjects in which the Chinese and Indians excelled—ought to be excluded in evaluating scientific achievement.

Technology—Perceptions of Backwardness; Qualified Praise

So much was new and strange in the overseas lands they visited in the early stages of expansion that Europeans often simply overlooked tools and methods of cultivation. Overall impressions of the material culture of a given society counted for a good deal more. Cities and housing,

[24]Clark, *Science and Social Welfare*, pp. 79–84; Alexander Koyré, "The Significance of the Newtonian Synthesis," in *Newtonian Studies* (Chicago, 1965), pp. 20–22; and W. C. Dampier, *A History of Science* (Cambridge, Eng., 1979), pp. 148–49, 172–23.

[25]Boas, *Scientific Renaissance*, chap. 7; Hall, *Revolution in Science*, pp. 84–91; Frank E. Manuel, *A Portrait of Isaac Newton* (Washington, D.C., 1979), chap. 8; and Betty Jo Dobbs, *The Foundations of Newton's Alchemy* (Cambridge, Eng., 1975). In recent decades a number of historians have argued that the magical-alchemical tradition played a critical role in the formulative stages of the Scientific Revolution. See, e.g., Francis Yates, "The Hermetic Tradition in Renaissance Science," in Charles Singleton, ed., *Art, Science and History in the Renaissance* (Baltimore, Md., 1967), or Yates, *The Rosicrucian Enlightenment* (London, 1972). For the application of this approach to the thought of Francis Bacon, see Paolo Rossi, *Francis Bacon: From Magic to Science* (London, 1968).

[26]Hall, *Revolution in Science*, pp. 26, 222, 233; Mandrou, *Modern France*, pp. 180–82. Centuries earlier the boundaries were even vaguer; one twelfth-century writer included hunting and theatrical performances among the "seven" sciences (Crombie, *Augustine to Galileo*, vol. 1, 186–87). For examples from overseas accounts from this period, see I. Grueber, *Voyage à la Chine* in *Relations de divers voyages curieux* (Paris, 1672), p. 8; and Gabriel Magaillans, *A New History of China* (London, 1688), p. 88.

public works and armies, and the way the inhabitants dressed and the products they offered in trade had a greater impact on European attitudes than their distinctive techniques of construction, the killing power of the weapons with which they fought their wars, and the quality of the looms on which they wove their textiles. But even the earliest voyagers were at least curious about African and Asian technology, and many compared what they saw overseas with what they themselves employed or knew to exist at home. Few observers sensed that western Europeans had gained an overall lead in technology, but many travelers identified specific areas in which they believed the Europeans to be markedly superior to the overseas peoples they contacted. In combination with general estimates of the level of a given society's material culture, technological achievement did much to determine the extent to which its people were esteemed or held in contempt. It also had considerable bearing on the levels allotted to different peoples in the hypothetical hierarchies of human development which were beginning to form in the European imagination. Machines were not yet the preferred measure of human worth, but their quality and complexity had begun to be associated with cultural advancement and creative potential.

Of the areas visited in the "known world" (the eastern hemisphere), Africa was perceived to be the most different from Europe and the most unsettling. Europe's pre-expansion image of Africa had been a bifurcated one: on the one hand the fabulously wealthy rulers and great cities that dotted the Saharan interior on medieval maps; on the other the reports of creatures that were half-man, half-beast, and tales of bizarre customs and rituals. What most explorers actually saw appeared for the most part to confirm the second and darker half of the African image. There was little evidence of the great empires and fabled cities such as Timbuctu which medieval scholars had read about in the accounts of Muslim scholars and Maghribi merchants. These busy emporiums were located in the savanna zone deep in the African interior, where Europeans rarely journeyed in this period. The legendary Prester John, a rich and powerful Christian king whom the Europeans hoped to find in Africa and enlist in their struggle against the infidel Muslims, was located in distant Ethopia only after decades of futile searching in western and southern Africa. Here again, the reality that corresponded to the positive side of the pre-expansion image proved disappointing. Not only were the Ethopians a very different sort of Christian from any variety found in western Europe, but their kingdom lacked great cities and fabulous wealth and, like Europe itself, was encircled and besieged by hostile Muslim peoples.[27]

[27] Hoffman, *Le nègre romantique*, p. 18.

Many aspects of the societies of coastal Africa, which had the most extensive contact with Europeans in this period, appeared to confirm unfavorable medieval impressions of the continent. Except for a few kingdoms, such as Benin and the Kongo, the peoples of the west African littoral were organized in stateless societies or loose political confederations. Even though the process of building nation-states was still in its early stages in Europe itself, and the populations of the Iberian kingdoms that led the way in overseas expansion were not large—about a million in Portugal; six to seven million in Spain—European observers found the states of west Africa so small that some asserted there were no states at all. When strong rulers were found, they were usually pictured as petty despots whose retainers groveled before them as if they were gods. The fact that Europe itself was embroiled in numerous wars of dynastic ambition and increasing civil and religious strife in the sixteenth and early seventeenth centuries did not deter numerous travelers from deploring the seemingly incessant but small-scale wars waged by the African coastal peoples. Many of the writers of this era conveyed the sense of political chaos, legal corruption, and social formlessness which remained a central feature of educated Europeans' attitudes toward sub-Saharan Africa until well into the twentieth century.[28]

There were some signs of wealth and what the Europeans considered social development along certain parts of the African coast. In the west there were the gold, ivory, and pepper that for decades provided the main motive for Portuguese exploration and the focus of their trading efforts.[29] On the east coast there were a complex trading system and a string of impressive commercial centers. But these were dominated by Arabs and Gujaratis, most of whom were Muslim and thus usually regarded as quite distinct from the peoples of "Negroland," "Nigritarum" or black Africa. Aside from a few palace complexes such as Benin and Luanda and the Muslim-controlled ports of east Africa, there was little about the material culture of coastal Africa that impressed European observers in the sixteenth and seventeenth centuries.[30]

[28]Many of the accounts written in this era deal with patterns of law and government. For predominant French views, see Roger Mercier, *L'Afrique noire dans la littérature française* (Dakar, 1962), p. 15; and William B. Cohen, *The French Encounter with Africans* (Bloomington, Ind., 1980), pp. 24–26. For Dutch responses, see the numerous discussions in Olfert Dapper, *Africa: Being an Accurate Description* (London, 1670), vol. 2; and Martin Ouwinga, "The Dutch Contribution to the European Knowledge of Africa in the Seventeenth Century" (Ph. D. dissertation, Indiana University, 1975), pp. 121, 144–45.

[29]Chaunu, *L'expansion européenne*, pp. 70–71.

[30]Dapper, *Africa*, vol. 2, pp. 470–71; J. Cuvelier and L. Jadin, *L'ancien Congo d'après les archives romaines* (Brussels, 1954), pp. 120, 134; Catherine Coquery, *La découverte de l'Afrique* (Paris, 1965), p. 118; Valentim Fernandes, *Description de la côte occidentale d'Afrique* (Bissau, 1951), pp. 38–39; Duarte Barbosa, *The Book of Duarte Barbosa* (London, 1918), vol. 1, pp. 19–23; and Carvajal Luys del Marmol, *L'Afrique* (Paris, 1667), p. 61.

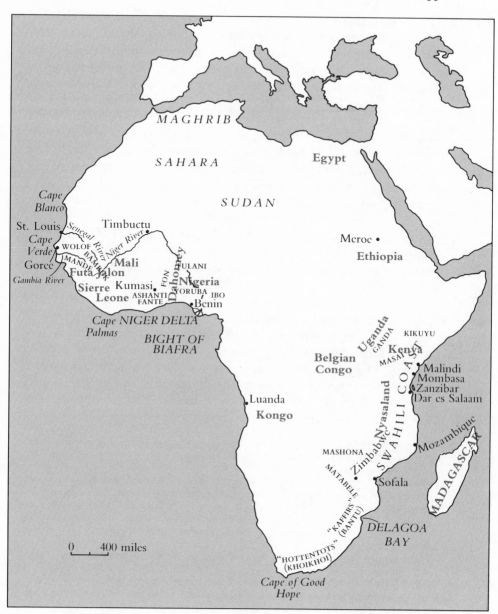

Map 1. Africa, with the location of places and peoples mentioned in the text

Numerous travelers noted the absence of walled towns, which for centuries had been prominent features of the European landscape. The size and quality—or the absence—of such fortifications in newly "discovered" areas took on added importance for European observers in these centuries when advances in firearms were forcing major innovations in fortress design throughout western Europe and heavy expenditures on intricate and extensive defensive works.[31] Travelers also noted the lack of construction in stone, which was either unavailable or much more difficult to work with than the abundant wood, leaf, and grass building materials that were preferred in coastal rainforest areas. When stone buildings such as those at Zimbabwe in south central Africa were reported, European observers refused to believe that Africans had built them or could even imagine building them.[32]

Housing, which was very different in size, form, and manner of construction from that found in Europe, was perhaps the most frequently discussed aspect of African material culture. Some travelers lauded the ingenuity and beauty of African dwellings, but most found them small, miserably built, misshapen, and poorly provided with doors, windows, and ventilation. Samuel Brun, a German physician who visited Africa many times early in the seventeenth century, remarked on the paucity of furniture in general and the lack of beds in particular. Olfert Dapper, a Dutchman who later in the century based a lengthy description of Africa on a wide selection of travel accounts, captured the majority sentiment when he dismissed African dwellings as mere hovels or pigsties.[33]

[31]Pierre Lavedan, *Histoire d'urbanisme: Renaissance et temps modernes* (Paris, 1941), pp. 14–21, 224–28; and Michael Howard, *War in European History* (Oxford, 1976), esp. pp. 34–37.

[32]Mercier, *L'Afrique noire*, p. 15; Gomes Eannes de Azurara, *The Chronicle of the Discovery and Conquest of Guinea* (London, 1946), p. 230; G. R. Crone, trans. and ed., *The Voyages of Cadamosto and Other Documents* (London, 1937), p. 29; Duarte Pacheco Pereira, *Esmeraldo de Situ Orbis* (London, 1937), p. 98; and William Towrson, "First Voyage to the Coast of Guinea," in Richard Hakluyt, comp., *The Principal Navigations . . . of the English Nation* (Edinburgh, 1889), vol. 11, p. 109.

[33]For sample positive views on African housing, see Pierre Bergeron, *Les voyages fameux du sieur Vincent le Blanc* (Paris, 1649), p. 23; and Sieur de Bellefonds Villault, *A Relation of the Coasts of Africa Called Guinee* (London, 1670), p. 162. Examples of negative impressions are included in J. Cuvelier, *Relations sur le Congo du Père Laurent de Lucgues* (Brussels, 1953), pp. 80–81; Crone, *Voyages of Cadamosto*, pp. 29, 37, 78; Dapper, *Africa*, vol. 2, pp. 347, 453; Claude Jannequinn, *Voyage de Lybie au royaume de Senega* (Paris, 1643), pp. 72–73; and Dierick Ruiters, *Toortse der Zee-vaert om te beseylen de Custen gheleghen bezuyden den Tropicus Cancri* (The Hague, 1913), pp. 50–51. Some authors made distinctions by areas, praising some peoples for their construction skills and dismissing others as inept. See Samuel Brun, *Schiffarten* (The Hague, 1913), pp. 6, 32, 39, 53–54; and Fernandes, *Description*, pp. 13, 93. On household furnishings, see Brun, *Schiffarten*, pp. 32, 40; and Ernest van den Boogaart, "Colour Prejudice and the Yardstick of Civility: The Initial Dutch Confrontation with Africans," in R. Ross, ed., *Racism and Colonialism* (The Hague, 1982), p. 47.

However small and vulnerable the caravels and *naos* that carried the Portuguese to India and the Spanish to the Americas, they were vastly superior to any vessels encountered along the west African coast or in the New World. The great empires of sub-Saharan Africa were oriented to the interior of the continent and the camel-borne Saharan trade on routes that ran to the Muslim kingdoms of the Maghrib. The coastal peoples traded primarily with other groups in the rainforest zone or with middlemen in the savanna region. They fished in the sea and traveled or transported goods along the many rivers and lagoons on the coast below Cape Verde in dugout canoes of varying sizes. Early Portuguese explorers were impressed with the size of these vessels, believing some could carry from 50 to 120 warriors or fishermen, but later Dutch and English voyagers claimed that African canoes accommodated much smaller numbers of passengers.[34] These contrasting impressions correspond in part to differences in the areas visited, but they also suggest that as time passed and European ships grew in size and technical proficiency, criticism of African vessels and navigational abilities grew decidedly harsher and more sweeping. The Dutch merchant Dierick Ruiters, who visited the Sierre Leone area just over a century after Valentim Fernandes had described the massive canoes found there, claimed that the people of the region were poor fishermen partly because they did not have craft large enough to enable them to fish at sea, where the catch would surely have been more bountiful.[35] Richard Jobson, who traveled along the African coast in the same years as Ruiters, stated categorically that the peoples he encountered had no canoes and no boats capable of sea travel.[36] As the narrative of the Venetian navigator Cadamosto vividly illustrates, from the very earliest contacts between Europeans and Africans in the mid-fifteenth century, the Europeans were well aware of the superiority of their ships and navigational skills. Cadamosto surmised that because the Africans had neither seagoing ships nor sailing instruments (he specifically mentions the compass and sea charts), they were "amazed" by the Portuguese voyages to their coasts and awed by the great sailing vessels, which they believed to be phantoms.[37]

Fear of disease and ambush by African warriors, who were known to use such insidious weapons as poisoned darts, kept the Europeans close to their ships and coastal forts. Thus, land engagements between Euro-

[34]Fernandes, *Description*, p. 95; Pereira, *Esmeraldo*, pp. 100, 132; Pieter de Marees, *Beschryvinghe ende historische Verhael van het Gout Koninckrijck van Gunea* (The Hague, 1912), pp. 89, 121–24; and Pieter van den Broecke, *Reizen naar West-Afrika 1605–1614* (The Hague, 1950), p. 23.

[35]Ruiters, *Zee-vaert*, p. 57.

[36]Richard Jobson, *The Golden Trade* (London, 1932), p. 20.

[37]Crone, *Voyages of Cadamosto*, pp. 16, 20–21, 34, 150–51.

peans and Africans were infrequent and small in scale during the six-teenth and seventeenth centuries. But the accounts of European travelers and soldier-adventurers leave little doubt that they believed their advantages over all African peoples in military technology and organization to be comparable to their superiority in ships and navigation. Duarte Lopez, who visited the Kongo region in the 1570s, boasted that a single Portuguese cavalryman was the equal of a hundred African warriors because the latter had poor weapons and were ignorant of firearms. Other observers supported Duarte's conviction that firearms, whether cannon or arquebuses, gave small numbers of Europeans a decisive advantage over much larger numbers of Africans. Some writers told of the awe and fear that European weapons instilled in coastal peoples, who could not believe that it was possible to kill at such great distances. These Africans also found that they were unable even to draw the strings of English longbows, which appeared similar to their own weapons but were much more powerful.[38]

Apart from passages dealing with ships and weapons, specific references to technology in early European accounts of Africa are remarkably sparse and usually disparaging. Most observers were aware that Africans, in contrast to the Amerindian peoples of the New World, were able to work iron, and some commented on the high quality of the weapons and utensils produced by peoples such as those of the Kongo and Sierra Leone.[39] But though European travelers might praise the final products of African blacksmiths, they also noted that their forges and bellows were small and primitive. Valentim Fernandes drew the clearest distinction between artisan skills and poor technology. If the peoples of Sierra Leone knew how to make proper "machines," he wrote, they could produce more iron than was then found in the entire Bay of Biscay region.[40] A similar pattern of disparagement dominated discussions of African textile production. The ingenuity of the peoples of the Kongo region, who wove clothing from palm leaves and grasses, was widely admitted, and there was high praise for the blue-colored textiles produced and sold along the Leeward Coast further north. But most travelers considered the tools of African weavers crude, and European merchants regarded cloth as an item to be traded *to* the Africans,

[38]Filippo Pigafetta, ed., *A Report on the Kingdom of the Congo . . . Drawn out of the Writings and Discourses of the Portuguese, Duarte Lopez* (London, 1881), p. 39; Towrson, "Second Voyage to the Coast of Guinea," and "Dutch Travellers to the East Indies," both in Hakluyt, *Principal Navigations*, vol. 11, p. 136, and vol. 10, p. 226.

[39]For comments pro and con on African metalworking, see Dapper, *Africa*, vol. 2, p. 434; Jacques le Maire, *Voyage to . . . the Coast of Africa* (Edinburgh, 1887), pp. 66–67; and J. Cuvelier, *Relations sur le Congo*, pp. 139–40.

[40]Fernandes, *Description*, p. 95.

not purchased from them—though it is worth remembering that the cloth in question was often of East Indian and sometimes African manufacture.[41]

Although African cultivation patterns were rarely described in any detail in this period, we can detect in European accounts the beginnings of the myth that tropical lands were lush and fertile but poorly developed. Some travelers provided detailed descriptions of the abundant plant and animal life of the coastal rainforest belt, and they contrasted this fecundity with the small portion of the land that was regularly cultivated. A few of these writers linked what they perceived to be low levels of agricultural production to the lack of draft animals or plows and other farming implements. These and many other observers also suggested that there was a connection between the "natural" indolence or idleness of the Africans and what the Europeans considered the backward state of African agriculture.[42]

Given the great advances in mining made by the Europeans in the centuries before and during the first phase of overseas expansion, and given Africa's initial importance as a source of gold, it is surprising that so little is said in travelers' descriptions about African mining equipment or techniques. As William Bosman's account indicates, the early travelers' neglect of this topic—which would elicit frequent and extensive commentaries from writers in the eighteenth century—can best be explained by the fact that African mines, whether in the west or the southeast, were located deep in the interior where disease and African resistance prevented European travel.[43]

In the early centuries of expansion the Europeans' generally low regard for African technological abilities was usually expressed indirectly through remarks on the poor quality and limited supply of African textiles and the primitive state of African warfare and agriculture. Tools and weapons, when mentioned, are more often merely described than

[41]Georges Balandier, *Daily Life in the Kingdom of the Congo* (New York, 1969), pp. 163–69; Bitterli, *Schwarzen Afrikaners*, p. 18; Cuvelier and Jadin, *L'ancien Congo*, pp. 116–17; Brun, *Schiffarten*, p. 30; Crone, *Voyages of Cadamosto*, p. 31; Dapper, *Africa*, vol. 2, p. 256; and Broecke, *Reizen*, pp. 17, 67.

[42]For the most explicit expressions of the lush tropics theme see Brun, *Schiffarten*, p. 34; Saint-Lô, *Voyage du Cap Verd*, pp. 54–55; or Dapper, *Africa*, vol. 2, pp. 378, 456–57, 560. On African agriculture more generally, see Jobson, *Golden Trade*, pp. 169–70; Cuvelier and Jadin, *L'ancien Congo*, pp. 117, 391; and Thomas Herbert, *Some Years Travel into Africa and Asia* (London, 1638), p. 22.

[43]William Bosman, *A New and Accurate Account of the Coast of Guinea* (London, 1705), pp. 72–73, 80. See also, Marmol, *L'Afrique*, p. 115; Jobson, *Golden Trade*, p. 22; and Villault's vague comments as late as the 1660s, *Relation*, 278–80. The most detailed account of the goldfields in this period is included in Marees, *Beschryvinghe van Gunea*, pp. 193–96, but Marees admits that what he relates was "told to him by some Negroes."

evaluated, though their absence is sometimes pointedly noted. As far as I am aware, no explicit overall comparison between levels of technological development in Europe and Africa was attempted. But it is clear that European observers considered the Africans deficient in the invention of tools and weapons and in their application to production and war. Low esteem for African material culture is reflected in the products—mostly raw materials at first; later human beings as slaves—that Europeans sought to obtain through trade with the coastal peoples. It is also indicated by the manufactured goods that European merchants and slavers peddled in exchange for African exports, including cheap textiles, obsolete firearms, bar iron, and even beads.[44]

The first stirrings of European contempt for African technological abilities can also be detected in tales of the dramatic effects of European firearms on "hostile natives" and the wonder shown by coastal peoples at even the simplest mechanical devices. No writer in the first centuries of expansion excelled the Portuguese explorer Alvise da Cadamosto in regaling the scholars and future explorers back home with narratives of such incidents. He told of wary African visitors to his ship who were terrified by the firing of a mortar; of how they found a burning candle a "beautiful and miraculous" object; and of their pleasure at hearing "the sweetest music ever" on common country bagpipes. Unlike many nineteenth-century travelers, Cadamosto did not turn African encounters with these unfamiliar devices into occasions for practical jokes, but like later observers he did report the awe evinced by the Africans for the Europeans who had created these wonders. Cadamosto claimed that he and his compatriots were thought to be "great wizards" who possessed "knowledge of everything."[45]

Finding little in the material culture of Africa that was likely to inform or amuse prospective readers, most travelers consistently focused on certain other themes, even before Bernard Varen in the mid-seventeenth century explicitly proposed a standardized agenda of inquiry.[46] Many authors provided detailed descriptions of African social patterns, including marriage (with special attention to polygamy), modes of child rearing, and male and female roles. The witchcraft craze that peaked in Europe in the sixteenth and early seventeenth centuries may well account for frequent accounts in this period of African sorcery and "devil-

[44]A. G. Hopkins, *An Economic History of West Africa* (New York, 1973), pp. 110–12. As Hopkins argues (pp. 120–21), these goods were not necessarily substandard, but they suggest a low estimate of African manufacturing capabilities.

[45]Crone, *Voyages of Cadamosto*, pp. 31, 41, 50–51, 58–59, 67–68.

[46]Annemarie de Waal Malefijt, *Images of Man* (New York, 1974), pp. 44–45.

worship."[47] These practices, like African religion more generally, were usually covered in only the vaguest and most garbled manner. Sorcery, superstition, and idolatry were linked by inference to reports of cannibalism, human sacrifice, and the mass slaughter of war prisoners.[48] Ignoring the obvious climatic explanations and their own discomfort, early travelers also deplored the scant dress or outright nudity of Africans of both sexes. Nudity in turn was associated with promiscuity, lasciviousness, and sexual excess, which came to be seen as characteristic African vices.[49] Though such positive qualities as generosity and hospitality were occasionally mentioned, the emphasis on practices that struck European travelers as bizarre and immoral fixed an early image of savagery that would shape later attitudes toward African peoples. Physical differences, especially color and facial features, and what was perceived to be a low level of material culture gave added credibility to this vision of a continent mired in a state of sin and savagery.

If European discoveries in Africa did little to confirm medieval legends of fabulous riches and powerful rulers, these visions appeared to be more than justified by what early travelers found in Asia, particularly in India and China. The spices and the delicately woven textiles of India, which had found a ready market in the West since ancient times, had been the ultimate goal of all early explorers.[50] When Vasco da Gama's fleet finally reached Calicut on India's Malabar coast in the spring of 1498, the Mughal empire that would dominate most of the subcontinent during the first centuries of European expansion had not yet been founded. Nonetheless, Da Gama's crews and those that came later found much to enchant and, quite literally, dazzle them in the large and beautiful trading cities that dotted the coasts of India and maritime southeast Asia. They were struck by the luxury and sophistication of Asian societies and by the variety and volume of products they exchanged in the great Indian Ocean trading complex. Before the rise of

[47]The literature on European witchcraft has proliferated rapidly in the past decade or so. For studies of the "craze" at its height, see Robert Mandrou, *Magistrats et sorciers en France au XVIIe siècle* (Paris, 1968); Keith Thomas, *Religion and the Decline of Magic* (New York, 1971), esp. chaps. 14–18; and Alan MacFarlane, *Witchcraft in Tudor and Stuart England* (London, 1970).

[48]See Brun, *Schiffarten*, pp. 7, 15–17, 30, 34, 38–40, 63; Dapper, *Africa*, vol. 2, pp. 347, 363–64, 368, 416–18, 477–80, 544, 568; Broeke, *Reizen*, pp. 13, 65, 68; Herbert, *Some Years Travel*, p. 11; Ruiters, *Zee-vaert*, pp. 48–49, 68–69, 82, 85; and Saint-Lô, *Voyage du Cap Verd*, pp. 29–30, 128–30.

[49]Brun, *Schiffarten*, pp. 13, 28–30, 35, 53–54; Dapper, *Africa*, vol. 2, pp. 360, 373, 390; Maire, *Voyage*, pp. 56, 59; and Villault, *Relation*, pp. 147–48.

[50]Most of the spices were actually located In southeast Asia, but before the sixteenth century the Europeans believed their source to be a vaguely defined India.

the Mughals in the middle decades of the sixteenth century, Portuguese chroniclers marveled at the size and opulence of the south Indian kingdom of Vijayanagar and its imposing capital on the Tungabhadra River.[51] From the 1550s onward European merchants, missionaries, and adventurers made their way in increasing numbers to the Mughal court centers of Lahore, Delhi, and Agra, where the rulers Akbar and later Jahangir presided over an empire far larger and richer than any European state. The Mughal emperors themselves and their splendid courts, replete with marble palaces, jewel-encrusted mosques, pleasure gardens, and hordes of richly attired courtiers, seemed to epitomize the wealth and power of Asian potentates that had been celebrated in medieval legends and romances.[52]

In contrast to Africa, the material culture of India was, to all but the most critical of observers, indeed impressive. There was an abundance of great walled cities (some of them deemed to be larger than either Paris or London),[53] massive stone or marble mosques and temples, and palatial residences for nobles, high officials, and, in the coastal enclaves, rich merchants.[54] Admittedly, some travelers noted that the houses of the peasants and urban poor were wretched hovels, and even the dwellings of well-to-do Indians had come in for criticism by the beginning of the seventeenth century. Their sparsely furnished interiors were unfavorably compared with those of the residences of the European nobility and bourgeoisie, which had begun to fill with the furnishings and Oriental carpets so wonderfully depicted by such Dutch painters as De Hooch and Vermeer.[55]

[51]For accounts of Vijayanagar, see Robert Sewell, *A Forgotten Empire* (New Delhi, 1962). On Portuguese reactions to the Malabar ports, see Donald Lach, *Asia in the Making of Europe* (Chicago, 1965), vol. 1, pt. 1, pp. 354–55.

[52]Lach, *Asia*, vol. 2, pt. 2, pp. 153, 265, 320; Hodgen, *Early Anthropology*, pp. 141–42; and Edward Oaten, *European Travellers in India* (London, 1909).

[53]Sections by Ralph Fitch and Nicholas Withington in William Foster, ed., *Early Travels in India, 1583–1619* (New Delhi, 1968), pp. 17–18, 206, 226–27.

[54]For representative samples of commentary on these aspects of Indian society by travelers of different nationalities, see Barbosa, *Book*, pp. 108, 140, 200, 226; Bergeron, *Les voyages fameux*, pp. 68–69, 85, 132–34; J. Albert Mandelso, *Voyages and Travels into the East-Indies* (London, 1662), pp. 30, 39–40, 45; Antony Monserrate, *Commentary on His Journey to the Court of Akbar* (Oxford, 1922), pp. 15, 30–32, 159; John Fryer, *A New Account of East India and Persia, 1672–1681* (London, 1909), vol. 1, p. 231, and vol. 2, pp. 119–20; Fitch in Foster, *Early Travels*, pp. 136–38, 144–46, 161, 173–74, 182–83; and J. H. Ryley, *Ralph Fitch: England's Pioneer to India and Burma* (London, 1899), pp. 97–99.

[55]For a discussion of the furnishing and decoration of the interiors of European residences in this period, see Braudel, *Structures of Everyday Life*, pp. 303–11. For criticisms of lower-caste Indian housing or upper-caste domestic furnishings, see John Careri's travel account in S. Sen, ed., *Indian Travels of Thevenot and Careri* (New Delhi, 1949), p. 163; Edward Terry, *A Voyage to East-India* (London, 1655), pp. 187–88, 191; Withington in Foster, *Early Travels*, p. 226; Thomas Bowrey, *A Geographical Account of Countries*

Early European travelers also disapproved of what they termed Hindu idolatry (after it became clear to them that the pantheon of Hindu
gods were not Christian saints), regarded Indian religious beliefs as little
more than superstition, and considered Indian rituals grotesque. They
were particularly appalled by such Indian practices as *sati* (widow-
burning), which they mistakenly believed to be a widespread practice
based on ancient Indian teachings.[56] Yet despite their reservations about
extensive poverty, excessive exactions by Indian lords, and what were
seen as bizarre customs, until the seventeenth century most European
travelers were captivated by India's princes and palaces and impressed
by its sheer size, its large population (which in 1500 equaled, perhaps
exceeded, that of all of western Europe),[57] the extent of its cultivated
lands, and the variety and quality of its manufactures. The scale and
splendor of India's material culture outweighed its exoticism and veiled
from all but the most astute observers the underlying weaknesses that
led to the collapse of the Mughal imperial edifice in the eighteenth century.[58]

If much of what European travelers found in India lived up to the
subcontinent's advance billing, what they encountered in China exceeded even the most effusive pre-expansion descriptions of Cathay or the
Land of the Great Khan. In the decades of the first overseas expeditions,
the hyperbolic descriptions of Marco Polo, who had actually visited
China, and John Mandeville, who had not, had come to be mistrusted
and regarded as excessive by geographers and sea captains involved in
the exploration efforts.[59] The early reports of Portuguese sailors, however, and later accounts by Jesuit missionaries of varying nationalities
indicated that, if anything, Polo and Mandeville had been too sparing in
their praise of China's achievements. As the French missionary Evariste
Huc pointed out in his much-cited nineteenth-century account, six-
teenth- and seventeenth-century merchants and missionaries went out
to China from a Europe ravaged by dynastic struggles, peasant re-

---

around the Bay of Bengal, 1669–1679 (Cambridge, Eng., 1905), p. 26; Lach, *Asia*, vol. 1,
pt. 1, p. 432; Francisco Pelsaert, *Jahangir's India: The Remonstrantie of Francisco Pelsaert*
(Cambridge, Eng., 1925), p. 67; and François Bernier, *Travels in the Mogul Empire* (Westminster, Eng., 1891), pp. 242–48.

[56]For sample reactions, see Barbosa, *Book*, vol. 1, pp. 115, 213–14, 220; Bergeron,
Voyages fameux, pp. 74–75, 86–89; Bernier, *Mogul Empire*, pp. 306–8; and Olfert Dapper,
Asia, of naukeurige beschryving van het Rijk des Grooten Mogols (Amsterdam, 1672), pp. 55–
58.

[57]Braudel, *Structures of Everyday Life*, 45–46; and Kingsley Davis, *The Population of
India and Pakistan* (New York, 1951), pp. 24–26.

[58]For typically effusive praise, see Thomas Coryat's account in Foster, *Early Travels*,
pp. 245–46.

[59]Parry, *Discovery*, pp. 51–52.

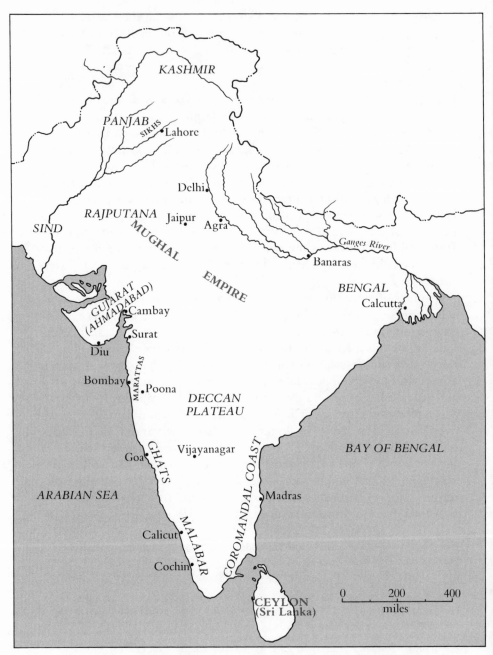

Map 2. India with cities and regions mentioned in the text

bellions, and religious wars. In east Asia they encountered an empire so vast that the largest of Europe's nascent nation-states could be tucked into several of its provinces.[60] They extolled the virtues of what they believed to be the absolute but paternal authority wielded by the Ming emperors and the training and dedication of the officials who ran the elaborate bureaucracy that made good the emperor's decrees and collected his apparently limitless revenues. Explorers and missionaries remarked on the extent to which tranquillity prevailed in a society where the power of fractious nobles had been broken, religious strife (in contrast to India) was unknown, crime was checked by strict laws and harsh penal codes, and vagabonds were rarely seen. Furthermore, China's material culture exceeded even that which had so greatly impressed visitors to India. The accounts of early travelers unanimously celebrate the great number of walled cities in China, whose size, broad and regularly laid-out streets, and crowded markets made Europe's largest urban centers seem like mere provincial towns. European visitors delighted in the number and variety of great bridges, the spectacle of the imperial city and palaces at Beijing, and the beauty of China's gardens and temples. Few actually visited the Great Wall, but reports of its antiquity and prodigious length served to underscore China's overall achievements in monumental architecture, perhaps the most important gauge of material development in this era.(Few visitors to China in this period appear to have attempted to compare the quality of the construction of the Great Wall or other fortifications with those found in Europe. The task of castigating the Chinese for their failure to produce military engineers of the caliber of Vauban or Coehorn would be left to the writers of a later, even more aggressive, phase of expansion.) European visitors also remarked on the extensive cultivation of a great variety of crops by a population larger than that of either western Europe or India, a population that the Europeans deemed frugal, industrious, and subservient to the will of the state.[61]

Given the great size of the Chinese empire, early European observers can be forgiven for failing to recognize the gravity of the weaknesses in the Ming regime and the signs of social unrest that in the mid-seventeenth century would lead to internal rebellions, the fall of the

[60]Evariste Huc, *The Chinese Empire* (London, 1855), p. xxii.

[61]The missionary Matteo Ricci found Chinese defensive works lacking in geometric sophistication and complexity. See Jonathan D. Spence, *The Memory Palace of Matteo Ricci* (New York, 1985), p. 43. For a valuable survey of many of these themes, see Lach, *Asia*, vol. 1, pt. 2, pp. 753, 764, 770, 803; and vol. 2, pt. 2, pp. 238–39. Some of the works that convey this very positive image include those of Mendoza, Da Cruz, Kircher, and Pereira (discussed below).

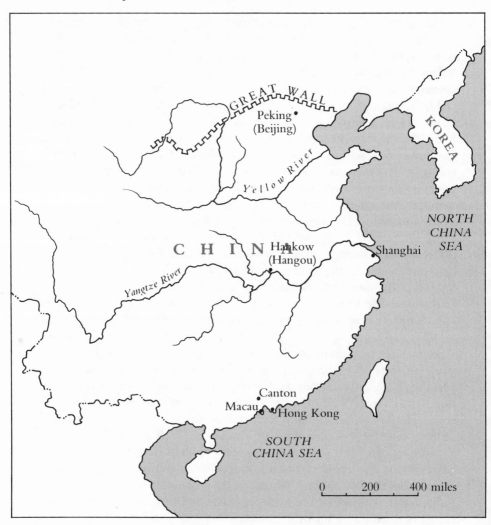

Map 3. China, with major centers visited by European travelers

dynasty, and the Manchu conquest. Europe's own difficulties in this period—the Thirty Years War throughout much of the continent from 1618 to 1648, the civil war in England in the 1640s, and the confusion and violence of the Fronde in France from 1649 to 1653—may well have reduced the shock of the disturbances in China. In any event, like so many previous "barbarian" conquerors, the Manchus, who had begun to be Sinified long before their conquest, simply occupied the throne and key positions at court and in the upper levels of the Confucian bureaucracy. A combination of collaborating Chinese administrators and their own banner armies allowed the Manchus to pacify and rule an even larger empire than that first encountered by the Europeans in the early 1500s.

The European "discovery" of India and China, not to mention the many kingdoms and trading states of southeast Asia whose size, wealth, and importance were often greatly inflated by travelers in this period, appeared to confirm the view of medieval geographers that the inhabited portions of the earth consisted of three continents, and of these, Asia was by far the largest, richest, and most powerful.[62]

The much greater attention given to material culture in European accounts of India and China, compared with an emphasis on social patterns in Africa, resulted in more frequent and considerably more detailed descriptions of Asian technology. From the time of the arrival of Da Gama's tiny fleet in Asian waters in 1498, the necessity of understanding and, wherever possible, matching Asian technology was readily apparent to European sailors, merchants, and military captains. The first challenges came while the Europeans were still at sea, where their advantages in ship design and navigational skills were a good deal less clear than they had been in African waters. The Arab *jahazis* and Indian *kotias* encountered in the Indian Ocean were equipped with lateen rigging similar to that of the Portuguese caravels. These vessels proved as swift and maneuverable as the ships of the European intruders, though not as well constructed or armed.[63] Along the coasts of India, southeast Asia, and farther east in the China seas, the Europeans came up against ships bigger than their own which accommodated much larger crews and numbers of passengers. Asian sailors had long used the compass, and their larger ships—especially those from China—had rudders, watertight compartments, as many as four masts, and numerous deck cabins.[64] Weaponry was the one area in which European seafarers were conscious of their decisive superiority from the first years of their involvement in Asia. Though Indian ships had small cannon,[65] and early European naval guns were slow-firing and inaccurate, the Europeans' steadily improving naval artillery proved critical to their efforts to main-

---

[62]See, e.g., R. W. Southern, *The Making of the Middle Ages* (New Haven, Conn., 1969), p. 71; and for a seventeenth-century expression of this view, see J. Ovington, *A Voyage to Surat* (Oxford, 1929), pp. 102–3.

[63]Parry, *Discovery*, chaps. 1 and 9.

[64]Needham, *Science and Civilization*, vol. 4, pt. 3, pp. 396–422, 477–86 and esp. 508–28; Henry Yule, *Cathay and the Way Hither* (London, 1913), vol. 1, p. 77n; White, *Medieval Technology*, p. 132; and J. C. van Leur, *Indonesian Trade and Society* (The Hague, 1955), pp. 86–87, 159–60, 212–14.

[65]Parry, *Discovery*, p. 174.

tain their Indian Ocean enterprises in the face of violent Muslim opposition.

Most early European observers were impressed with the size and great number of Asian ships. Nicolo Conti, a Venetian merchant who journeyed overland through much of Asia for nearly twenty-five years in the mid-fifteenth century, praised the quality of Indian ship construction and remarked on the many masts and sails that Gujarati ships could carry.[66] But closer inspection in the sixteenth century led to much less flattering estimates of the quality of Indian shipbuilding and navigational skills. Edward Terry, who served as chaplain with the 1615 mission of Thomas Roe to the Mughal court, admitted that the "junks [*sic*]" that carried goods between India and the Middle East were very large, but he considered them "ill-built"—broad-beamed and slow, too short, and likely to break up in bad weather. Terry's views were supported by later visitors to India, who also concurred in his opinion that the guns aboard Indian vessels, though often numerous, were either of poor quality or manned by sailors who had not been trained to use them effectively.[67]

Views on Chinese maritime technology were divided. Some travelers favored the position of the Spanish friar Juan Mendoza, who marveled at the number of Chinese vessels and considered the best of them equal to Europe's finest in size and construction.[68] Other observers shared the reservations of Martin da Rada, also a Spanish missionary, who reported that Chinese ships were slow, ill made, and poorly armed, and who claimed that Chinese navigators lacked sea charts.[69]

Although early travel accounts convey a sense that European ships and navigational instruments were more advanced than any but those of the Chinese (who, not purely coincidentally, possessed the only craft able to best the Portuguese in sea fights) naval technology was seldom cited as an indicator of European superiority over non-Western peoples.[70] This is not surprising, for despite great naval successes in this era,

---

[66]Nicolo Conti in R. H. Major, ed., *India in the Fifteenth Century* (London, 1857), p. 27.

[67]Terry, *Voyage to East-India*, pp. 137–38. Terry's views were widely disseminated by the republication of portions of his account in a compilation by Johannes Da Laet in 1631. See also Mandelso, *Voyages and Travels*, p. 87, and Fryer, *New Account*, vol. 1, p. 267.

[68]Juan Gonzalez de Mendoza, *The History of the Great and Mighty Kingdom of China* (London, 1853–54), vol. 1, pp. 148–52; John Speed, *A Prospect of the Most Famous Parts of the World* (London, 1662), p. 38; John Nieuhoff, *An Embassy to the Emperor of China . . .* (London, 1669), p. 232; and Athanasius Kircher, "*Antiquities of China*," appended to Nieuhoff, p. 99.

[69]Cipolla, *Guns, Sails, and Empires* (n.p., 1965), pp. 124–26; Martin da Rada, "Narrative of His Mission to Fukien," in C. R. Boxer, ed., *South China in the Sixteenth Century* (London, 1953), pp. 294–95; F. Alvarez Semedo, *The History of the Great and Renowned Monarchy of China* (London, 1655), p. 99; and Grueber, *Voyage à la Chine*, p. 15.

[70]For a comparison of the size and structure of Chinese war junks and Portuguese ships in the first century of overseas expansion, see Needham, *Science and Civilization*, vol. 4, pt. 3, pp. 508–9.

European ships remained vulnerable and their instruments crude and fallible. Above all, the average seaman's or traveler's ignorance of many of the advances that had been made in European navigational instruments and ship design, together with limited firsthand observation of Asian ships, rendered far from obvious the extent of the edge that European ships had gained over Asian rivals.[71]

Their rather modest estimate of their own naval capacity relative to that of the Indians or Chinese can be contrasted with the Europeans' frequently disparaging assessments of Asian methods and weapons for waging land warfare. This contrast is all the more striking in view of the fact that the conquest of even small Asian kingdoms was well beyond the means of expansive European nations in this period. The European acceptance of the limits of their capacity to project their power in mainland Asia is reflected both by the fact that land expeditions were rarely attempted or even contemplated and by the very deferential posture adopted by agents of European trading companies toward Asian potentates.[72] Before the eighteenth century, European conquests were possible only in island or coastal areas where seapower could be brought to bear. Nonetheless, a number of travelers commented on the vulnerability of even the most powerful of the Asian empires. European writers admitted that the armies of the Ming or Mughal rulers were much larger than any found in Europe but considered them poorly trained, organized, and armed. Asian artillery was singled out for special criticism. It was thought to be badly cast and too large and bulky to be used effectively in battle. These defects, plus less demonstrable allegations that Asian soldiers were cowards likely to panic and flee at the first sign of reverses in the field, led as early as the mid-sixteenth century to boasting on the part of visitors such as the trader-adventurer Galeate Pereira that a handful of European conquistadores could "make headway" against the whole of China.[73]

To the acute embarrassment of the Portuguese, Da Gama's initial contacts with both the Muslim merchants of east Africa and the Hindu and Muslim traders of India's Malabar coast revealed several key areas in handicraft manufacturing where Asia's technological lead over Europe remained considerable. Displaying a surprising disregard for the sophistication of Asian production and commerce, which had long been re-

---

[71]Parry, *Discovery*, pp. 148–54.

[72]For superb illustrations of this deference, see Van Leur, *Indonesian Trade*, pp. 241–45; and for an extreme example of early groveling, Donald Keene, *The Japanese Discovery of Europe* (New York, 1954), pp. 3–4.

[73]Galeate Pereira, "Certain Reports of China," in Boxer, *South China*, p. 28; Mendoza, *Mighty Kingdom*, pp. 128–30; Semedo, *Renowned Monarchy*, pp. 96–100; Terry, *Voyage to East-India*, pp. 158–62; Mandelso, *Voyages and Travels*, p. 51; and Pietro della Valle, *Travels in India* (London, 1892), vol. 1, p. 147.

ported by European overland travelers, Da Gama's party arrived in Asian waters with little more than coarse cloth, crude hardware, and assorted beads to trade for fine cotton and silk textiles, spices, and other much-sought Asian products. The response of Asian merchants was cool, even contemptuous, and Da Gama's ships returned to Portugal with little tangible evidence that the centuries-long quest for the Indies had been worth the great effort. Therefore, the Portuguese were forced to make use of another critical advantage that their ability to wage war at sea gave them: they took by force what they could not gain by fair trade. Warships became the cement of the succession of Portuguese, Dutch, and English trading-post empires that formed the basis of European enterprise in Asia from the sixteenth to the eighteenth century and made it possible for the Europeans to obtain a limited range of products, most notably spices, by force.[74] Some of the spices were shipped home to be sold at inflated prices, but others provided the means by which Europeans participated peacefully in the Asian interport trading system, where they continued to find few markets for European exports.

The European demand for large quantities of Indian and Chinese manufactured goods obscured from most observers the overall lead that Europe had gained in machine-assisted production by the sixteenth century. Small numbers of merchants traveled from Europe to draw products from the far-flung trading network that had linked the Mediterranean to the great civilizations of Asia since ancient times. In India, finely woven cotton textiles were in great demand for reshipment to other trading areas as diverse as the spice islands of the Indonesian archipelago and the slave coasts of Africa, or for export to Europe itself. From the seventeenth century Indian cotton cloth was in great demand in England among people at all social levels, from the ladies of Queen Mary's court to her more humble subjects in search of a washable shirt or pair of breeches.[75] Their importance in Western commerce is strikingly illustrated by the names—calico, muslin, chintz, gingham—still in use for different sorts of cotton materials. Not surprisingly, descriptions of the stages of Indian textile production from spinning to dyeing, where the use of indigo was a topic of continuing interest, abound in European

[74]See the fine surveys of the Dutch and Portuguese seaborne empires by C. R. Boxer; Cipolla, *Guns, Sails, and Empires*; M.A.P. Meilink-Roelofsz, *Asian Trade and European Influence in the Indonesian Archipelago between 1500 and about 1630*, (The Hague, 1962); and Michael Pearson, *Merchants and Rulers in Gujarat* (Berkeley, Calif., 1976).

[75]P. J. Thomas, *Mercantilism and the East India Trade* (New York, 1963), pp. 25–47; Holden Furber, *John Company at Work* (Cambridge, Mass., 1948), pp. 3, 14–17; Meilink-Roelofsz, *Asian Trade*, pp. 208–10; and Hopkins, *Economic History*, p. 110. Hopkins points out that there was also a considerable re-export of Indian manufactures of other sorts to the slaving areas in West Africa.

travel accounts of India, and this interest was sustained by British and French writers until late in the eighteenth century.[76]

European observers also commended Indian craftsmen for their skill in jewelrymaking, the working of precious and semiprecious metals, papermaking, and silk production.[77] But from the seventeenth century on there were significant reservations about Indian tools and techniques. These criticisms increased as Europeans became more familiar with Indian manufacturing processes and as the Europeans, partly on the basis of what they had learned from the Indians, improved and expanded their own production of comparable goods. In the 1660s, for example, the French physician François Bernier drew attention to the poor quality of Indian artisans' tools in general and to the lack of a proper system of apprenticeship among Indian craftsmen. Bernier also complained of the indolence of Indian workmen, which he attributed in part to their low status and pay.[78]

In this same period, key gaps in India's technological endowment as a whole also began to be noticed. John Ovington, a chaplain in the service of the East India Company, noted that he had seen no printing presses during his travels about the subcontinent in the 1680s.[79] Some decades earlier, the Flemish geographer Da Laet had included in his compilation of Indian travel accounts strong criticisms of Indian mining techniques based on the descriptions of William Methold, who wrote in the early 1600s. It is significant that though Methold was one of the first travelers actually to visit an African or Asian mine and thus fully grasp how advanced Europe was in this area of technology, the language of Da Laet's version of Methold's description is much harsher than the original. Da Laet dismissed Indian tools and techniques as "clumsy" and "arduous" and called the use of tens of thousands of laborers in the mines typical of a people who are "almost ignorant of machinery."[80]

Despite the criticisms of the late seventeenth century, however, except for Bernier's generally critical appraisal of the state of manufacturing in India, there is little in the early descriptions of the subcontinent to

[76]John Irwin and P. R. Schwartz, *Studies in Indo-European Textile History* (Ahmedabad, 1966), esp. p. 34.

[77]Terry, *Voyage to East-India*, pp. 113, 115–16, 134; Mandelso, *Voyages and Travels*, p. 83; Monserrate, *Commentary*, p. 160; Peter Mundy, *Travels in Europe and Asia* (London, 1914), vol. 2, pp. 98, 221–23; and Ovington, *Voyage to Surat*, pp. 166–67.

[78]Bernier, *Mogul Empire*, pp. 202, 225, 254–55. See also John Ogilby, *Asia: An Accurate Description of Persia and the Vast Empire of the Grand Mogol* (London, 1673), p. 157.

[79]Ovington, *Voyage to Surat*, pp. 149–50.

[80]Johannes Da Laet, *Description of India and a Fragment of Indian History* (Bombay, 1928), p. 76; and William Methold, "Relations of the Kingdom of Golchonda and other Neighboring Nations," in W. H. Moreland, ed., *Relations of Golconda in the Early Seventeenth Century* (London, 1931), pp. 31–32.

indicate that European travelers believed that *in general* the West was technologically more advanced. In fact, in key areas of production, especially textiles, sixteenth- and seventeenth-century writers stressed how much Europeans had to learn about manufacturing from India and how little they had to offer in exchange for the Indian products they so avidly sought. This was, after all, an era in which European technology only trickled to Asia in driblets of contraband firearms, clocks, and mechanical toys intended as presents for Asian potentates, while the Europeans crammed the holds of their galleons and merchantmen with spices and such manufactured goods as textiles and porcelain for export to Africa, the New World, and Europe itself.

For most European visitors, China was not just the premier source of silk and porcelain; it was recognized as the civilization that had invented much of mankind's basic technology. Travelers frequently commented on the ingenuity and industriousness of the Chinese and praised them for inventing printing, paper, gunpowder, the compass, and other devices such as wagons with sails, which are often mentioned in early accounts but strike us today as quaint contrivances rather than fundamental discoveries. In part because the Europeans themselves had acquired the processes of paper and silkmaking even before overseas exploration had begun and later mastered the manufacture of porcelain,[81] these products, despite the great demand for them in Europe, were less critical in determining a generally high European regard for Chinese technology than textiles were in evaluations of India. European visitors were most impressed with the inventiveness of the Chinese, with the engineering skills amply manifested in their public works, and with the massive irrigation systems that dominated the most heavily populated areas of the Middle Kingdom. As in the case of India, however, a few travelers found some aspects of Chinese technological development clearly inferior to European achievements. Perhaps the most portentous comments were made by such writers as Kircher and Bothero, who conceded China's ancient technological contributions but faulted the Chinese for not developing the full potential of the tools and techniques they had devised.[82] Others mentioned a more positive characteristic of the Chinese: their capacity to imitate the production of furniture, artwork, and some of the mechanical devices imported from the West.

On the whole, the dominant European view of Chinese technological achievement was that expressed by Alvarez Semedo, who resided in China for more than twenty years in the mid-seventeenth century. One

[81]Braudel, *Structures of Everyday Life*, p. 326; Crombie, *Augustine to Galileo*, vol. 1, p. 200.

[82]See Kircher, "Antiquities," p. 103; and Lach, *Asia*, vol. 2, pt. 2, p. 246.

of the few observers in the first centuries of overseas expansion to single out technological capacity as a key gauge of the level of development attained by a non-European people, Semedo chided fellow Europeans who dared regard the Chinese as barbarians, "as if they spoke of the Negroes of Guynea or Capuyi of Brazil." The skill of the Chinese in manufacturing alone, Semedo declared, was sufficient proof of the high level of civilization they had achieved.[83]

"Natural Philosophy"—Illiteracy and Faulty Calendars

If detailed discussions of African technology in the early centuries of expansion were rare, references to African science—or natural philosophy, as it was usually called in this period—were virtually nonexistent. Most accounts contained rather extensive descriptions of African religious beliefs, rituals, and practice of sorcery, all of which tended to be dismissed as superstition. But other than scattered asides about their rudimentary knowledge of medicine and arithmetic,[84] the only assessments made of the Africans' understanding of the natural world were generalizations such as that offered by Peter Heylyn, who declared in 1639 that the inhabitants of "Nigritarum . . . doe almost want the use of reason, most alienate from the dexterity of wit; and all arts and sciences."[85] Except in the Muslim areas in the western savanna and on the east coast, there were no indigenous written sources that travelers could use to study African thought at any level. Some Europeans apparently assumed that the absence of literacy meant an absence of both scientific learning and philosophy or "higher" religion.[86] This assumption proved to be a major liability for the Africans, especially when European writers contrasted their lack of written languages with the existence of ancient Indian, Chinese and other Asian texts. Literacy itself came to be regarded as a major attribute of civilized societies, and educated Europeans increasingly viewed illiterate black Africans as peoples sunk in ignorance and superstition, devoid of any learning that might be instructive for the Europeans themselves.[87] Yet there were many early observers who did not necessarily equate intellectual potential with

---

[83]Semedo, *Renowned Monarchy*, pp. 27–28.

[84]See Saint-Lô, *Voyage du Cap Verd*, pp. 27, 29–30; Maire, *Voyage*, pp. 59–60; Herbert, *Some Years Travel*, p. 22; Marees, *Beschryvinghe van Gunea*, pp. 180–81; Jannequinn, *Voyage de Lybie*, pp. 120–21; and Villault, *Relation*, pp. 45–48, 116.

[85]Peter Heylyn, *Microcosmos: A Little Description of the Great World* (Oxford, 1639), 719.

[86]Dapper, *Africa*, vol. 2, pp. 458–59; Boogaart, "Colour Prejudice," pp. 52–53; "Letter of Antoine Malfante," in Crone, *Voyages of Cadamosto*, p. 89; Herbert, *Some Years Travel*, p. 22.

[87]For a discussion of the links between writing and civilization in the eighteenth century, see Mercier, *L'Afrique noire*, pp. 80–82.

literacy and learned writings. Though some, like Peter Heylyn, dismissed the Africans as brutes without "sense or reason,"[88] numerous travelers commented on the cleverness, lively intellect, and capacity for reasoning of various sub-Saharan African peoples.[89]

In contrast to their sparse comments regarding the sciences in Africa, European writers in the early centuries of expansion had a fair amount to say about the state of natural philosophy in India and particularly China. It is ironic that the strongest criticisms were directed against Chinese scientific learning, which was, along with Indian and Islamic, the most diversified and advanced among the civilizations beyond Europe. To a large degree, Chinese scientific thinking came in for the heaviest criticism because Europeans knew, or thought they knew, the most about it. Unlike the Africans, the Chinese had built up a large corpus of written works on astronomy, mathematics, physics, medicine, and other areas of natural philosophy. These were accessible to European visitors once the Chinese language had been mastered, because the Chinese—unlike the Indians, who also possessed an extensive scientific literature—made their treatises available to educated Westerners. Beginning with Matteo Ricci in the 1590s, a succession of brilliant Jesuit missionaries, who resided for long periods at the Ming and Qing courts at Beijing, mastered the Chinese language, and studied and translated Chinese writings, including those in a variety of scientific fields.[90]

Though Jesuits and missionaries from other orders also made their way to the Mughal court centers, and some acquired familiarity with Persian and Urdu—the court and camp languages of the Mughal elite—Indian treatises on scientific subjects were much less well known by Europeans until the last decades of the eighteenth century. In part this was because Indian, as distinct from Islamic, scientific writings were predominantly Hindu and written in Sanskrit rather than the court languages. The Brahmin scholars who preserved and studied the Sanskrit texts, whether on science or yogic mysticism, were wary of, if not

[88]Marmol, *L'Afrique*, p. 313; Mercier, *L'Afrique noire*, p. 15; and Dapper, *Africa*, vol. 2, p. 590.

[89]Saint-Lô, *Voyage du Cap Verd*, p. 32; Cuvelier and Jadın, *L'ancien Congo*, p. 119; Ouwinga, "Dutch Contribution," pp. 122–23, 126; Joao de Barros, *Asia*, in Crone, *Voyages of Cadamosto*, p. 120; and Villault, *Relation*, pp. 140–41.

[90]The best general accounts of the Jesuits and their work in China remain Arnold Rowbotham, *Missionary and Mandarin* (Berkeley, Calif., 1942); and Virgile Pinot, *La Chine et la formation de l'esprit philosophique en France (1640-1740)* (Paris, 1932). The recent study by Jonathan Spence, *The Memory Palace of Matteo Ricci*, provides a fine account of the background to and the first phase of the Jesuit mission.

hostile to, the prospect of sharing India's ancient wisdom with outsiders. Their resistance was, of course, heightened by the attitude of the
Christian missionaries, who usually made no secret of their abhorrence
of Indian rituals and contempt for Indian religious beliefs. The more
tactful and tolerant Christian missionaries, most notably Roberto Di
Nobili, did study Indian languages and gain access to Indian texts (at
times provided by Indian converts or sympathetic pundits), but they
concentrated their efforts on religious works.[91] Consequently, most of
what Europeans thought they knew about Indian science until the late
eighteenth century they had learned through the perusal of the astronomical diagrams and computational tables of Hindu astrologers and
in conversation with Hindu or Muslim scholars, physicians, and—
regarding numbers, arithmetic and Indian methods of calculation—
merchants.[92]

Of all of the travelers to India in the early centuries of European
overseas expansion, none saw as clearly as François Bernier beyond
the marble palaces and richly caparisoned elephants of the massive Mughal armies to the underlying weaknesses of the empire which would lead
to its collapse in the first decades of the eighteenth century. Born in 1620
into a family of peasant-leaseholders in Anjou, Bernier proved to
be a clever student with remarkably varied interests. Capping a distinguished scholarly career, which included study with the famed
philosopher-scientist Pierre Gassendi, Bernier received a doctorate in
medicine in 1652. Six years later he embarked on a thirteen-year journey
that took him to much of the vast Indian subcontinent. He witnessed the
costly war among the four sons of Shah Jahan for the Mughal throne.
With a prophetic gift equal to that of the great nineteenth-century
French traveler Alexis de Tocqueville, Bernier commented on the
extremes of wealth and poverty, the political fissures and dynastic
squabbles, the stifling effects of despotic rule on commerce and manufacturing, and the religious divisions that were already undermining the
empire during Aurangzeb's reign (1658–1707). During this period it
reached its greatest size and, in the eyes of most contemporary observers, the apex of its power and wealth. Bernier's strictures of Mughal
political policies and economic practices were complemented by a critique of Indian scientific learning that was confined largely to medicine,
mathematics, and astronomy. It was, however—together with the account of another man of medicine, the surgeon John Fryer, whose years
of service with the English East India Company overlapped the last

---

[91]Vincent Cronin, *Pearl to India: The Life of Robert Di Nobili* (New York, 1959).
[92]Some of the Jesuits working in India were exceptions to this rule. See the *Lettres
édifiantes* citations in Chapter 2.

years of Bernier's travels in India—the most extensive European commentary on Indian science before the end of the eighteenth century.

Bernier and Fryer had strong reservations about earlier travelers' brief but generally positive assessments of Indian medical knowledge and practices. Bernier conceded that Hindu physicians had achieved some success with cures that stressed dietary restrictions, but he regretted the fact that they never employed bleeding as part of their treatment of illness. He claimed that Indian physicians knew nothing about anatomy—which the Europeans had studied systematically since the twelfth century[93]—and were unlikely to learn anything about it, since the Hindus did not permit dissection. He added mockingly that their ignorance did not prevent the Indians from pontificating on questions like the number of veins in the body, which they set at 5,000—"no more, no less."[94] Fryer was even more extreme in his censure, finding Hindu doctors poorly trained, untested, and unskilled in pharmacy, anatomy, bleeding, and surgery. Fryer claimed that Indian physicians only "pretended" to take the pulse of their patients and would have nothing to do with urine samples, because bodily excretions were seen as polluting.[95]

Bernier had nothing to say about Hindu arithmetic, which perhaps more than any other aspect of Indian learning had impressed earlier European visitors; they had marveled at the ability of Indian merchants and pundits to do complex calculations in their heads. Fryer commented briefly on this skill and observed sardonically that the Indians did in fact know arithmetic the best because after all it was the "most profitable" of all disciplines.[96] Neither Fryer nor any other European observer appeared to be aware of the Indians' considerable mathematical achievements or Europe's debt to India in this field, beginning with the numerals, which had facilitated mathematical advances in Europe and thus proved critical to the Scientific Revolution.[97]

Bernier and Fryer both had a good deal more to say about Hindu astronomy, which, they concurred, was sorely deficient compared to European learning in this area. Once again, their views ran counter to those of earlier travelers, who had stressed the great antiquity of Indian astronomy and commented on the accuracy of the elaborate tables used

[93]Hall, *Revolution in Science*, p. 44.
[94]Bernier, *Mogul Empire*, pp. 338–39.
[95]Fryer, *New Account*, vol. 1, pp. 285–88.
[96]Ibid., pp. 89–90; vol. 2, p. 103. For other expressions of this view, see Bowrey, *Geographical Account*, p. 24; and Jean Tavernier, *Travels in India* (London, 1889), vol. 1, p. 161.
[97]Crombie, *Augustine to Galileo*, vol. 2, pp. 131–35.

by Hindu pundits to chart planetary movements and predict eclipses.[98] Bernier claimed that Indian predictions and calculations were much less accurate than European ones; Fryer pointed out that most Indian scholars were interested less in astronomy than in "judicial astrology" (the casting of individual horoscopes) and magic. Both writers drew attention to the "ridiculous" myths by which Hindu and Muslim alike explained the occurrence of such phenomena as eclipses. Both delighted in describing the panic and waves of pot-banging and ritual ablutions that eclipses touched off among the Indian populace. But it is worth noting that Bernier, displaying the penchant for relativism that became so pronounced in the works of the eighteenth-century philosophes, also heaped scorn on his fellow Frenchmen for a similar reaction, which he had witnessed several years earlier.[99]

As Bernier's account perhaps best illustrates, early European ambivalence toward Indian scientific learning had turned decidedly negative by the last half of the seventeenth century. The compiler Olfert Dapper, who put together a volume on India similar to his earlier one on Africa, caught the mood when he stated categorically, and without a shred of evidence, that the Hindus were "very ignorant in natural philosophy and astronomy." He added that the Hindus had utter contempt for European astronomy, preferring instead to adhere "blindly to their ridiculous fables."[100] Dapper's blanket dismissal of Indian scientific learning is all the more striking because his description of India as a whole retained much of the sense of awe inspired in earlier writers by the size and splendor of the Mughal empire.

Matteo Ricci, the Jesuit missionary who entered into the service of the Ming emperors in the last years of the sixteenth century, recounted his amazement at being regarded by the Chinese *literati* as a "prodigy of science."[101] Their esteem "quite made him laugh," for despite his work in mathematics under the famed astronomer Christopher Klau, Ricci

[98]Pelsaert, *Jahangir's India*, p. 77; Terry, *Voyage to East-India*, p. 235; Ovington, *Voyage to Surat*, p. 206.

[99]Bernier, *Mogul Empire*, p. 301–4, 339–40; Fryer, *New Account*, vol. 1, pp. 275–76; vol. 2, pp. 102–3.

[100]Dapper, *Asia*, p. 116. Dapper's volume was quickly translated into several European languages. The English version, acknowledging Dapper and other authors, appeared under the name of John Ogilby.

[101]Background on Ricci and the Jesuit mission is based primarily on Henri Bernard, *Matteo Ricci's Scientific Contribution to China* (Peking, 1935); Spence, *Memory Palace*; George Harris, "The Mission of Matteo Ricci, S.J.," *Monumenta Serica* 25 (1966), 1–168; and Rowbotham, *Missionary and Mandarin*.

felt himself to be poorly educated in the sciences. He had missed Archimedes and algebra but had studied arithmetic, geography, geometry, astronomy, and perspective at the Jesuit Collegio Romano, where these subjects had been a standard part of the curriculum since the 1560s. He confessed that he knew little of astronomy beyond the techniques of charting the paths of the planets, predicting eclipses, and correcting calendars; and he played down his aptitude for mathematics, even though his fellow students had dubbed him "the mathematician." Yet the Chinese, he mused, persisted in treating him as an authority on all sorts of scientific questions, from those dealing with astronomy and calendar adjustment to the particulars of mapmaking and clock repair.

Like almost all early European visitors to China, Ricci greatly admired Chinese society and was highly impressed by the size and power of the Ming empire, where he lived from 1582 until his death nearly thirty years later. Initially he too formed a high estimate of Chinese scientific abilities; after witnessing court scholars correctly predict the timing of two eclipses in 1583, he exclaimed that the Chinese were "very learned" in medicine, mathematics, astronomy, the "science" of morals, and the mechanical arts. Ricci found it remarkable that Chinese views on many scientific subjects were similar to those of European thinkers, even though there had been little contact between the two civilizations. His subsequent studies of Chinese learning, however, both during the long years he waited anxiously at Macau to be called to Beijing and after he had arrived at the Ming court, led to a gradual but definite revision of his earlier views on the state of the sciences in China. But like virtually all his contemporaries, Ricci retained a high estimate of Chinese civilization as a whole.

After mastering Chinese and exploring Chinese learning in texts and through long discussions with Chinese scholars, Ricci arrived at a view of Chinese scientific knowledge that concurred with many of the earlier criticisms offered by the Spanish friars Martin da Rada and Gaspar da Cruz. Though one might well question their claims to familiarity with the necessary Chinese texts, both Da Rada and Da Cruz, after generally praising what they had seen in their travels through China, charged that its scientific works and education contained (in the words of Da Rada) "nothing to get hold of."[102] Like Da Rada, Da Cruz, and most early writers, Ricci retained a favorable view of Chinese medicine.[103] He also

[102]Da Rada, "Narrative," pp. 261, 295–96; Gaspar da Cruz, "Treatise in Which the Things of China Are Related . . . ," in Boxer, *South China*, pp. 212–16; and Lach, *Asia*, vol. 1, pt. 2, pp. 782–83.

[103]Semedo, *Renowned Monarchy*, pp. 56–57; Kircher, "Antiquities," p. 83; Nieuhoff, *Embassy*, p. 162.

recognized that Chinese astronomy was ancient and not without value. But Ricci's astronomical calculations and his work on the official Chinese calendar, which were so critical to his and later Jesuits' positions at the court, made him increasingly aware of the mistakes and areas of complete ignorance in the texts championed by the Chinese scholars who were vying with Ricci and his successors for the emperor's favor. Ricci and later Jesuit scholars discovered that similar errors in Chinese thinking and lacunae in Chinese knowledge were to be found in mathematics, geography, physics, and other fields. The Jesuit scholars also considered the instruments the Chinese had devised for astronomical observation, timekeeping, and navigation clearly inferior to those developed in Europe. By the mid-seventeenth century, as Alvarez Semedo's *History of the Great and Renowned Monarchy of China* demonstrates, Chinese grammar, logic, and rhetoric were also under assault.[104]

In contrast to the Europeans' view of Indian responses and later stereotypes of xenophobic Chinese mandarins, there was praise from a number of travelers and missionaries in the early centuries of expansion for Chinese receptivity to Western scientific learning. Indeed, the Jesuits were well aware that it was their scientific skills that gave them whatever influence they enjoyed with a succession of Ming and Manchu emperors. That influence in turn allowed them to keep alive the hope that the interest of the Chinese in European ideas about the natural world would eventually lead to their acceptance of Christian beliefs about the supernatural.

Scientific and Technological Convergence and the First Hierarchies of Humankind

Although the exchanges in this period between scientists and the artisans who were responsible for many of Europe's early technological advances were much more haphazard and far less frequent than in the eighteenth century and later, there were many instances of collaboration between the two. It is not my intent to become involved in the extended and sometimes heated debate that has developed over the magnitude and significance of these exchanges.[105] But the correspondence between

---

[104]Semedo, *Renowned Monarchy*, p. 51; Louis J. Gallagher, ed., *China in the Sixteenth Century: The Journals of Matthew Ricci, 1583–1610* (New York, 1953), pp. 325–26; Spence, *Memory Palace*, pp. 145–49; and Rachel Attwater, *Adam Schall: A Jesuit at the Court of China, 1592–1666* (London, 1963), pp. 16–20, 43–47, 53–56, 60–61, 121, 149–50.

[105]For discussions of differing positions in the debate, see Peter Mathias, "Who Unbound Prometheus? Science and Technical Change, 1600–1800," and A. R. Hall, "Sci-

areas of endeavor where science and technology converged in the first centuries of expansion and fields in which Europeans perceived major differences between their own acomplishments and those of non-Western peoples suggests a pattern of importance to the issues explored in this book. As we have already seen, early European travelers repeatedly noted the superiority of their instruments, whether nautical or astronomical, over those of any of the peoples they encountered overseas. It is not coincidental that some of the most critical innovations during the early centuries of expansion were in instrumentation and toolmaking. Equally noteworthy is the fact that many of these breakthroughs involved exchanges between artisans and scientists, or inventions by men whose work, like Galileo's, included both scientific investigation and technological innovation. Except for advances in navigational devices, which were vital to exploration, innovations in instrumentation were modest until the late fifteenth and early sixteenth centuries, when watches came into use, and lathes and refined screws were developed. Beginning with the odometer in the 1520s, a wide range of key instruments appeared: the single and later the compound microscope (1590s and 1650s respectively), the telescope (1608), the thermometer (1592), the barometer (1643), the air pump (1650s), and devices for reading the pulse rate (early 1600s).[106] By the seventeenth century the better informed travelers were clearly aware of this flurry of innovation and the advantages it had given Europe in scientific exploration and in the production of key mechanical devices ranging from firearms to clocks.

The Europeans' recognition of the superiority of their precision tools and instruments was connected to a further and more fundamental revelation. European travelers and missionaries began to sense that even their ways of thinking and perceiving the world were fundamentally different from those of any of the peoples they encountered overseas. Perhaps the most striking examples of this realization were linked to changing European approaches to time and space. As David Landes has recently shown, calendar inaccuracies and the need to regulate monastic routines had led medieval scholars and artisans to take a strong

ence, Technology, and Utopia in the Seventeenth Century," both in Mathias, ed., *Science and Society 1600–1900* (Cambridge, Eng., 1972); Boas, *Scientific Renaissance*, chap. 7; Nef, *Material World*, pp. 35–40, 278–81, 318–25; Crombie, *Augustine to Galileo*, vol. 1, pp. 183–84, vol. 2, pp. 133–34, 298; Clark, *Science and Social Welfare*, pp. 9–11, 14, 19–22, 76–82; and Margaret Jacob, *The Cultural Meaning of the Scientific Revolution* (New York, 1987).

[106]Hall, *Revolution in Science*, pp. 248–54; and Silvio Berdini and Derek de Solla Price, "Instrumentation," in Melvin Kranzberg and Carroll J. Pursell, Jr., eds., *Technology in Western Civilisation* (London, 1967), vol. 1, pp. 168–87.

interest in the measurement of time.[107] The effort to measure and record accurately the passage of time generated a variety of devices to perform these tasks. Of these, clocks proved the most efficient. By the fifteenth and sixteenth centuries considerable numbers of Europeans, particularly those living in urban areas and engaged in textile production or commerce, had undergone the profound reorientation of time perception that the spread of public and, later, private clocks had effected. Personal and social activities, especially those related to work and market exchanges, were increasingly regulated by manufactured, secularly oriented machines rather than natural rhythms or religious and ritual cycles.[108]

A surprising number of European travelers commented on the lack of a similar machine-regulated time sense among the peoples they contacted overseas. European writers stressed that the Africans had no instruments for measuring time and only the most primitive of natural gauges, such as planting and harvest cycles or the phases of the moon.[109] Merchants and missionaries reported that both the Chinese and the Indians had abstract units and mechanical devices by which to measure the passage of time, but most writers found them crude and inaccurate in comparison with European gauges and clocks. The most detailed critiques of Asian time-measuring devices can be found in the writings of Ricci and his successor missionaries at the Chinese court, who judged Chinese instruments for both astronomy and timekeeping excessively complex, hopelessly antiquated, and exceedingly unreliable. Jesuit scholars and other visitors also complained that the Chinese could not run or repair, much less manufacture, Western timepieces, samples of which had been sent to China as gifts for prominent officials or for the emperors themselves.[110] Though the lack of accurate timepieces in civilizations as ancient and as highly developed as China and India suggested a very different approach to time from that becoming dominant in urban Europe, travelers in the early centuries of expansion made a good deal less of this contrast than writers in the nineteenth century, when the

[107]David Landes, *Revolution in Time* (Cambridge, Mass. 1983), pp. 58–66.

[108]Ibid, pp. 89–90; Le Goff, *Time, Work, and Culture*, pp. 35–37, 43–52. For a perceptive comparison of Europe and China, see Louis Dermigny, *La Chine et l'Occident: Le commerce à Canton au XVIIIe siècle, 1719–1833* (Paris, 1964), vol. 1, pp. 45–47.

[109]Fernandes, *Description*, p. 101; Pigafetta, *Kingdom of the Congo*, p. 61; and Dapper, *Africa*, vol. 2, p. 398.

[110]Landes, *Revolution in Time*, chap. 2; Gallagher, *China*, pp. 23, 168; Carlo Cipolla, *Clocks and Culture, 1300–1700* (New York, 1967), pp. 80–92; Nieuhoff, *Embassy*, pp. 166–68; Kircher, "Antiquities," p. 65; and Lewis Le Comte, *Memoirs and Remarks Made in Above Ten Years Travels through the Empire of China* (London, 1737), pp. 117–18, 304. For Indian parallels, see Ogilby, *Asia*, p. 119; Ovington, *Voyage to Surat*, p. 166; and Fryer, *New Account*, vol. 2, p. 92.

Industrial Revolution had further transformed Western perceptions of time.

Leonardo da Vinci once declared that painting, like all the other *sciences*, must be based on mathematics.[111] His view of the interdependence of art, mathematics, and the study of anatomy reflects the degree to which artistic production was shaped by developments in a number of scientific fields. Much has been written for a popular audience about the quest for visual accuracy that drove Michelangelo and Leonardo to dissect cadavers. In fact, a broader science-derived stress on careful empirical observation and realistic representation was a major concern of virtually all major Renaissance artists. Renaissance painters were also heavily dependent on mathematical formulas and concepts in their efforts to produce works reflecting three diminsions, to render depth and perspective convincingly. They drew on geometry in their attempts to capture a true sense of space, volume, and mass and on optics to highlight the contrasts between light and shadow.[112]

These developments appear to have had little effect on travelers to Africa, whose rare comments on African art appear only with reference to African idolatry or religious rituals. Perhaps even more than later European observers, they considered African art too primitive, grotesque, and alien to be compared seriously with their own?[113] But European visitors to Asia devoted considerable attention to the painting and sculpture of India, China, Japan, and other societies. In fact, the interest in Western art on the part of some of the Mughal as well as the Ming and Qing emperors led to the dispatch of painters associated with the missionary effort at each court. Many of the Jesuit scholars who resided in Delhi or Beijing also dabbled in art (or music), in addition to their work in optics, astronomy, and mechanics.[114] The responses of such men as Ricci and Giovanni Ghirardini, an Italian artist who visited China at the end of the seventeenth century, indicate an early perception of fundamental differences in European and Asian approaches to art and space itself. Europeans found the Indians and Chinese talented at decoration

[111]M. Kline, "Painting and Perspective," in Hugh Kearney, ed., *Science and Change, 1500–1700* (London, 1971), pp. 24–25.

[112]Ibid.; Herbert Butterfield, "Renaissance Art and Modern Science," in Kearney, *Science and Change,* pp. 7–15; Crombie, *Augustine to Galileo,* vol. 2, pp. 273–75; and Giorgio de Santillana, "The Role of Art in the Scientific Renaissance," in Marshall Claget, ed., *Critical Problems in the History of Science* (Madison, Wis., 1959), pp. 33–65.

[113]On the neglect of African art by Dutch travelers in general, see Boogaart, "Colour Prejudice," p. 53.

[114]Bernard, *Ricci's Scientific Contribution,* pp. 31, 74; F. S. Feuillet de Conches, "Les peintres européens en Chine et les peintres chinois," *Revue contemporaine* 25 (1856), 219–20; Pinot, *La Chine,* pp. 21–24; and Giovanni Ghirardini, *Relation du voyage fait à la Chine. . .* (Paris, 1700), p. 14.

and design, expert at rendering plants and animals, and often imaginative in their use of colors. But they felt that none of the Asians had a sense of perspective or proportion. They believed that the Indians and Chinese were ignorant of the technique of chiaroscuro and that Asians could not begin to match European artists in the realistic depiction of human anatomy.[115]

An incident related by Ghirardini perhaps best illustrates the chasm of incomprehension that was opening ever wider between Chinese and European approaches to art and underlying conceptions of space. The emperor had commissioned Ghirardini to paint a grand mural with colonnaded buildings. When the court officials viewed the finished product, they were at first "stupefied" and concluded that Ghirardini had summoned up diabolical forces to create a scene that looked so real. On discovering that the mural was but a *trompe-l'oeil*, Ghirardini reported, the officials exclaimed that it was "contrary to nature" to depict distances where none really existed. However much the artist may have exaggerated the Chinese scholar-officials' reaction, apparently the emperor shared their disapproval, for Western painters at the Qing court were increasingly forced to adopt Chinese techniques. Chiaroscuro, perspective, and other European innovations exercised little influence on Chinese painting for the remainder of the Manchu period.[116]

The Europeans' awareness of differences in space perception was also evidenced in their reactions to African and Asian modes of measuring distances and their approach to geography in general. Most Europeans continued like the rest of humankind to measure distance on the basis of human dimensions (hand, foot, and the like) or activities (a day's journey) and were oblivious to standards beyond those specific to their home parish or county.[117] But educated Europeans, drawing on advances in mathematics, instrumentation, and astronomy, had begun to make maps and charts and to impose standards of measurement that would radically transform the Western concept of space. An awareness of these developments is reflected in the view of early travelers that the Africans were ignorant of geography and that they lacked maps, direction-finding instruments, and standard units by which to measure distances.[118]

In contrast to observers in India, who made little mention of map-

[115]Gallagher, *China*, pp. 21–22; Ghirardini, *Relation du voyage*, p. 79; Semedo, *Renowned Monarchy*, p. 56; and Grueber, *Voyage à la Chine*, p. 8.

[116]This incident and its long-term effects on Chinese painting are discussed in Feuillet, "Les peintres," pp. 224–26.

[117]Mandrou, *Modern France*, pp. 66–67.

[118]Pigafetta, *Kingdom of the Congo*, p. 111; Crone, *Voyages of Cadamosto*, p. 51; Pereira, *Esmeraldo*, p. 79.

making or geographical learning in this era,[119] the Jesuit scholars in China were disconcerted to find such an advanced civilization so backward in basic geography, despite its long-established tradition of mapmaking, overseas navigation, and astronomical study. As a number of writers observed, the Chinese possessed rather accurate maps of the provinces of their own kingdom, but their conceptions of the world beyond China bordered on the absurd. According to the Jesuit observers, the Chinese held to the notion—then rejected by all but the most unenlightened of Europe's scholars—that the earth was flat and square. They positioned China in the very center of this square, viewing it as a large land mass surrounded by much smaller islands representing the other lands that were known to the Chinese. Matteo Ricci did add that though they were originally deeply disturbed by China's much reduced and peripheral position on European world maps and globes, some court scholars admitted after careful study that these were superior to their own ideographic representations (comparable to the so-called "TO" maps of medieval Europe).[120] Significantly, Ricci also noted that European maps appeared to reassure inquisitive Chinese officials because they showed such a great distance separating China from Europe.[121]

Although science and technology were far less important than religion or sociopolitical organization in shaping European attitudes toward African and Asian peoples in the early centuries of overseas expansion, as aspects of material culture they contributed significantly to varying European perceptions of the level of development attained by different non-Western cultures. Scale and degree of complexity, insofar as European travelers were able to determine the latter, counted for a great deal in European assessments, whether these attributes were manifested in far-flung imperial bureaucracies or walled cities and irrigation works. Religion, as we have seen, was a vital point of reference for virtually all European observers. But, with such rare exceptions as the Ethopians, all overseas peoples shared in European eyes the onus of being non-Christian. European writers did identify differing degrees of heathen depravity, but these were more difficult to substantiate and compare than the quality of housing, the size of ships, and the volume of trade. Idolatry was idolatry; the devil allegedly being worshipped was the same in Benin and Gujarat.

Thus, it was differences in material culture—which included science

---

[119]Though there are brief references to Indian precision in measuring distances in Mundy, *Travels*, pp. 66–67; and Ovington, *Voyage to Surat*, p. 116.

[120]Lloyd Brown, *The Story of Maps* (Boston, 1949), esp. pp. 94–100.

[121]Gallagher, *China*, pp. 165–69; Bernard, *Ricci's Scientific Contribution*, p. 63; Kircher, "Antiquities", pp. 65–66.

and technology in the view of many travelers and missionaries—that had the most to do with the emergence of a hierarchy of non-Western peoples that began to take shape in the minds of European observers from the very first decades of expansion. This hierarchy was not as explicitly delineated as it would be in the eighteenth century, when classification was the rage; nor was it as hard and fixed as it would become in the writings of racist writers, especially in the nineteenth century. Nonetheless, the basis for the arguments and schemas of later writers had begun to be formulated. India and especially China had clearly impressed European visitors more than the cultures they contacted in Africa. Though all would subsequently be devalued in relation to Europe, this ranking of non-Western societies by level of development—China, India, Africa—persisted, with rare exceptions, well into the twentieth century.

The differences in early overall assessments of African, Indian, and Chinese societies also have bearing on the long-standing debate over the origins of European racist attitudes toward the Africans. In the early centuries of expansion Europeans rarely resorted to racial explanations for differences in social or cultural development between themselves and the Africans, or between the Chinese or Indians and the Africans.[122] If early writers sought to account for these differences (and often they were content merely to describe them), they favored environmental explanations and those that stressed such perceived cultural differences as the presence or absence of Christianity. The latter were unabashedly ethnocentric but not racist. Early European travelers often commented on the color or described the physical appearance of the peoples they contacted. But my reading of a large number of accounts by individuals of varying nationalities indicates that color and physical features have been unduly stressed by some of the authors engaged in the debate over the origins of racism.[123] It is true that some writers equated blackness

[122]Proto-racial speculations can be found in the suggestions of such thinkers as Isaac de la Peyère (1594–1676), who proposed multiple creations–including a pre-Adamite one–to explain extreme differences in human types. See Hodgen, *Early Anthropology*, pp. 230, 273, 275–77.

[123]See esp. Winthrop D. Jordan, *White over Black: American Attitudes towards the Negro, 1550–1812* (New York, 1968), chap. 1, in which religion and nudity are discussed, but the centrality of color and other physical differences is stressed. Jordan limits himself to English sources and within those to Shakespeare's writings and a handful of travel accounts drawn mainly (and selectively) from Hakluyt's collection. For other recent authors who share Jordan's views, again largely on the basis of their reading of Shakespeare rather than the travel literature, see James Walvin, *Black and White* (London, 1973), pp. 19–26; Anthony J. Barker, *The African Link* (London, 1978), pp. 42–44; Philip Mason, *Prospero's Magic: Some Thoughts on Class and Race* (London, 1962), pp. 54–74, 77, 99, 122–23; G. K. Hunter, "Elizabethans and Foreigners," *Shakespeare Survey* 17 (1964), 37–52; and Leslie A. Fiedler, *The Stranger in Shakespeare* (London, 1972).

or, less commonly, wiry hair and flat noses with savagery and bestiality,[124] but a far larger number simply observed that the Africans were dark-skinned, just as travelers in Asia noted that the Indians had "dusky" or "tawny" skin. In the latter case there is no suggestion that complexion had anything to do with the level of development of the peoples in question.[125] On the contrary, on the basis of the accounts I have consulted, I would argue that there are at least as many travelers who praise the Africans' physical appearance or link mention of their dark skin to positive assessments of African intelligence and other abilities as those who associate dark skin with degradation.[126]

A number of authors found black preferable to brown or "olive." Cadamosto, for example, spoke more highly of the "well-formed and large" bodies of the peoples on the "black side" of the Senegal River than of the "small, lean, brownish" peoples of the desert, and Duarte Pereira observed that the lighter-skinned peoples of the Cape of Good Hope region (the Khoikhoi?) were savages compared to the darker Wolofs and Mandingos.[127] Though it has been established that in medieval Europe generally and Elizabethan England in particular the color black was associated with unpleasant things like dirt, death, and Satan,[128] early travelers rarely expressed these sorts of connections. In fact, travelers of allegedly unremitting ethnocentrism[129] displayed a much more heightened sensitivity to African disdain for their pale or pink skin than have the present-day authors who argue for a universal preference for the color white.[130] Some travelers reported meeting Africans

[124]For examples, see Bergeron, Les voyages fameux, pp. 13, 16, 22; Herbert, Some Years Travel, pp. 9, 16; and Pigafetta, Kingdom of the Congo, pp. 118, 120. Some travelers, (e.g., Jacques le Maire, Voyage, pp. 36, 51–52), found the Africans comely but lacking in intelligence. Among recent studies, see David Brian Davis, The Problem of Slavery in Western Culture (Ithaca, 1966), p. 447; and A. R. Russell-Wood, "Iberian Expansion and the Issue of Black Slavery," American Historical Review 83/1 (1978), 38–39.

[125]Barbosa, Book, p. 8; Jannequinn, Voyage de Lybie, p. 19; Jobson, Golden Trade, p. 51; and Fernandes, Description, p. 17.

[126]See Fernandes, Description, p. 47; Jobson, Golden Trade, p. 45; Maire, Voyage, p. 36; Dapper, Africa, p. 598; Marees, Beschryvinghe van Gunea, pp. 28–29; Ovington, Voyage to Surat, p. 47; Ouwinga, "Dutch Contribution," pp. 113, 134–36; Boogaart, "Colour Prejudice," p. 46; Broeke, Reizen, pp. 13, 66; and Davis, Problem of Slavery, pp. 448–50.

[127]Pereira, Esmeraldo, p. 154; Crone, Voyages of Cadamosto, p. 28.

[128]Jordan, White over Black, pp. 7–8; Walvin, Black and White, pp. 24–25; Don Allen, "Symbolic Color," Philological Quarterly 15 (1936), 81–92; and Caroline E. Spurgeon, Shakespeare's Imagery and What It Tells Us (Boston, 1958), esp. pp. 64–66. As Hoffman points out (Le nègre romantique, pp. 16–18) and I have tried to show, these vague negative associations were counterbalanced by positive images of the black Africans linked to the legendary Prester John and the "Ethiopian" among the Magi.

[129]Katherine George, "The Civilized West Looks at Primitive Africa, 1400–1800: A Study in Ethnocentrism," Isis (1958), 62–72.

[130]Harold R. Isaacs, "Blackness and Whiteness," Encounter 21/2 (1963), 8–21; P. J. Heather, "Color Symbolism," Folk-Lore 59 (1948), 165–83; and Kenneth Gergen, "The Significance of Skin Color in Human Relations," Daedalus 96/2 (1969), 390–406.

who depicted the devil as white and believed the light-skinned Europeans to be monsters, while Jan Linschoten related that the Africans believed themselves to be the proper color and whites counterfeit copies of humans.[131] Early European observers did not know African languages or customs sufficiently to understand that the Ibo, for example, referred to leprosy in polite speech as "the white skin",[132] and no writer, as far as I am aware, realized before the late eighteenth century that white is associated with death and mourning in many Asian cultures.[133] Nonetheless, only the most arrogant and insensitive of European travelers had illusions about African and Asian admiration for their physical appearance.

Although many authors in recent decades have stressed color or physical differences in accounting for the origins of racial prejudice, my reading of early travel narratives suggests that interpretations emphasizing the social and historical contexts in which racist attitudes emerge may be nearer the mark. One side of this contextual approach, which has been ably studied, focuses on the slave trade and the conditions under which Africans were transported and employed as forced labor in the New World or enslaved within parts of Africa itself.[134] Another side, which has received less attention, involves European perceptions of Africans living in their own societies and cultures. It is clear that early European travelers found African cultures much less developed than the Indian or Chinese in virtually all institutions and endeavors, from political systems and marriage customs to handicraft manufacturing and conceptions of the supernatural or of time and space. The fact that the peoples of Africa and the New World came to be perceived as sav-

[131]Despite his emphasis on physical differences as the key to unfavorable English views of the Africans in this period, Jordan concedes that many travelers were aware of the Africans' preference for their own dark skin and some visitors averred that the Africans associated white skin with the devil (*White over Black*, pp. 9–11). For evidence from sixteenth- and seventeenth-century accounts, see "Letter of Antoine Malfante," in Crone, *Voyages of Cadamosto*, p. 89; Heylyn, *Microcosmos*, p. 678; Jan Huyghen von Linschoten, *Voyage to the East Indies* (London, 1885), vol. 1, p. 271; Boogaart, "Colour Prejudice," p. 45; and Bitterli, *Schwarzen Africaners*, p. 50. For negative Indian responses to pale Europeans, see Careri in Sen, *Indian Travels*, p. 246; and Edward Terry, "A Relation of a Voyage to the Eastern Indies," in S. Purchas, *Purchas, His Pilgrims* (London, 1625), pt. 2, p. 1473.

[132]Chinua Achebe, *Things Fall Apart* (Greenwich, Conn., 1959), p. 71.

[133]Abbé Grosier, *A General Description of China* (London, 1788), vol. 2, p. 301.

[134]For an overview of this approach, see George M. Fredrickson, "Toward a Social Interpretation of the Development of American Racism," in Nathan I. Huggins, Martin Kilson, and Daniel M. Fox, *Key Issues in the Afro-American Experience* (New York, 1971), pp. 240–54; and for an application of it to South Africa and the United States, Fredrickson, *White Supremacy* (Oxford, 1981). See also earlier essays by Carl Degler in *Comparative Studies in Society and History* 2 (1959); Oscar and Mary Handlin in *William and Mary Quarterly* 7 (1950); and Edmund S. Morgan in *American Slavery: American Freedom* (New York, 1975), chap. 16.

ages,[135] while those of India and China were considered as civilized as the Europeans themselves, is indicative of the gulf that had already opened, not just between Europeans and Africans but also between Africans and Asians in the European mind. European categorizations of some peoples as savages or barbarians and others as civilized had much less to do with narcissistic disdain for extreme differences in physical appearance than with ethnocentric perceptions of levels of sophistication in social organization and cultural development generally. The roles of science and technology in shaping these perceptions were secondary; they were facets of assessments of the material culture of non-Western peoples as a whole. But tools and cannons and conceptions of space and time were for many early European observers among the most tangible means of distinguishing civilized peoples from savages and barbarians.

[135]W. G. L. Randles, *L'image du sud-est Africain dans la littérature européenne au XVIe siècle* (Lisbon, 1959), pp. 151–54; Hodgen, *Early Anthropology*, pp. 195–201, 315, 361–65; and Boogaart, "Colour Prejudice," pp. 46, 53–54.

CHAPTER 2

The Ascendancy of Science:
Shifting Views of Non-Western Peoples
in the Era of the Enlightenment

IN THE last decades of the seventeenth and the first half of the eighteenth century, the curiosity of small circles of missionaries and scholars, sea captains and merchants about the new worlds discovered by European overseas exploration swelled into a passion for news of expeditions to "exotic" lands and information about the institutions, customs, and beliefs of non-Western peoples. Travel accounts, imaginary as well as real,[1] enjoyed a popularity among western Europe's steadily growing educated classes unequaled in any other era. As Robin Hallett has observed, only theological works were in greater demand; he estimates, for example, that four times as many books were published on Africa between 1700 and 1750 as in all of the seventeenth century.[2] But more than quantitative increase distinguished the travel and missionary accounts of this period from earlier works of this genre. They were on the whole more informed and detailed, more accepting of cultural differences between Europeans and overseas peoples, and less judgmental than earlier writings. In addition, the meaning of travel

---

[1]Daniel Defoe's *Robinson Crusoe* (1719) is perhaps the most famous work of this imaginative genre. For a survey of its French counterparts, see Geoffroy Atkinson, *The Extraordinary Voyage in French Literature from 1700 to 1720* (Paris, 1922).

[2]Robin Hallet, *The Penetration of Africa* (New York, 1965), pp. 39, 137, 146. For more detailed discussions of the boom in travel literature in various national settings, see Pierre Martino, *L'Orient dans la littérature française au XVIIe et au XVIIIe siècle* (Paris, 1906), pp. 39–40, 53–60; and Norman Hampson, *The Enlightenment* (New York, 1982), pp. 71, 106–7. The cosmopolitan spirit of the age was nicely expressed by Samuel Johnson in his 1759 fantasy account of Ethopia, *The History of Rasselas, Prince of Abyssinia* (in *Shorter Novels: Eighteenth Century*, London, 1930, p. 25). In response to a question from the prince, the poet-philosopher Imlac, replies, "We grow more happy as our minds take a wider range."

literature and the uses to which it was put were transformed in major ways in the era of the Enlightenment. The earlier emphasis (understandable, given the ignorance of the Europeans and the accelerating pace at which new lands were discovered) on gathering and disseminating information gave way to a focus on interpretation and application of the accounts of overseas societies.

Although Jesuit letters and numerous travelers' accounts were mined for data, the eighteenth-century vision of non-Western societies was molded, disputed, and revised primarily by writers in Europe itself. These authors made selective and often highly questionable use of materials on overseas societies to construct models for Europeans to emulate, to buttress critiques of European beliefs and institutions, and to provide ammunition for the philosophical and policy debates that raged throughout the eighteenth century. Views of African culture, favorable and unfavorable, fueled the abolitionist assault on the slave trade and the often equally vigorous counterthrusts of the pro-slavery party. Noble savages, whether Amerindian, Tahitian, or (less commonly) African, living an imagined idyllic existence in communion with nature and their natural instincts, were fabricated to expose the decadence and corruption of an allegedly overdeveloped, suffocatingly mannered and refined European civilization. Imaginary visitors from rival civilizations, such as Montesquieu's Persian travelers and Horace Walpole's Chinese philosopher, provided superb vehicles for social satire and political commentary.

Voltaire, whose wide-ranging genius epitomized the age, made extensive use of accounts of Chinese and to a lesser extent Indian culture, freely acknowledging his debt to Jesuit writings on both civilizations. Though these writings were perhaps the best informed of the period, they were heavily influenced by the Jesuit missionaries' need to defend their strategies of proselytization against the bitter denunciations of their Franciscan and especially their Dominican rivals.[3] Voltaire freely refashioned the already distorted images of China and India that emerged from the *Lettres édifiantes* (see below) and J. B. Du Halde's monumental history of China to suit his own designs as polemicist, philosopher, historian, and man of letters. On the basis of highly suspect evidence he exaggerated the antiquity of the Indian and Chinese civilizations in order to cast doubt on Judeo-Christian chronologies. He also refashioned the

---

[3]Virgile Pinot, *La Chine et la formation de l'esprit philosophique en France, 1640–1740* (Paris, 1932), remains the best account of these struggles in China and Siam. For India, see Vincent Cronin, *Pearl to India: The Life of Robert Di Nobili* (New York, 1958).

monistic Hindu concept of the divinity into a God very like that of the Deists.[4]

Voltaire's practices were widely adopted by lesser writers such as N. Le Clerc, who went so far as to "invent" a work on physics by an imaginary disciple of Confucius, intended to spur on the reform efforts of a Russian nobleman.[5] In the last decades of the century, images of China and India varied widely, depending on whether they were being championed as model states by the Physiocrats and defended by the Orientalists or excoriated for excessive bureaucracy and commercial restrictions by Captain Anson and Laurent Lange and for corrupt institutions and wicked superstitions by the Utilitarians and Evangelicals. Whatever side of a debate they were on, philosophes and government officials alike reshaped the available facts to suit their own arguments and policies.

Whether fashioned by philosophes in Europe or travelers and missionaries in foreign lands, the varied and often contradictory images of African and Asian peoples and cultures that vied for acceptance by the educated classes in the eighteenth century were much more influenced by scientific standards than European attitudes had been in the first centuries of expansion. The many scientific breakthroughs that culminated in Isaac Newton's experiments and writings on optics, mechanics, and mathematics in the last decades of the seventeenth century left little doubt among the educated that a decisive break with the past had occurred. This realization appeared to be repeatedly confirmed by the less spectacular but numerous and important scientific discoveries of the eighteenth century.[6]

Widely discussed and debated in England soon after their publication, Newton's works were also known to small circles of French, German, and Dutch thinkers, including such prominent figures as Huygens and Leibniz, from the 1670s onward. During the first decades of the eighteenth century, the power of the Newtonian combination of empiricism and the application of mathematics to scientific questions became in-

[4]Antonin Debidour, "L'indianisme de Voltaire," *Revue de littérature comparée* 4 (1924), 28, 35–40; Pinot, *La Chine*, pp. 314–27; Basil Guy, *The French Image of China before and after Voltaire* (Geneva, 1963), pp. 242–44.

[5]Etiemble, "De la pensée chinoise aux 'philosophes' français," *Revue de littérature comparée* 30 (1956), 468. For additional examples, see Guy, *French Image of China*, pp. 431–32.

[6]For a veritable catalogue of scientific discoveries in the eighteenth century, see A. Wolf, *A History of Science, Technology, and Philosophy in the 18th Century*, vol. 1 (New York, 1952). For a discussion of the philosophes' awareness of the radical transformations wrought by scientific developments, see Ernst Cassirer, *The Philosophy of the Enlightenment* (Princeton, N.J., 1968), esp. pp. 3–5, 9–12.

creasingly apparent to scholars both in England and on the Continent. From the first years of the century, French savants had traveled to England to observe first-hand the nation that had produced Newton as well as Locke and Boyle. The implications of Newton's work for the Cartesian system of natural philosophy were widely argued, and such scholars as Malebranche and his disciples struggled to reconcile the theories of Descartes with Newtonian findings. Voltaire's *Letters on the English Nation* (1734) and *Elements of Newton's Philosophy* (1738), along with Maupertuis's *Discourse on the Shape of the Stars* (1732), led to Continental recognition of Newton's work beyond the circles of scientists and philosophes where it was already familiar. By the late 1730s Francesco Algarotti's *Newtonianism for the Ladies* and similar works were popularizing the empirical and mathematical approach to scientific thinking and disseminating a sense of the unprecedented nature of Europe's achievements in the sciences.[7]

The philosophes' awareness of the magnitude of the changes set in motion by discoveries in the sciences and their centrality to intellectual discourse in the eighteenth century is vividly summarized in the following passage from the *Elements of Philosophy* by Jean d'Alembert. As one of the original editors of the *Encyclopedie*, that mammoth compendium of eighteenth-century knowledge, D'Alembert was in a superb position to judge.

> Natural science from day to day accumulates new riches. Geometry, by extending its limits, has borne its torch into the regions of physical science which lay nearest at hand. The true system of the world has been recognized. . . . In short, from the earth to Saturn, from the history of the heavens to that of insects, natural philosophy has been revolutionized; and nearly all fields of knowledge have assumed new forms. . . . the discovery and application of a new method of philosophizing, the kind of enthusiasm which accompanies discoveries, a certain exaltation of ideas which the spectacle of the universe produces in us; all these causes have brought about a lively fermentation of minds. Spreading throughout nature in all directions, this fermentation has swept with a sort of violence everything before it which stood in its way, like a river which has burst its dams.[8]

---

[7]In recent years the early views of Pierre Brunet on the dissemination of Newton's ideas and methodology have been considerably revised. This summary of Newton's reception on the continent is based primarily on Henry Guerlac, *Newton on the Continent* (Ithaca, 1981); A. R. Hall, "Newton in France: A New View," *History of Science* 13 (1975), 233–50.

[8]Quoted in Cassirer, *Philosophy of the Enlightenment*, pp. 46–47.

The circle of potential readers for works dealing with the new discoveries was admittedly small, as evidenced by Edmund Burke's famous estimate that as late as 1789 "serious readers" in England totaled only about 80,000 persons.[9] Limited access to education, especially advanced education, and the high cost of books kept the audience for scholarly scientific works much smaller than the readership of satirical pamphlets or low-brow romances. Yet the publication of relatively inexpensive popular encyclopedias that sold in the hundreds of thousands, the establishment of lending libraries, and the proliferation of literary societies as well as salons and the better sorts of coffeehouses, which became centers for intellectual exchange, spread the new learning to a readership that expanded remarkably over the course of the century. The high level of participation by amateur experimenters and collectors in the scientific endeavors of this period and the persistence of a common culture that encompassed both the sciences and the fine arts meant that scientific knowledge was, as D. G. Charlton put it, "a subject of deep interest to the educated public as a whole, not an esoteric mystery as it has increasingly become."[10] Physicians, for example, who made up a substantial portion of the nonspecialist investigators who had dominated scientific inquiry for centuries, conversed in the same salons and debated in the same societies as the leading philosophers and writers of the day. Often, as in the case of Quesnay, Jaucourt, and La Mettrie, physicians were themselves influential philosophes.

Lawyers, physicians, bureaucrats, merchants, clerks, and even skilled artisans joined the nobility and clergy as avid readers of both popularized accounts of scientific investigations and narratives of the discovery of the world beyond Europe. The two kinds of works combined to reinforce the rather impressionistic claims of writers like Bernier and Fryer that the Europeans had advanced far beyond all other peoples and civilizations in the sciences. They had explored realms, posed questions, and gained knowledge of the natural world that was unimaginable for the priests and scholar-administrators of other civilizations. The improvements that such popularizers as Jean Fernel and Jan Stradanus had celebrated in the mid-sixteenth century were increasingly viewed as

[9]Peter Gay, *The Enlightenment: An Interpretation* (London, 1979), vol. 2, pp. 59–60. The following discussion on the development of the channels through which ideas spread in the eighteenth century is based on ibid., pp. 58–69, and vol. 1, pp. 176–78; John Lough, *An Introduction to Eighteenth Century France* (New York, 1960), chap. 7; Robert Darnton, *The Literary Underground of the Old Regime* (Cambridge, Mass., 1982), pp. 25, 135–47, 173–83; Hampson, *Enlightenment*, pp. 152–53; and A. R. Hall, *The Revolution in Science, 1500–1750* (London, 1983), pp. 230–35.

[10]D. G. Charlton, *New Images of the Natural in France* (Cambridge, Eng., 1984), pp. 66–67, 79.

unique to Europe.[11] This sense of Europe's singularity is vividly captured in the response of the pilgrim-poet Imlac to the questions of the Ethiopian prince who is the protagonist of Samuel Johnson's 1759 novella, *The History of Rasselas*:

> In enumerating the particular comforts of life, we shall find many advantages on the side of the Europeans. They cure wounds and diseases with which we languish and perish. We suffer inclemencies of weather which they can obviate. They have engines for the dispatch of many laborious works, which we must perform by manual industry. There is such communication between distant places, that one friend can hardly be said to be absent from another. Their policy removes all public inconveniences; they have roads cut through their mountains, and bridges laid upon their rivers. And, if we descend to the privacies of life, their habitations are more commodious, and their possessions are more secure.[12]

More significantly perhaps for the thinkers of an age that so revered classical learning, especially that of the Romans,[13] eighteenth-century Europeans were confident that largely because of advances in the sciences they had surpassed all other civilizations, past or present. The controversy between the defenders of ancient learning and the advocates of modern thought, which had been a central intellectual concern in the seventeenth century, ended with a decisive victory for the "Moderns." The widely held scholastic conviction, that medieval scholars were no more than dwarfs who stood on the shoulders of the classical giants, was revived but with a crucial difference: eighteenth-century thinkers such as Bernard Fontenelle, the great French popularizer of scientific ideas and long-time secretary of the Académie des Sciences, emphasized the greater distance that the Moderns could see from their perch on the shoulders of the colossi of antiquity, particularly in the realm of natural philosophy.[14]

[11]See the discussion of Fernel's *De Abditis Rerum Causis* (1548) in Charles Sherrington, *The Endeavour of Jean Fernel* (Cambridge, Eng., 1946), pp. 16–17; and the engravings of Stradanus's *Nova Reperta* (c. 1590) that were reprinted by the Burndy Library (Norwalk, Conn.) in 1954.

[12]Johnson, *Shorter Novels*, p. 25.

[13]Gay, *The Enlightenment*, vol. 1, chap. 2.

[14]Hans Baron ("The Querelle of the Ancients and Moderns as a Problem for Renaissance Scholarship," in P. O. Kristeller and P. P. Weiner, eds., *Renaissance Essays* [New York, 1968], pp. 95–114), emphasizes the earlier shift to this view on the part of some Renaissance thinkers who were impressed with the quickening pace of invention and overseas discovery as well as the beginnings of the Scientific Revolution. On Fontenelle, see L. Marsak, *Bernard de Fontenelle: The Idea of Science in the French Enlightenment* (Philadelphia, 1959), esp. pp. 10–12, 46–49; and Gay, *The Enlightenment*, vol. 2, pp. 124–25.

The influence of scientific thinking on the writers who shaped European attitudes toward non-Western cultures in the eighteenth century was manifested in a variety of ways. The accounts of overseas travelers took on added importance as one form of the empirical evidence that eighteenth-century thinkers were convinced would enable them to undertake the "scientific" study of human societies.[15] Travelers such as G. de la Galaisière le Gentil and Michael Adanson prided themselves on their membership in or contacts with prominent members of the Académie des Sciences, and participants in the Jesuit mission to China were often corresponding members. Galaisière traveled primarily to make astronomical observations from various locations overseas; Adanson (more typically) filled the account of his trip to Senegal with detailed descriptions of geology, and flora and fauna, astronomical observations, and temperature readings. In the letters of savants such as J. S. Bailly and Voltaire and missionaries such as Dominique Parennin, frequent comparisons were made between the scientific achievements of the Europeans and those of the Indians and Chinese. Sir William Jones, who played a pivotal role in the European discovery of Asian learning in the late eighteenth century, was considered a "man disciplined in the school of science" by his early nineteenth-century biographer, Lord Teignmouth. Jones, whose father was a tutor in mathematics and a friend of Newton and Halley, read widely in chemistry and medicine throughout his life. Both in England and while he was in India in the service of the East India Company, Jones assiduously collected botanical specimens and recorded astronomical observations, in addition to displaying proficiency in several branches of mathematics.[16]

References to the specific findings and general achievements of European scientists abound in the works of the French philosophes who evaluated Indian and Chinese scientific learning (still broadly defined to include such subjects as grammar and ethics)[17] throughout the early

[15]For discussions of the Enlightenment emphasis on empiricism and experiment, see Gay, *The Enlightenment*, vol. 1, pp. 135, 141-42; Gladys Bryson, *Man and Society: The Scottish Inquiry of the Eighteenth Century* (Princeton, N.J., 1945), pp. 21–22, 27; Cassirer, *Philosophy of the Enlightenment*, pp. 10–11, 42, 46; and Hampson, *Enlightenment*, pp. 84–85.

[16]G. de la Galaisière le Gentil, *Voyage dans les mers de l'Inde, 1761–1769* (Paris, 1779), title page and intro.; Michael Adanson, *A Voyage to Senegal* (London, 1759), pp. 4, 19–20, 154–55, 235, 253; M. D. Mairan, *Lettres au R. P. Parennin contenant diverses questions sur la Chine* (Paris, 1770), esp. 7–13, 60–61; Jean-Sylvain Bailly, *Lettres sur l'origine des sciences et sur celle des peuples de l'Asie* (Paris, 1777); Lord Teignmouth's biography of Sir William Jones in Jones, *Works* (London, 1807), esp. vol. 1, pp. 204–52, and vol. 2, 295ff.; S. N. Mukherjee, *Sir William Jones* (Cambridge, Eng., 1968), pp. 17–18; and Guerlac, *Newton*, pp. 104–5.

[17]Bryson, *Man and Society*, pp. 15–17; G. N. Clark, *Science and Social Welfare in the Age of Newton* (Oxford, 1937), pp. 118–19. For examples from the Enlightenment era, see R.

decades of the eighteenth century. Comparisons of ideas in astronomy, mathematics, and medicine in Europe and Asia also held a prominent place in the writings of the British Orientalists, who sought to make Indian scientific learning known to Western scholars at the end of the century. Some travelers challenged the findings of European scientists on the basis of their overseas observations. Paul Isert, for example, was certain that the specimens of flora and fauna he had examined in Africa would pose serious problems for the classification schemes proposed by biologists and zoologists who had not traveled outside of Europe.[18]

Often, topics of special concern to overseas observers were influenced by recent advances in Europe. The increased length of many travelers' discussions of medical diagnosis and treatment in non-European lands provides a case in point. Though mortality rates at all social levels in Europe remained appallingly high by modern standards, and quacks of many varieties continued to pass as healers, the eighteenth century did witness significant advances in medicine. These included the brilliant work done in anatomy and physiology by such physicians as Giovanni Morgagni, the liberation of surgeons from the barbers' guilds in England and France in the 1740s, and increased efforts to legislate proper training and licensing for practicing physicians. The century-long struggle to rid Europe of the scourge of smallpox was reflected in detailed accounts of Asian immunization techniques. The attention given by European travelers in this era to Indian and Chinese learning in mathematics also increased dramatically, reflecting the essential roles that the various branches of mathematics had played in the Scientific Revolution of the seventeenth century. European commentary on Asian physics and chemistry, by contrast, in which advances had been more modest and theoretical propositions were less secure (as evidenced by the widespread acceptance of Stahl's phlogiston explanation for fire until late in the century)[19] was for the most part vague and superficial. Travelers' observations about these nascent disciplines usually amounted to little more than bald assertions that non-European peoples knew little or nothing about them.

P. Intorcetta's *La science des chinois* (Paris, 1673), which is largely on politics and moral philosophy; the discussions of Asian "sciences" in the *Lettres édifiantes* cited in n. 30 below; and Sir William Jones's lectures on the sciences in Asia (see nn. 78, 88).

[18]Paul Isert, *Voyages en Guinée et dans les Iles Caraïbes en Amerique* (Paris, 1793), pp. 53, 60. For other references to scientific work, see Dominique Lamiral, *L'Afrique et le peuple affriquain* (Paris, 1789), p. 101; and Thomas Winterbottom, *An Account of the Native Africans in the Neighborhood of Sierre Leone* (London, 1803), vol. 1, pp. 198–99, 207, and vol. 2, pp. 254–56, 275–77.

[19]Stephen Mason, *A History of the Sciences* (New York, 1962), chap. 26; and D. S. L. Cardwell, *The Organisation of Science in England* (London, 1972), pp. 16–17.

Many eighteenth-century writers on non-Western societies claimed that their observations and conclusions had been based on research conducted according to scientific standards. Voltaire, for example, aspired to scientific accuracy in his writing of history and his research into the antiquity of Indian astronomy and Chinese civilization.[20] Edward Long and Charles White sought to give scientific respectability to their discourses on African or "Negro" inferiority through references to the purported laws of breeding and propagation, the use of such (elementary) scientific terminology as genus and species, and exercises (of differing degrees of sophistication) in comparative anatomy. Though what were held to be scientific techniques for ranking the varieties of humankind were not widely employed or accepted until the nineteenth century, such writers as Long and White demonstrated what potent weapons allegedly scientific investigations and findings might be in arguing the case for white superiority.

Even though the pace of invention accelerated rapidly in the eighteenth century, technological achievement was far less important than scientific advance in shaping European attitudes toward African and Asian societies. The "cluster of innovations"[21] central to the process that has come to be known as the Industrial Revolution—from Newcomen's steam engine (1712) and Kay's flying shuttle (1733) to Arkwright's "water frame" spinning jenny (1769) and Watt's improved steam engine (1776)—was devised for the most part by eighteenth-century artisans and engineers. New machines and sources of power made it both possible and increasingly profitable to concentrate production in factories, though the bulk of the nonagrarian laboring force even in nations as advanced as Britain and France was employed in households and small workshops until well into the nineteenth century.[22] The tools employed by these artisans changed little in most forms of manufacturing in the eighteenth century, and except in parts of England there was little improvement in agricultural implements and cropping techniques.[23] In

[20]Debidour, "L'indianisme de Voltaire," pp. 30, 34–35, 38–39.

[21]The term was coined by J. A. Schumpeter and is applied by Phyllis Deane in her fine discussion of technological changes: *The First Industrial Revolution* (Cambridge, Eng., 1965), chaps. 7 and 8.

[22]Eric P. Lampard, "The Social Impact of the Industrial Revolution," in Melvin Kranzberg and Carroll Pursell, eds., *Technology in Western Civilization* (London, 1967), vol. 1, pp. 317–18. For more detailed discussions of the persistence of artisans in the British working class, see I. J. Prothero, *Artisans and Politics in Early Nineteenth-Century London* (Folkestone, Kent, Eng., 1979); and Raphael Samuel, "Workshop of the World: Steampower and Hand Technology in Mid-Victorian Britain," *History Workshop* 3 (1977), 6–72.

[23]Hermann Kellenbenz, "Technology in the Age of the Scientific Revolution, 1500–1700," in Carlo Cipolla, ed., *The Fontana Economic History of Europe* (Glasgow, 1974),

addition, many eighteenth-century inventions especially in textile and iron production amounted to improvements on technology that had been evolving since medieval times, rather than radically new departures. Water mills, for example, remained the major nonanimate source of energy until the nineteenth century. The steam engine, which did represent a genuine leap—both conceptually and operationally—in power-generating technology, was used for only a very narrow range of tasks (chiefly for pumping water out of coal mines) through most of the eighteenth century. The improvements of Watt and others which greatly expanded the applications for which steam power could be harnessed came only in the last decades of the century.[24]

Therefore, "the sciences," in which dramatic advances had been occurring for nearly two centuries, loomed much larger than "the arts" or technology in the thinking of such authors as Voltaire, Sir William Jones, and David Hume, whose writings strongly influenced European views on Asia and Africa at various points in the century. Technology was of less consequence for most eighteenth-century writers because breakthoughs in production and communications were largely contemporary and had yet to make their impact on French and British societies as a whole. The industrial transformations that would sharply reduce the European demand for Indian and Chinese textiles and other handicraft products and establish the West as the manufacturing center of the world were under way, but these momentous shifts did not become obvious until the first decades of the nineteenth century. Except in seapower, the advantages in weaponry and communications enjoyed by eighteenth-century Europeans over African and Asian adversaries were marginal at best. The land conquests that were extended in Java and begun in India at midcentury owed much more to superior leadership and organization and to the quarrels of the Europeans' Asian rivals than to the speed or firepower of the Europeans' armies.[25] Thus, the Euro-

---

vol. 2, pp. 178–79, 218, 264–65; and Robert Muchembled, *Popular Culture and Elite Culture in France, 1400–1750* (Baton Rouge, La., 1985), p. 15. On persistence and change in British agriculture in the eighteenth century, see G. E. Fussell, "Science and Practice in Eighteenth-Century British Agriculture," *Agricultural History* 43/1 (1969), 7–18; Deane, *First Industrial Revolution*, chap. 3; and Peter Mathias, *The First Industrial Nation* (New York, 1969), pp. 64–80.

[24]Samuel Lilley, "Technological Progress and the Industrial Revolution, 1700–1914," in Cipolla, *Fontana Economic History*, vol. 3, esp. pp. 190, 203. For an extended discussion of improvements in waterpower generation, see D. S. L. Cardwell, "Power Technologies and the Advance of Science, 1700–1825," *Technology and Culture* 6/2 (1965), 188–207, esp. 192–93.

[25]Clark, *Science and Social Welfare*, pp. 109–111; H. J. De Graaf, *Geschiedenis van Indonesië* (The Hague, 1949), sec. 3; Geoffrey Parker, *The Military Revolution* (Cambridge, Eng., 1988), pp. 128–36; Philip Mason, *A Matter of Honour: An Account of the Indian Army*

peans' assumption that they had gained an unprecedented understanding of the workings of the natural world, rather than the conviction that they had as yet put these discoveries to practical use, vied with or joined religion as the main proof of their own superiority and the chief measure of the achievements and potential of non-Western peoples.

Model of Clay: The Rise and Decline of Sinophilism in Enlightenment Thought

Although it mattered little to the Chinese what the "hairy barbarians" from western Europe thought one way or the other, no culture or civilization has been as lavishly praised or as widely acclaimed as a model to be emulated as was Qing China in the first half of the eighteenth century.[26] Gardens in the Chinese style complete with ornate pagodas, such as those at Kew and the Tivoli, were à la mode. Chinese porcelains, both the imported real thing and imitations made at Delft, Meissen, and Saint-Cloud, graced the lacquered cabinets and cupboards of the rich and trend-conscious. In the early decades of the century such leading artists as Watteau and Boucher employed Chinese motifs in their paintings and engravings, and the influence of the Chinese was apparent in the furnishings and wall hangings of palaces and fashionable homes and the ornamental flourishes of rococo architecture. But the rage for *chinoiserie* went far beyond latticed garden houses and themes for theatrical works. Some of the most prominent thinkers of the age, including Leibniz, Voltaire, and Quesnay, looked to China for moral instruction, guidance in institutional development, and supporting evidence for their advocacy of causes as varied as benevolent absolutism, meritocracy, and an agriculturally based national economy.

The China that inspired the vogue of *chinoiserie* and the admiration of the philosophes was largely a creation—partly real, partly imagined—of generations of Jesuit missionaries, whose writings remained the main

---

(Harmondsworth, Eng., 1976), pp. 39–40. For a good overview of military advances which stresses organization and discipline, see William H. McNeill, *The Pursuit of Power* (Chicago, 1984), chap. 4.

[26]There is a large literature on the vogue of China and *chinoiserie* in the eighteenth century. Adolf Reichwein's *China and Europe: Intellectual and Artistic Contacts in the Eighteenth Century* (London, 1925), remains a useful introduction despite some factual problems; cf. R. C. Bald, "Sir William Chambers and the Chinese Garden," *Journal of the History of Ideas* 11 (1950). A more recent survey of artistic and decorative influences is provided by Madelaine Jarry, *Chinoiserie* (New York, 1981), and there are fine insights in Raymond Koechlin "La Chine en France au XVIIIᵉ siècle," *Gazette des Beaux-Arts* ser. 4, 4 (1910), 89–103. For the impact of China on literature, esp. that of France, see Martino, *L'Orient dans la littérature française;* and Guy, *French Image of China*, which supersedes all other works.

source of information on the "Celestial Empire" until the late eighteenth century. As the servants of a succession of Chinese emperors, the Jesuits came to view Chinese society in a generally positive, at times even excessively flattering, light. Their criticisms of China tended to be muted or repressed in the face of challenges by Dominicans and Jansenists to their strategy of proselytization through accommodation. These rivals saw a good deal less in China to praise and preserve and much more to condemn and uproot in the name of Christian conversion. Jesuit writings on China, as well as those of their adversaries, became particularly prominent in the late seventeenth and early eighteenth centuries as the controversy intensified over whether Chinese ancestral rites ought to be considered an admirable social custom (as the Jesuits argued) or condemned as a pagan religious ritual (as their rivals insisted).[27] The overall Jesuit view of China, which shared much with the awestruck impressions of early visitors, was exuberantly summarized by Gabriel Magaillans, who served as a missionary in China in the 1640s and whose "New History" of the Qing empire was published in several European languages in the late seventeenth century: "China is a country so vast, so rich, so fertile, and so temperate; the Multitude of the People so infinite, their Industry in Manufacture, and their Policy in Government so extraordinary, that it may truly be said, that ever since the undertaking of Long Voyages, there was never any Discovery made, that might stand in Competition with this."[28]

Such plaudits were given added credibility in the last decades of the seventeenth century by reports of the flourishing state of China under the great Qing emperor Kang-xi, who ruled from 1662 to 1722.[29] Though Qing China would reach the peak of its power and prosperity during the equally long reign of the Qien-Long emperor (1736–95), in the late seventeenth century its size and wealth, massive bureaucracy and splendid court, bustling cities and well-cultivated fields continued to evoke admiration on the part of most European visitors. China's reputation was greatly enhanced by the emperor Kang-xi himself, who

[27]See Arnold Rowbotham, *Missionary and Mandarin* (Berkeley, Calif., 1942), chaps. 9–12; and Pinot, *La Chine*, esp. chap. 2.

[28]Gabriel Magaillans, *A New History of China* (London, 1688), p. 172. For similar views from the first decades of the eighteenth century, see E. de Silhouette, *Idée générale du gouvernement et de la morale des chinois* (Paris, 1731); and J. B. Du Halde, *The General History of China*, 4 vols. (London, 1736).

[29]For recent studies of this era, see Jonathan Spence, *Emperor of China: Self-Portrait of K'ang-hsi* (New York, 1974); Lawrence Kessler, *K'ang-Hsi and the Consolidation of Ch'ing Rule, 1661–1684* (Chicago, 1976); and Frederic Wakeman, Jr., *The Great Enterprise: The Manchu Reconstruction of Imperial Order in the Seventeenth-Century*, 2 vols. (Berkeley, Calif., 1985).

was depicted by Jesuit writers as a veritable philosopher-king, devoted to his subjects' welfare and deeply interested in the fine arts and sciences, both Chinese and Western.[30] Kang-xi's so-called edict of toleration in 1692 particularly caught the attention of Bayle, Leibniz, and Voltaire, who like virtually all the philosophes deeply detested religious bigotry and persecution. Even though few dared to make the comparison explicit, the contrast between Kang-xi's religious policies and Louis XIV's revocation of the Edict of Nantes in 1685—with the consequent renewal of religious strife in France and neighboring states—strengthened the arguments of those who sought to defend Chinese political wisdom and ethical probity.[31]

Yet the praise of even the most enthusiastic of Sinophiles was nearly always qualified. Leibniz, for example, repeatedly urged his fellow Europeans to learn from China's example in matters of law, government, and morality, but he left no doubt of his conviction that the West was far ahead of China in theoretical thinking, whether in philosophy or in the sciences. Leibniz dreamed of a scientific mission to China and firmly believed that once the superiority of the Christian West in speculative thought had been demonstrated, the way would be open for the conversion of the Chinese to Christianity.[32] He saw Europe and China as equals with much to exchange, but other writers were less generous. Though recent works on the impact of China on Europe in the seventeenth and eighteenth centuries have either scarcely noted or altogether ignored them,[33] criticisms of Chinese attainments in mathematics, astronomy, and anatomy, for example, increasingly shaped the attitudes of Enlightenment thinkers toward Chinese civilization as a whole.

Louis Le Comte's controversial *Nouveaux mémoires sur la Chine* (1696) contained one of the earliest published critiques of Chinese scientific learning and one of the most detailed to appear before the nineteenth century. The attention that Le Comte's work received at the time of its publication was largely due to his outspoken opinions on religious issues

[30]*Lettres édifiantes et curieuses écrites des missions étrangères par quelques missionnaires de la Compagnie des Jésus*, ed. J. B. Du Halde (Paris, 1707–73), vol. 7, pp. 186–92; vol. 14, pp. xii–xiii; and vol. 16, pp. 346–49.

[31]For a discussion of the importance of tolerance in the thinking of Pierre Bayle and other early philosophes, see Paul Hazard, *The European Mind, 1680–1715* (Cleveland, 1967), esp. chaps. 4 and 5; on this theme in the "high" Enlightenment, see Daniel Mornet, *Les origines intellectuelles de la révolution française, 1715–1787* (Paris, 1967), pp. 22–23, 30, 39–41. For the impact of Kang-xi's attitudes in particular, see R. F. Merkel, *Leibniz und China* (Berlin, 1952), pp. 5–6, 11, 15 , 22; and Donald Lach, "Leibniz and China," *Journal of the History of Ideas* 6/4 (1945), 440–41.

[32]Merkel, *Leibniz und China*, esp. pp. 14–16, 21–26, 29; and Lach, "Leibniz and China," pp. 441–43, 446–47, 454–55.

[33]Except David Landes, *Revolution in Time* (Cambridge, Mass., 1983), chaps. 1 and 2.

with bearing on the controversy over Chinese ancestral rites.[34] But his sweeping and often harsh appraisal of the science and to a lesser extent the technology of the Chinese had a much more lasting impact on the European image of the Middle Kingdom. Le Comte, who was a member of the first party of French Jesuits sent to China as well as one of Louis XIV's court mathematicians, repeated the customary accolades to China's size and wealth, large cities, and admirable bureaucracy. From the outset, however, he openly displayed his annoyance at what he considered the conceit of the Chinese, which he charged was based on a highly inflated estimate of their own achievements and a tendency to devalue those of other peoples. He complained that even after his Jesuit colleagues had again and again demonstrated the superiority of European astronomical calculations and mathematical techniques, the Chinese continued to adhere to their antiquated procedures and instruments.[35] Like virtually all earlier observers, Le Comte had a high regard for Chinese manufactures, such as porcelain and silk textiles. But he included in his account lengthy critiques of Chinese instruments, especially horological and astronomical, and he made no secret of his low regard for Chinese tools in general. He asserted that the Chinese had not begun to develop mining techniques worthy of their empire's great mineral wealth. He also made clear his preference for European—specifically, Parisian—artisans over those of China. He believed the latter incapable, even with European supervision and conceding their propensity for hard work, of producing instruments or tools that would measure up to European standards.[36]

Le Comte's criticisms of certain aspects of Chinese technology were rather reserved compared to his blanket dismissal of whole categories of their scientific endeavor. Leveling a series of charges that would become a standard feature of later works, he declared that the Chinese knew little physics, that their anatomy was mostly wrong and their geometry superficial, and that their astronomy was ancient but deeply flawed. He conceded that the Chinese were skilled in practical arithmetic calculations. He also lauded their techniques of medical diagnosis, specifically citing their knowledge of pulse-taking. However, he questioned the efficacy of their modes of treatment, noting, for example, that they regarded the Western practice of bloodletting as barbaric.[37] But his

[34]Rowbotham, *Missionary and Mandarin*, pp. 141–47; Martino, *L'Orient dans la littérature française*, pp. 126–27.

[35]Louis Le Comte, *Memoirs and Remarks Made in Above Ten Years Travels through the Empire of China* (London, 1737), pp. 65, 69, 120, 213.

[36]Ibid., pp. 63, 69, 120–22, 138–42, 150–56, 229, 240. Le Comte was also critical of Chinese architecture, though he admitted that it had a "rude" kind of beauty.

[37]Ibid., pp. 213–19.

comments on specific areas of Chinese scientific learning were far less scathing than his general appraisal of the state of the sciences in the Qing empire. However well educated and ingenious the Chinese had shown themselves to be, he wrote, they had not produced "one single man, of great achievement in the speculative sciences." In fact, he observed (with the *femmes savantes* of the Parisian salons clearly in mind), even in the sciences, educated French women were more knowledgeable than all the great doctors of China.

Le Comte felt that the Chinese lacked the qualities of precision and "penetration" that he considered essential to those who "devoted themselves to the study of nature." He believed them incapable of the logical thought that his French countrymen had brought to such a degree of perfection. Still, in his overall assessment of Chinese civilization, Le Comte concurred with the favorable impression of most French writers in the first half of the eighteenth century. He concluded that despite their "mean" achievements in the speculative sciences, the Chinese had more than demonstrated their capacity for civilized development in their superb laws and administration, their good sense in commerce, and their "genius" for ethics and reason.[38]

All these observations were of importance to the philosophes who made their judgments about China on the basis of such works as Le Comte's. Some decades after his memoirs appeared, for example, Jean-Baptiste Dortous de Mairan, then the director of the Académie des Sciences, puzzled over a contradiction that had been noted in numerous early assessments of China, beginning with the first reports of Ricci and his fellow missionaries. In a letter written in the early 1730s to Dominique Parennin, a Jesuit who had served with distinction for decades at Kang-xi's court, Mairan conceded the great genius of the Chinese for government, lawmaking, commerce, and invention and their attachment to order, authority, and hard work. How then, Mairan asked, could a people who had distinguished themselves in so many areas of civilized activity progress so little in the sciences, despite thousands of years of application to fields they considered as critical as astronomy and medicine?[39] Parennin's response summarizes the views of most contemporary Jesuit observers as they were expressed in the *Lettres édifiantes*, which J. B. Du Halde selected, compiled, and published between 1707 and the 1770s. Parennin's response also represents a veritable catalogue of the defects in Chinese character and society that eighteenth-century

[38]Ibid., pp. 122, 212–13.
[39]This discussion is based on the Mairan-Parennin exchanges published in *Lettres édifiantes*, vol. 21, pp. 79–138; vol. 24, pp. 24–28; and vol. 26, pp. 62, 138, 147ff.; and in Mairan, *Lettres au R. P. Parennin*, esp. pp. 2–16, 33–35, 60–61.

European authors thought responsible for China's failure to advance in the sciences and that nineteenth-century writers viewed as the cause of China's backwardness in both science and technology.

Other than complaints about the poor quality of Chinese paper, Parennin had little to say about Chinese technology. He made no attempt to refute Mairan's suggestion—which would become an article of faith for nineteenth-century observers—that the Chinese were incapable of improving what they had invented. In fact, Parennin concurred with Mairan's claim by noting the Chinese failure to develop artillery despite their early invention of gunpowder, and their lack of printing presses even though they had a centuries-old tradition of block printing.[40] He discussed these failings at some length but focused his analysis of Chinese shortcomings on their work in the sciences. He conceded that the Chinese were far behind Europe in virtually all the sciences. He believed their work in astronomy and geometry, for example, vastly inferior to their writings on history or law. Parennin found their instruments crude and reminded Mairan that Western assistance had been essential to Chinese efforts to correct their calendars and planetary tables, despite the fact that these were central to the imperial ritual cycle and thus to the well-being of the entire kingdom. Parennin disputed what he viewed as inflated Chinese claims about the antiquity of their scientific discoveries and the depth of their understanding. He could not find much that was "solid" in their works on the diagnosis and treatment of disease and thought it absurd that some writers had attempted to equate Leibniz's binary theorem with the teachings of the ancient scholar "Fo Hi."[41]

Expressing a view that would often be set forth in discussions of Indian sciences at the end of the eighteenth century, Parennin argued that the level of scientific understanding in China had actually declined. He asserted that the capacity of ancient Chinese scholars to combine theory and practice in the study of such vital questions as the circulation of blood had degenerated in later centuries into an unthinking and mechanical application of prescriptions derived from earlier investigations. Again and again in his correspondence with Mairan, Parennin returned to speculations concerning the causes of this decline. Anticipating arguments that would be frequently repeated in later works, he faulted the Chinese for failing to probe beyond specific findings to underlying principles and for disregarding the necessity of "exactitude" in observation

[40]For a contemporary reference to this failure to follow up on inventions, see L. de la Barbinais Gentil, *Nouveau voyage autour du monde* (Paris, 1728), vol. 2, pp. 155–56.

[41]*Lettres édifiantes*, vol. 26, p. 62. It is interesting to note that Leibniz himself drew attention to the "striking connections" between his thinking and that of "Fo Hi" in his 1716 *Discours (Lettre) sur la philosophie chinoise*. See Merkel, *Leibniz und China*, p. 28.

and investigation. Above all, he stressed the Chinese attachment to ancient customs and teachings and hostility toward innovation as key impediments to advances in the sciences.[42] He argued that in their education and their examinations for bureaucratic positions, the Chinese valued and rewarded veneration for tradition, while original thinking and attempts to challenge long-accepted views in astronomy or geography met with disapproval, dismissal, or, in extreme cases, official repression.

Parennin also linked the degeneration of the sciences in China to a system in which intellectual pursuits were too dependent on the whims of individual rulers. Advances under learned emperors like Kang-xi could be lost by the neglect or outright hostility of less enlightened successors. In one of his last letters Parennin capped his speculations on the causes of the presumed backwardness of the sciences in China with a summary judgment on Chinese scholar-bureaucrats which one might expect to find in a diatribe by a late eighteenth-century Sinophobe or a Victorian missionary. Scientific learning stagnated in China, he charged, because in both ancient and contemporary times it was in the hands of indolent men with superficial minds who were more concerned to pursue "immediate and concrete" matters that would bring personal reward than to probe fundamental questions that might lead to new understandings.[43]

In the face of so thorough a dismissal of Chinese attainments in the sciences, it is important to point out that both Mairan and Parennin professed the highest regard for the civilization of China as a whole. In fact, Mairan explicitly sought to assure Parennin that Chinese deficiencies in the sciences had not diminished his respect for their achievements in other areas, including politics and ethics. He contrasted the peace and contentment of the Qing empire with the social strife and unceasing warfare of Europe and observed that Europe's success in the abstract sciences had not produced stable governments or tranquil lives. Despite these concessions and high praise for Chinese technology and medicine by other missionaries,[44] the writings of the Jesuits, who have generally been depicted as staunch Sinophiles, contained much information and opinion that later critics could draw upon in their efforts to discredit the

[42]These weaknesses were given particular stress by David Hume in the 1740s (see "On Progress," in *Essays, Moral, Political and Literary* [London, 1875], p. 183) and by Jean-Sylvain Bailly later in the century (*Lettres sur l'origine des sciences*, pp. 21–25).

[43]Parennin's criticisms were later repeated virtually verbatim by J. S. Bailly in his letters to Voltaire. Parennin rejected Mairan's suggestion that the nature of the Chinese language might be at fault, but this explanation of Chinese backwardness would become quite popular in the nineteenth century.

[44]*Lettres édifiantes*, vol. 20, pp. 305–46; vol. 22, pp. 427ff.

image of China as a prosperous and well-governed empire. Even though China was arguably the most advanced civilization that Europeans had encountered in their ventures overseas, a reading of their letters leaves little doubt that by the first decades of the eighteenth century the Jesuits believed Europe to be clearly superior in virtually all scientific endeavors. Missionary recollections of conversations with such learned and intelligent individuals as the emperor Kang-xi also alleged that the Chinese themselves acknowledged the lead that Europeans had gained.[45]

The picture of China conveyed by J. B. Du Halde's monumental *Description géographique, historique, chronologique, et physique de l'Empire de la Chine . . .* (first published in 1735), which was considered the most authoritative source on China until well into the nineteenth century, was very much influenced by the ambivalence expressed by earlier Jesuit writers. Du Halde, who had not actually served in China himself, mixed the general admiration of Jesuit observers with their reservations about Chinese attainments in the sciences and (to an extent unparalleled by earlier observers) their technological achievement. Though he reckoned that a single province of the vast and productive Qing empire would "flatter the ambition of no mean Prince,"[46] he was highly critical of the military technology of the Chinese, which he labeled backward, and Chinese armies, which he considered undisciplined and poorly armed. He stressed the obsolescence of Chinese artillery and their dependence on Jesuit priests for the casting of serviceable cannon. Du Halde repeated the praise of other observers for Chinese silks and porcelains, which he considered proof of Chinese "ingenuity," but he judged their tools on the whole "simple" in comparison with those in use in Europe. He acknowledged China's ancient contribution to humankind's technological advance but believed that the Chinese craftsmen of his own day were less inventive than European mechanics, even though they were superb at imitating the manufacture of European devices from watches to muskets.[47]

Du Halde's estimate of Chinese scientific achievements was a good deal more generous than either Mairan's or Parennin's. He extolled the long-standing commitment of the Chinese to scientific investigation and learning in general, which he found reflected in their "magnificent libraries," their "vast numbers" of doctors and colleges, and the extensive education of their elite classes. Yet he concurred with the view of Mair-

[45]Ibid., vol. 14, pp. xii–xiii; vol. 16, pp. 346, 348.
[46]Du Halde, *General History of China*, vol. 1, p. 236.
[47]Ibid., vol. 2, pp. 75–80 (on the military); pp. 108–13; 123–26 (on tools and craftsmen).

an and Parennin that despite their dedication to the sciences for such a long period of time, they had not "brought to perfection" any of the "speculative sciences," which required "penetrating" and "subtle" minds. His evaluations of Chinese learning in various areas followed roughly those of the Jesuit correspondents with mixed reviews for Chinese astronomy, arithmetic, and medicine and sweeping assertions that they knew nothing about anatomy, geometry, and more advanced forms of mathematics. But he countered these blanket dismissals with high praise for Chinese work in such "sciences" as logic and law, praise that may have influenced Voltaire's thinking on the Middle Kingdom.[48] Like earlier Jesuit writers, Du Halde delighted in the superior skills in instrumentation and scientific learning demonstrated by the succession of Jesuit missionaries who served at the Qing court. Noting with concern Chinese resistance to the adoption of European ideas and techniques, he expressed the hope that the talents of the Jesuit fathers in "all these inventions of Human Wit, till then unknown to the Chinese, [have] abated something of their natural Pride, and taught them not to have too contemptible an Opinion of Foreigners."[49]

Reflecting the general sentiment of eighteenth-century authors, Du Halde was convinced that if they chose to do so, the Chinese could master Western learning and recover their ancient aptitudes for the sciences and invention. He argued that their success in such areas as government and agriculture, which he believed required intellectual "penetration" equal to that of the "speculative sciences," was indisputable proof of the Chinese capacity for scientific work. Though he recognized that drastic shifts in values would be required of the Chinese if they were to adopt Western scientific approaches, he appears to have assumed that these changes would surely come, given the increasing exposure of the Chinese to Western ideas and their inevitable recognition of the obvious benefits to be garnered by accepting the European approach to the natural world.[50]

Voltaire, who did more to advance China as a model for Europe than any other writer of the Enlightenment, tended to concentrate on the positive side of the Jesuit assessments of the Middle Kingdom. In his early or *chinoiserie* phase,[51] he made use of motifs that he imagined to be Chinese to buttress his unrelenting attacks on intolerance, to argue that morality was possible without Christianity, and to provide exotic settings for such dramatic works as *L'orphelin de la Chine* (1755) and such

[48]Ibid., vol. 3, esp. pp. 63–75, 78–80, 356ff.
[49]Ibid., vol. 2. pp. 1, 75–78 (quoted portions), 135–36.
[50]Ibid., vol. 3, pp. 63–64.
[51]Guy, *French Image of China*, pp. 219ff.

satiric tales as *Zadig* (1748). Drawing heavily on the works of Jesuit authors, including Semedo, Kircher, and Du Halde, however, Voltaire soon adopted a radically different approach to China, which first became apparent with the publication of his *Essai sur les moeurs* in 1756. From a mere embellishment, China was elevated to the status of a civilization deserving serious and lengthy historical consideration. This transformation was evidenced both by Voltaire's ridicule of Bossuet's pretensions to writing an *histoire universelle*, despite his ignorance and neglect of China, and by the fact that sections on China begin and end Voltaire's *Essai*, his own pioneering ten-volume venture into the writing of global history. Often repeating virtually verbatim the effusive praise of early travelers and Jesuit scholars, Voltaire celebrated in the *Essai* and later in his letters and the *Dictionnaire philosophique* the size and splendor of the Chinese empire.[52] He extolled its government and its laws; its multitude of great cities, canals, bridges; its many inventions and manufactures. He pointedly referred again and again to the antiquity of Chinese civilization, and he contrasted its high level in early times with that of the contemporary Celts who were struggling to master the wild forests covering most of France. He lauded the Chinese for their honesty, integrity, industry, and respect for authority, whether governmental or paternal. At times Voltaire's enthusiasm for things Chinese appears to have been boundless. For example, after informing his readers that criminals in China could not be executed until their cases had been reviewed three times by the "grand council," he exclaimed that this "law" alone was sufficient basis for concluding that the Chinese possessed the "most just and humane" civilization in the world.

And yet, though Voltaire's advocacy of Chinese civilization has invariably been the focus of works on the influence of China on Enlightenment thinking, he was quite critical of Chinese attainments in one area—science. Like Mairan, he confessed his perplexity at encountering an ancient civilization that had so long been devoted to scientific study and yet was centuries behind Europe and had only recently embarked on serious work in astronomy, mathematics, and chemistry. Also like Mairan, Voltaire linked scientific and technological trends. He noted that the Chinese were the originators of key inventions—printing, the compass, gunpowder—but that these devices had remained "mere curiosities" until the Europeans obtained them and carried them to a "high

---

[52]This discussion of Voltaire's views is based on his *Essai sur l'histoire générale et sur les moeurs et l'esprit des nations* (Paris, 1756), vol. 1, pp. 9–18, 20–21, and vol. 4, pp. 308, 310; his *Philosophical Dictionary* (New York, 1835), s.v. "China"; and his *Lettres chinoises, indiennes et tartares à monsieur Paw (un Bénédictin)* (London, 1776), pp. 27–28, 30–31, 35–36.

level of perfection." Voltaire drew attention to the marked inferiority of Chinese instruments—especially those for timekeeping and astronomy—to those devised by European craftsmen. He had much less than Parennin to say about the causes of China's backwardness, but he made some rather vague references to the excessive respect shown by the Chinese for tradition and their indolent "natures" (notwithstanding his earlier praise of their penchant for hard work). Voltaire also thought it important to distinguish between the rulers, whom he pictured as highly educated sage-scholars, and the mass of the people, whom he considered as credulous and superstition-ridden as their European counterparts.

However harsh Voltaire's criticisms, however, their impact was buffered by the clear sense conveyed in his writings that on balance China was an impressive civilization with much to teach even the increasingly arrogant Europeans. Though he thought Chinese failings in the sciences a matter for serious concern, he argued that they were more than outweighed by Chinese excellence in other areas, particularly in moral philosophy—which, after all, he considered the most exalted of the "sciences."[53]

In his qualified but nonetheless enthusiastic praise of China as a civilization with much to teach Europe, Voltaire was struggling against a rising tide of criticism directed against early impressions of the Qing empire. The overall Jesuit assessment of China had long been contested, especially by Jansenist and Dominican rivals. An even more negative appraisal, based primarily on commercial rather than religious interest in the Qing empire, was set forth in the works of a number of English writers, beginning with Daniel Defoe's *Farther Adventures of Robinson Crusoe* in 1719. Defoe's caustic commentary on current affairs, thinly disguised as fiction, takes his intrepid protagonist to China in search of trade and profit. Crusoe's tribulations there permit Defoe to expose the general poverty, government oppression, corruption, and dishonesty that he thought were the chief attributes of the Middle Kingdom. The sharply contrasting Jesuit view of the Chinese is as much a target of Defoe's barbs as China itself. Crusoe confesses that he finds it difficult to compare China and Europe because the latter is so much more advanced in virtually all fields.[54] He surmises that the high praise lavished on China resulted largely from the surprise of European travelers and missionaries at finding this reasonably civilized people amid the "rude"

[53]Guy, *French Image of China*, pp. 262–63.

[54]In the same period Samuel Johnson also refused to consider the comparison, believing the Europeans too far superior. See Bald, "Chambers and the Chinese Garden," p. 289, for discussion of the introduction Johnson is thought to have written for landscape designer William Chambers.

barbarians who inhabited most of the rest of the known world beyond Europe. Though Defoe's critique emphasized Chinese commercial restrictions and bureaucratic oppression and the weaknesses of Qing armies and war fleets, he also assured his readers that Chinese learning in the sciences was far behind that of the western Europeans. He dwelt at some length on the often described Chinese superstitions about eclipses, claiming that they knew nothing about the motions of the planets and had no more than a "smatch" of mathematical understanding.[55]

A number of less influential English writers, including Francis Lockier and William Wotton in works that appeared in the first decades of the eighteenth century, attached even greater importance than Defoe to Chinese scientific and technological shortcomings.[56] But Defoe's fictionalized criticisms received their most convincing eyewitness validation in a 1748 travel account, *A Voyage around the World*, attributed to Captain George Anson. Given the brevity of the sections devoted to China and Anson's obvious ignorance of that country, the work had an impact out of all proportion to its merits. Anson not only rendered Defoe and lesser writers credible; his judgments helped shape the thinking on China of such prominent French philosophes as Montesquieu, Diderot, and Rousseau.[57] The sheer force of his charges, their apparent validation by his personal experience, and the future First Lord of the Admiralty's official status all help to explain why his account was so influential. But even more important was Anson's skill in expressing the complaints and frustrations of merchants and seamen from many European nations who were attempting to engage in commerce in China. Decades earlier the Swedish engineer Laurent Lange, whose account may have shaped Defoe's views, had described at some length the bureaucratic and cultural obstacles facing those who hoped to trade in China.[58] Anson repeated many of Lange's charges and added to them a sweeping indictment of Chinese character, knowledge, and artisan

[55]Daniel Defoe, *The Farther Adventures of Robinson Crusoe* (London, 1925), pp. 259–70, esp. 267–68.

[56]See Shou-yi Ch'en, "Daniel Defoe, China's Severe Critic," *Nankai Social and Economic Quarterly* 8/3 (1935), 537–9, 541.

[57]Guy, *French Image of China*, pp. 210–11; Dean B. Coen, "The Encyclopédie and China" (Ph.D. diss., Indiana University, 1962), p. 198. Glyndwr Williams has concluded that the account was actually written by one of Anson's subordinates, but he has no doubt that the opinions and underlying perspective are those of Anson; see Williams's introduction to the 1974 edition of *A Voyage around the World* (London).

[58]Laurent Lange, "Journal of the Residence of M. de Lange at the Court of Pekin, 1721 & 1722," in John Bell, *Travels* (Edinburgh, 1806), pp. 423–26, 433–38, 447–49, 473, 476, 495. My discussion of changing attitudes toward China is strongly influenced by the brilliant introductory essay in Louis Dermigny, *La Chine et l'Occident: Le commerce à Canton au XVIIIe siècle, 1719–1833*, 4 vols. (Paris, 1964); for a discussion of the importance of Anson and Lange, see vol. 1, pp. 28–29.

skills. Harping on a theme that was becoming increasingly popular and would later be taken up by writers of the stature of Montesquieu and the popularity of the Abbé Raynal, Anson accused the Chinese of greed, deceit, dishonesty, and outright thievery.[59] He related a series of incidents involving confrontations with Chinese officials to back up these accusations and additionally to illustrate both the cowardice of the Chinese and their backwardness in military organization and technology. In a boast reminiscent of that made by the Portuguese adventurer Pereira in the sixteenth century, Anson reckoned that his warship alone was a match for the entire Chinese navy.[60]

Perhaps even more damaging than these criticisms in the long run, however, were Anson's comments on Chinese aptitude for manufacturing and scientific thinking. In a few pages he pronounced a series of judgments that would be repeated and elaborated upon for centuries to come. The Chinese reputation for industry and ingenuity, evidenced by the large foreign demand for their many "curious" manufactures, he argued, far exceeded what they actually deserved. He judged Chinese skills in the handicrafts inferior to those of the Japanese in goods that both countries manufactured, and to those of European workmen in virtually all areas of production. The Chinese were good at imitating, Anson concluded, but like all "servile" peoples they lacked innovative genius and the sense of "truth and accuracy" required for first-rate precision work on such devices as clocks and firearms. Although he put them in a very different context, Anson cited missionary critiques to show that even the most favorable observers found Chinese scientific thought far inferior to that of the Europeans. He dismissed the long-standing notion that the Chinese compensated for their deficiencies in mathematics and astronomy by their highly developed ethical sense and their work in the moral "sciences." The fraud, deception, and corruption he had encountered in his dealings with Chinese officials, Anson assured his readers, made a mockery of these well-meaning claims by missionaries and prominent essayists.[61] These brief, arrogant, and ill-informed charges created a series of remarkably durable impressions for the gener-

---

[59]But see Barbinais's attempts to rebut this charge in his *Nouveau voyage*, vol. 2, p. 147.

[60]George Anson, *A Voyage around the World* (London, 1748), pp. 386–97, 414–15. For expressions of similar views by Montesquieu, see *Spirit of the Laws* (London, 1949), pp. 297, 304; by Raynal, *A Philosophical and Political History of Settlements and Trade of the Europeans in the East and West Indies* (London, 1776), vol. 2, pp. 71–72; and by the Abbé Grosier, *A General Description of China* (London, 1888), vol. 2, pp. 519–24. William Falconer followed Montesquieu in ascribing these Chinese vices to overpopulation; see *Remarks on the Influence of Climate . . . on the Disposition, Temper . . . of Mankind* (London, 1781), pp. 206–9.

[61]Anson, *Voyage around the World*, pp. 411–13.

ations that followed. It was as if all the centuries of Jesuit learning and actual experience in China had been undone in an instant.

In contrast to Voltaire, many French philosophes—including Montesquieu, Diderot, and d'Holbach—adopted Anson's stress on the baleful effects of despotic authority on Chinese character and commerce.[62] But in the face of growing doubts about China as a model for Europe, the idea was briefly revived in the 1750s and 1760s. Large portions of the popular and influential *Encyclopédie* (which began to appear in 1751), edited by Diderot and d'Alembert, were devoted to lengthy descriptions of Chinese techniques of handicraft manufacturing as well as to their engineering feats in irrigation, fortification, and bridgebuilding.[63] The long-vaunted Chinese genius for agricultural production was emphasized by members of the circle of French thinkers who adhered to the Physiocratic approach to political economy. In the 1760s Mirabeau and Quesnay in particular sought to restore China to favor among French intellectuals and courtiers. They urged that Bourbon France, then entering into a period of prolonged financial crisis, reaffirm and strengthen the agrarian orientation that had supported civilization in China for millennia. Quesnay countered the charges of Montesquieu and d'Holbach that China's despotic regimes had blocked material progress by arguing that the Qing emperors and their predecessors were "legal despots," by which he meant enlightened absolute monarchs. He asserted that the emperors' participation in an annual ritual of spring plowing was proof of their concern to promote agriculture in their realm, thereby ensuring the well-being of their many subjects.[64]

Quesnay's claims for the success of Chinese agriculture appeared to be confirmed by the eyewitness report of Pierre Poivre, which appeared in 1768. In his *Voyages d'un philosophe*, Poivre, who was of the Physiocratic persuasion, pointedly contrasted the populous and well-cultivated Qing empire with the poorly (or not at all) cultivated areas he had visited in Africa, India, and southeast Asia. He lavished praise on the same emperors and bureaucracy that Lange and Anson had excoriated for their oppression and corruption. Poivre viewed their promotion of land reclamation, their canal and irrigation projects, and their hard work as the

[62]Montesquieu, *Spirit the of Laws*, pp. 102, 122–25, 268; Guy, *French Image of China*, pp. 308–39; Lewis A. Maverick, *China: A Model for Europe* (San Antonio, Tex., 1946), p. 35.

[63]Coen, "Encyclopédie and China," pp. 3, 55, 94–97, 104–6, 199.

[64]Maverick, *China*, pp. 125–26, 206–7, 213–19, 246–47; Lewis A. Maverick, "Chinese Influences upon Quesnay and Turgot," *Claremont Oriental Studies* 4 (1942), esp. 7–8; and Virgile Pinot, "Les physiocrates et la Chine au XVIIIe siècle," *Revue d'histoire moderne et contemporaine* 8 (1906–7), esp. 204–11. Quesnay admitted that China was overpopulated and that poverty was widespread as a result, but in contrast to Anson or Defoe he saw overpopulation as a sign of a beneficent, not an oppressive, government.

keys to China's flourishing agrarian economy, which he judged superior to Europe's own.[65] He strongly recommended that monarchs who wished their lands to prosper use China and its emperor, who "sat on the throne with reason at his side," as models.[66] But the Physiocrats' advocacy of China as the greatest of agrarian civilizations rendered their counsel anachronistic in an age when commerce and manufacturing were rapidly assuming predominant roles in the more advanced economies of western Europe. Within decades, commerce and industrialization would elevate tiny Great Britain to a position of global preeminence and give it the strength to overwhelm the once mighty Chinese empire both economically and militarily.

Physiocratic attempts to resuscitate the China model were short-lived. Though China was a peripheral concern in their writings, philosophes such as d'Holbach and Helvétius renewed the assault on Qing despotism in the years before the coming of the French Revolution. Even the once revered sage Confucius came under assault, and Chinese bureaucracy and education came to be viewed as major barriers to progressive social development.[67] Voltaire himself began to have doubts, which he confided to the Abbé Pauw, whose gloomy view of the flora and fauna of the New World had prompted a spirited rebuttal by Thomas Jefferson. Voltaire suggested that though Pauw was too ready to dismiss China, he himself had perhaps exalted it too much.[68]

These varying strains of French criticism of China were brought together in Pierre Sonnerat's multivolume account of his travels in Asia, first published in 1782. The son of a wholesale merchant, Sonnerat shared Anson's hostility toward the Qing bureaucracy and its restrictive policies regarding foreign commerce. He stridently dismissed notions of China as a benevolent despotism, declaring that taxes in the empire were oppressive, the mandarin bureaucrats harsh and corrupt, and the populace as a whole poverty-stricken, superstitious, and servile.[69] Sonnerat confirmed Anson and Lange's charges of merchant harassment and Groslier's image of a cowardly people that could be conquered by a handful of European soldiers. He then launched into the most detailed critique of Chinese scientific thought and manufacturing techniques to appear prior to John Barrow's pivotal account in the first years of the nineteenth century.

[65]Pierre Poivre, *Voyages d'un philosophe* (Paris, 1768), pp. 106–19, 133.
[66]Ibid., pp. 138–39.
[67]Guy, *French Image of China*, pp. 314–31.
[68]Voltaire, *Lettres chinoises*, p. 3.
[69]This discussion of Pierre Sonnerat's views is based upon his account in *A Voyage to the East Indies and China between 1774 and 1781* (Calcutta, 1788), vol. 2, pp. 184–94, 200–11, 216–21, 231–34.

Sonnerat's long service in the French Ministry of the Navy and his extensive travels made him acutely aware of the profound technological changes that were under way in Europe and an experienced observer of overseas societies. Like so many author-travelers of the age, he had acquired some knowledge of the sciences and viewed his expeditions as fact-gathering forays that would advance learning in Europe. In the 1770s and 1780s he visited parts of Africa and much of maritime Asia, collecting and sending back to Paris numerous specimens for the cabinet of the royal garden. He was also a corresponding member of the Académie des Sciences during the period of his travels, a fact that evidently fed his penchant for lengthy disquisitions on all manner of subjects, from eating habits and dress to the treatment of widows. This may also explain his intense interest in the tools and machines and scientific thinking of the peoples and societies he visited, particularly those of India and China.

In Sonnerat's account the usual litany of Chinese inadequacies is supplemented by critiques of Chinese silk manufacturing, which he found far inferior to the French, and of their "boasted agriculture," which he felt was praiseworthy only in the cultivation of rice. He even considered Chinese gardens, which had so strongly influenced William Chambers, Rococo Butlin, and other prominent eighteenth-century landscapers, unworthy of comparison to European designs. Sonnerat concluded that the sources of Chinese failings in virtually all fields of endeavor were rooted in Chinese culture and character. He thought there could be no progress under a regime so despotic and so utterly opposed to innovation. Nor was there hope for improvement on the part of a people so devoid of "genius, activity, and imagination" and so utterly bound by ancient custom.[70]

The general decline in interest in China and especially the increasing turbulence in France on the eve of of the Revolution may explain why Sonnerat's harsh appraisal of Chinese civilization received so little attention, despite its considerable detail and eyewitness validation. How far China had fallen in the esteem of the majority of European intellectuals by the last years of the eighteenth century and the importance to that decline of its alleged failings in the sciences is perhaps best illustrated by Condorcet's brief but devastating references to China in his *Sketch for a Historical Picture of the Progress of the Human Mind*. Written in 1793 and 1794 when he had gone into hiding to avoid arrest by the Jacobin enemies who would eventually bring about his death, Condorcet's *Sketch* emphasized the roles of technology and science in the human ascent

[70]Ibid., pp. 212–13.

from savagery to civilization. In his view the despotic rule of "priestly castes" (the use of these terms reflects Condorcet's very limited knowledge of Chinese society and history) had prevented China from advancing beyond the third of the ten stages he outlined for the history of human development. After a promising beginning in which the Chinese had "outstripped all the other nations in the arts and sciences," they had been overtaken because despotism and superstition had stifled progress. China became a helpless giant, subject to conquest by Asian barbarians and unable to realize its full potential because its sciences, "being absurd prejudices, are condemned to eternal mediocrity . . . even the invention of printing has remained entirely useless for the progress of the human mind."[71] Thus in effect ended a century of shifting attitudes toward China—a century that had begun with the court of Louis XIV celebrating an imagined Chinese culture with New Year's festivities at Versailles done in what was thought to be the Chinese style.[72]

Ancient Glories, Modern Ruins: The Orientalist Discovery of Indian Learning

The enthusiasm for aspects of Indian civilization displayed by small circles of late eighteenth-century European intellectuals is a good deal more difficult to explain than the earlier fascination with China in the heyday of the Qing dynasty. British and French savants such as Charles Wilkins and Anquetil Duperron struggled to decipher ancient Sanskrit texts, and philosophers such as Johann Friedrich Herder found inspiration in Hindu epics and literature in an age when Indian civilization was in disarray. Overextended and sapped from within by the administrative and economic weaknesses that François Bernier had so relentlessly exposed, the Mughal empire slowly disintegrated beginning in the first decades of the eighteenth century.[73] Dynastic struggles and religious strife, foreign invasions and regional revolts shattered the massive but fragile Mughal imperial edifice. As the empire crumbled, the leaders of numerous splinter states and kingdoms breaking free from the Mughal yoke fought to establish their claims to paramountcy in different regions of India. Decades of bloody but indecisive warfare exacerbated natural

[71]Condorcet, *Sketch for a Historical Picture of the Progress of the Human Mind* (London, 1955), p. 38.

[72]Reichwein, *China and Europe*, p. 22.

[73]The best studies of the causes of the Mughal decline remain Irfan Habib, *The Agrarian System of Mughal India, 1556–1707* (Bombay, 1963); and M. Athar Ali, *The Mughal Nobility under Aurangzeb* (Bombay, 1966). For detailed chronological accounts of the Mughal decline and the rise of British power, see *History and Culture of the Indian People*, vols. 7 and 8, ed. R. C. Majumdar (Bombay, 1974, 1977).

calamities; famine, disease, and banditry spread across the subcontinent. These scourges led to the decline of once prosperous urban centers and the further impoverishment of an already hard-pressed peasantry.

Internal divisions and prolonged social strife also provided the openings that enabled the British and French to expand their trading-post enclaves on the Indian coast into full-fledged colonial possessions from the 1740s onward. The rivalry between the two European powers—in ever shifting alliances with Indian princes—ended with the rout of the French during the period of the Seven Years War (1756–63) and the beginnings of Britain's empire in India. The corruption and harsh exactions of East India Company officials, who assumed the responsibility for governing ever larger portions of the subcontinent in the last decades of the eighteenth century, contributed to further disasters, most notably the great Bengal Famine of 1769–70.

Paradoxically, just as Indian civilization reached the nadir of its slow and painful decline, it was "discovered" by Orientalists such as Wilkins and Sir William Jones. In a number of ways political chaos and the rise of British power contributed to late eighteenth-century interest in India, especially Hindu India. As it became clear to the British that they, whether by conscious design or out of "fits of absent-mindedness," had to take over the administration of large parts of India for some time to come, the study of the languages, beliefs, and customs of the diverse and sophisticated peoples they were to rule was transformed from an enriching pastime for gentlemen with a taste for the exotic into a vital concern for India's new rulers.

Warren Hastings, who had played a leading role in the foundation of the British empire in India, considered the study of Indian languages, ideas, and social organization essential to effective British rule in the subcontinent. In 1784 he asserted that "every accumulation of knowledge, and especially such as is obtained by social communication with people over whom we exercise a dominion founded on the right of conquest, is useful to the state."[74] During his term as the first governor-general of British India (1773–85), Hastings initiated the policy of giving preference in promotion and offering monetary incentives to those East India Company officials who were willing to undertake the serious study of Indian languages.

The greatly enhanced power and influence of Company officials such as Hastings and William Jones gave them access to Hindu and Muslim writings that had earlier been denied to alien and infidel European visitors (with the exception of Jesuit missionaries such as Di Nobili who

[74]Quoted in Penderel Moon, *Warren Hastings and British India* (New York, 1949), p. 352.

were willing to assmiliate to the host culture). The willingness and ability of Hastings and Jones and others like them to establish bonds of friendship with Hindu and Muslim princes and pundits further facilitated the European discovery of Indian learning.[75] The Company's policy of opposition to proselytization by Christian missionaries in its dominions, which was maintained until 1813,[76] and its subsidies for Hindu and Muslim education and the publication of classical Indian writings also predisposed Indian scholars and teachers to share their ancient texts with the conquerors from overseas. Heated and prolonged disputes between "Westernizers" (such as Charles Grant and James Mill) and Orientalists over the manner in which India ought to be governed gave added impetus to the serious study of Indian culture. These quarrels, which raged intermittently for fifty years, brought Indian writings in translation to the attention of the largest Western audience they would reach until the twentieth century.

Like the efforts of late medieval and Renaissance scholars to recover the writings of classical Greece and Rome, the European exploration of Indian learning was (and is today) concentrated on literary and philosophical works, particularly those of Hindu rather than Muslim origin. The lyric poetry of *Shakuntala* and "The Cloud Messenger" and the mysticism of the profound religious writings of ancient Indian civilizations strongly influenced leading writers of the German and English romantic movements, including Schlegel, Hölderlin, Shelley, and Wordsworth.[77] However, just as the search for classical sources eventually led to the recovery of the scientific learning of the ancients, eighteenth-century authors such as the Frenchman Jean-Sylvain Bailly and especially British scholars Jones and H. T. Colebrooke sparked a lively if rather short-lived interest in Indian scientific writings. Discussions of ancient Hindu contributions to mathematics, astronomy, and medicine supplemented the arguments of Orientalist defenders of Indian civilization, which rested primarily on its literary, artistic and philosophical achievements. In contrast to the results of investigations in China earlier in the century, the study of Indian scientific learning generally enhanced In-

[75]David Kopf, *British Orientalism and the Bengal Renaissance* (Berkeley, Calif., 1969), pp. 13–21; Moon, *Hastings*, pp. 350–54; P. J. Marshall and Glyndwr Williams, *The Great Map of Mankind* (Cambridge, Mass., 1982), pp. 12–13; Mukherjee, *Sir William Jones*, p. 124.

[76]John Kaye, *The Administration of the East India Company* (Allahabad, 1966), pt. 5, chap. 2.

[77]Ronald Taylor, "The East and German Romanticism," in R. Iyer, ed., *The Glass Curtain between Asia and Europe* (London, 1965), pp. 188–200; H. G. Rawlinson, "India in European Literature and Thought," in G. T. Garratt, ed., *The Legacy of India* (Oxford, 1962), pp. 30–37.

dia's standing in the eyes of European thinkers. But the learning that was admired was associated with ancient Hindu civilizations, not with contemporary Indian society.

Like the writings on Chinese civilization in the age of the Enlightenment, European appraisals of the level of Indian technological development played a far less important role in shaping attitudes toward Indian peoples and society than did estimates of their scientific capabilities. In the last years of the eighteenth century, factory-produced British textiles had only begun to capture the overseas markets in Europe and elsewhere that had long been dominated by Indian silk and cotton cloth exports. Until the 1790s, government restrictions on the importation of Indian cloth into Britain, not technological factors, were mainly responsible for the gains made by British manufactures at the expense of Indian handicrafts. As late as 1786 Sir William Jones could declare with confidence that in the production of cotton textiles the Indians "still surpass all the world."[78] Until the end of the century, travelers such as the French engineer Cossigny Charpentier felt obliged to devote considerable portions of their accounts of India to rather detailed descriptions of dyeing and weaving techniques.[79] Yet though favorable views on other branches of Indian manufacturing from iron smelting and refining to shoemaking were expressed well into the first decades of the nineteenth century,[80] Bernier's stress on the simplicity of Indian tools and their marked inferiority to those of European workers had become a familiar refrain in British and French accounts of India by the mid-eighteenth century.

No writer in the late eighteenth century surveyed the state of Indian technology in greater detail than the French traveler and naturalist Pierre Sonnerat. As we have seen, Sonnerat's harsh appraisals of Chinese scientific learning capped the intellectual assault on the idea of China as a model civilization. On his way to China in the 1770s he had also visited many areas on the Indian coast and devoted a large portion of the account of his Asian travels to discussion of Indian religions, social practices, and material culture. He offered his readers the most exten-

---

[78]William Jones, "Third Anniversary Discourse, On the Hindus, February 2, 1786," in *Eleven Discourses* (Calcutta, 1873), p. 17. Jones added that in their development of textile dyes the Indians had made important contributions to the advancement of chemistry.

[79]Cossigny Charpentier, *Voyage à Canton* (Paris, 1799), pp. 473, 490ff. For similar accolades in this same period, see William Robertson, *An Historical Disquisition concerning the Knowledge Which the Ancients Had of India. . .* (London, 1791), pp. 277–87. See also, earlier in the century, *Lettres édifiantes*, vol. 9, pp. 419–23; vol. 15, p. 34; and vol. 26, the entire fifty-page letter (dated 1742) by Father Coeurdoux.

[80]See the selections on various branches of manufacture in India in Dharampal, *Indian Science and Technology in the Eighteenth Century* (Delhi, 1971).

sive, though one-sided, critique of Indian technology available before the appearance of James Mill's *History of British India* in the second decade of the nineteenth century.

Like Bernier, Sonnerat conceded that Indian craftsmen were capable of manufacturing beautiful cloth and intricate filigree jewelry, but he dismissed the tools they worked with as simple, even primitive, and few in number. He observed that Indian carpenters or blacksmiths made do with two or three tools at tasks for which European workmen might use as many as a hundred. Not surprisingly, Sonnerat concluded that the average Indian workman was much less productive that his European counterpart. He calculated, for example, that a single European miller could grind a thousand pounds of flour in a day, whereas two Indian laborers could produce a mere fifty pounds with their "rude" handmill. "It is the same with all of their machines," he assured his readers. Blacksmiths' forges were so simple that they could be carried about from village to village; shoemakers had only awls and knives to work their leather; and the excellence of the jewelry prepared by Indian craftsmen was the product of their incredible patience and owed little to their crude tools.[81] Sonnerat's account also suggests some of the adverse effects that a low level of technological development had had on India's material culture as a whole. Aside from imposing temples, Sonnerat found the Indians' architecture plain and uninspiring, their housing lacking in the "magnificence" he had been led to expect in the "Orient," and their household furnishings meager and crudely manufactured.[82]

Unlike most travelers, Sonnerat, who had earlier worked under Pierre Poivre and clearly aspired to the status of a philosophe, not only criticized Indian technology but speculated at some length on the causes of what he deemed its lowly state. He judged that the Indians, like all the peoples of Asia, had made considerable technological advance in ancient times when the peoples of Europe were still wallowing in barbarism. But in contrast to the backward Europeans, who struggled through the centuries to improve their technological endowment and eventually emerged as the most advanced of all civilizations, the Indians and Asians in general had progressed little in the manufacturing arts. Sonnerat's causal explanations for India's technological stagnation are expressive of some of the central intellectual preoccupations of his era. He identified despotism, excessive respect for tradition, and climate as the chief culprits. Despotism and traditionalism stifled the free expression of ideas and rewarded those who were skilled at maintaining stability and state

[81]Sonnerat, *Voyage*, vol. 2, pp. 125–34.
[82]Ibid., pp. 16–18, 123. For similar views, see *Lettres édifiantes*, vol. 12, p. 60.

control rather than those capable of thinking creatively and introducing innovations. Climate, in this instance the enervating heat of the Indian tropical environment, sapped the energies of even the brightest and most creative individuals, thus reducing the chance that inventions, however ingenious, would actually be effectively applied to production or to resource extraction.

At one point Sonnerat suggested that the great European innovations in machines and tools were tied to "progress" in civilized development as a whole. Unlike the nineteenth-century writers who singled out technological advance as one of several defining aspects in what it means to be civilized, Sonnerat did not elaborate on this connection. But in an aside that was perhaps the most explicit expression in this era of the belief that a society's level of development could be gauged by its technological attainments, Sonnerat observed that it was in their vastly inferior application of machine power that the Indians appeared to be "most distant from the Europeans."[83]

The pattern of early development, stagnation, and decline that Sonnerat had outlined for Indian technological history was mirrored, according to earlier and contemporary French authors, by Indian scientific thinking. In the first decades of the eighteenth century the letters of the Jesuit fathers, which played a somewhat less important role in shaping attitudes toward India than toward China, established the view that in ancient times the Hindus had excelled in the sciences, particularly astronomy, medicine, and mathematics. But the early promise of Indian scientific discoveries had not been fulfilled. Like Indian technological innovation, Hindu scientific thinking had stagnated, become mired in superstition and mythology. As time passed, astrology was stressed at the expense of astronomy, and sophisticated mathematical thinking degenerated into the simple arithmetic sums of shopkeepers. Most Jesuit observers believed that only traces remained of the splendid work of the ancient Hindus. Jesuit writers claimed that contemporary Brahmin pundits knew little of the underlying principles that informed the discoveries of their distinguished ancestors. The pundits employed mathematical procedures and planetary tables that had been passed down mechanically and uncritically for millennia. They added nothing to a body of knowledge that had been scarcely increased or improved for centuries. The Brahmins were oblivious to relationships between natural phenomena and ignorant of theory. They explained eclipses and other natural occurrences with childish myths, and they had forgotten how to

[83]Sonnerat, *Voyage*, vol. 2, pp. 125 (quoted portion), 120–21, 126.

use many of the surprisingly sophisticated instruments that had been devised in ancient times.[84]

Some decades later the highly regarded astronomers G. de la Galaisière le Gentil and Jean-Sylvain Bailly arrived at similar estimates of Indian scientific learning on the basis of more thorough but also more narrowly focused investigations. Galaisière, who received a royal commission to travel to the East Indies in the 1760s for the express purpose of making astronomical observations, had a good deal to say about Indian astronomical instruments and techniques. As a student of one of the astronomers of the famed Gassini family and a member of the Académie des Sciences, Galaisière was deemed highly qualified to evaluate Indian capabilities in this area. His opinions were widely noted in the decades following the appearance of the account of his travels, by writers as influential as William Robertson and James Mill. Though Galaisière admired the accuracy of some of the Hindu calculations of planetary movements, he felt that Indian astronomy in general had little other than its antiquity to recommend it. He noted that Hindu scholars were oblivious to the need to correct or to improve upon observations and calculations made millennia earlier and that they were contemptuous of European claims to superior knowledge. He added that with very few exceptions the Hindus showed little interest in European techniques or findings, a complaint also voiced by Bailly.

Bailly's appraisal as a whole was similar to Galaisière's, but the future mayor of revolutionary Paris made a good deal more of India's ancient contribution to the study of astronomy.[85] Bailly argued that the task of refining the initial discoveries of peoples living in the tropics had been left to the more vigorous and persevering inhabitants of the temperate zone. Both writers suggested that indolence engendered by excessive heat had molded the Hindu character in ways that were incompatible with scientific investigation. Bailly charged that the Indians were incapable of sustained experimentation, a failing that made advances in anatomy, botany, or chemistry impossible. He claimed that like all peoples of the tropics the Indians were so soft and so attached to ancient beliefs that they simply preserved existing ideas without understanding them or adding anything new. Galaisière also asserted that the Hindus were incapable of prolonged research and added that they showed little inter-

---

[84]*Lettres édifiantes*, vol. 10, pp. 30–41; vol. 13, pp. 22–23; vol. 16, pp. 137–39; and E. Souciet, *Observations mathématiques, astronomiques, géographiques, chronologiques, et physiques* (Paris, 1729), pp. 6–7.

[85]For a fine study of Bailly as scientist and political activist, see E. B. Smith, *Jean-Sylvain Bailly: Astronomer, Mystic, Revolutionary, 1736–1793* (Philadelphia, 1954).

est in the cumulative growth of scientific knowledge. He denounced the secretiveness of the higher castes and their propensity to use their knowledge to maintain their control over the superstitious masses rather than to foster social improvement. Bailly saw the Indians' failings in astronomy as symptomatic of their general failure to match European advances in the "arts" and sciences as a whole. The "genius of discovery," he concluded, belonged to the Europeans alone, and this explained why they had made more rapid "progress" than any other people in understanding the natural world.[86]

Although Voltaire was certainly aware of the criticisms of the Jesuits and of writers such as Galaisière—many of which Bailly had explored at length in letters to Voltaire himself—he chose to emphasize India's ancient scientific achievements and play down its modern deficiencies. Indian civilization was a good deal less central to Voltaire's work than China, but he made much of the former's antiquity to buttress his challenges to the primacy of Judeo-Christian chronology. At his most exuberant, he pictured India as the mother of all civilizations and the font of learning in a variety of fields, including the sciences. But while he celebrated the mathematical and astronomical calculations of the ancient Hindus, Voltaire derided the lethargy and backwardness of contemporary Indian scholars and the superstition of the Indian masses. He confessed that he found it difficult to believe that the glorious civilization that had contributed so much to human development had fallen in scientific learning to the dismal state reported by François Bernier a century earlier. He conceded, however, that it had in fact done so, and his sweeping censure of the level of scientific inquiry in India in his own day differed only in its vagueness and its lack of empirical grounding from the judgments of his better informed and more specialized French contemporaries.[87]

The British authors and translators who dominated the European study of Indian learning in the last decades of the eighteenth century were on the whole considerably better informed than French writers such as Voltaire and even more serious students of Indian texts such as Bailly. Having routed their Muslim and French rivals for economic and political control of India and become the rulers of large portions of the subcontinent, British officials of the East India Company greatly in-

[86]Galaisière, *Les mers de l'Inde*, pp. 207, 213–17, 236–37; Jean-Sylvain Bailly, *Histoire de l'astronomie ancienne* (Paris, 1775), pp. 104–28; Bailly, *Traité de l'astronomie indienne et orientale* (Paris, 1787), pp. xxiii–xxv, xlix, lxxiii–lxxvi, cxlv–cxlvi; Bailly, *Lettres sur l'origine des sciences*, pp. 18–23, 73, 87–88.

[87]Voltaire, *Fragments on India* (Lahore, 1937), pp. 24–25; Voltaire, *Essai sur les moeurs*, vol. 1, pp. 28–32; Bailly, *Lettres sur l'origine des sciences*, pp. 3–11; Debidour, "L'indianisme de Voltaire," esp. pp. 34–35.

creased their contacts with local Hindu lords and the Brahmin literati. Many of the latter preferred the British to their former Muslim over-lords and eagerly entered the service of the new rulers. In the 1770s a group of highly educated and remarkably versatile British officials be-gan to form around a shared interest in Indian literature. These Oriental-ists readily acknowledged the essential contributions of Indian scholars and religious leaders to their work of discovery, translation, and inter-pretation. Indian gurus taught them the languages they needed to master the ancient Hindu and (more rarely) Muslim texts that were the focus of their studies; they corrected the Englishmen's translations and answered endless questions about Indian philosophy, mythology, and astronomy. Sir William Jones, for example, who was the leader of the Orientalists through most of the 1880s and into the 1890s, spent many of his days in India—in rooms shuttered from the glare and heat of the deltaic plains of Bengal—conversing with learned Brahmins about Indian epic poetry or Hindu techniques of setting algebraic equations. At the same time, Jones regularly informed his Indian colleagues of the latest scientific discoveries in Europe. Samuel Davis, who specialized in the study of Indian astronomy, remarked on the willingness of the Brahmins to explain the teachings contained in the ancient Sanskrit treatises; H. T. Colebrooke admitted how much the locating of texts and the mastering of their context depended upon Indian informants and teachers.[88]

The collaboration between British scholar-officials and learned Indian Brahmins rendered the classic works of Indian civilization in a wide range of fields available to Western intellectuals for the first time. Trans-lations of great religious works, such as the *Bhagavad- Gita* and portions of the *Vedas*, impressed European readers with the sophistication and profundity of Indian philosophy, the complexity of Sanskrit and other languages of ancient intellectual discourse, and the imaginative capacity of Indian writers. The Orientalists shared these discoveries with their colleagues at the meetings of the Asiatic Society of Bengal and in essays published in *Asiatick Reseaches* or as occasional papers sponsored by the society. Though the Orientalist emphasis was decidedly philosophical and literary, prominent scholar-officials such as H. T. Colebrooke pub-lished translations of Indian works on astronomy, mathematics, and

---

[88]Mukherjee, Sir *William Jones*, pp. 113–14; Jones, "Tenth Anniversary Discourse," in *Eleven Discourses*, p. 120; Jones, *Works*, vol. 2, p. 134; Jones, "Eleventh Anniversary Discourse" in *Discourses Delivered before the Asiatic Society* (London, 1821), pp. 41, 51; Samuel Davis, "On the Astronomical Computations of the Hindus," in Sir William Jones, ed., *Supplemental Volumes to the Works of Sir William Jones*, (London, 1801), vol. 1, pp. 264–65, 285; T. E. Colebrooke, *The Life of H. T. Colebrooke* (London, 1873), vol. 1, p. 304; Kopf, *British Orientalism*, p. 29.

medicine with valuable and often lengthy introductions. Less distinguished members of the Orientalist coterie, such as John Bentley and Reuben Burrow, wrote essays on subjects ranging from the Hindus' astronomical instruments to their techniques of arithmetic calculation.[89]

In contrast to the Europeans' increasingly negative view of Chinese scientific attainments in this period, there was much that was positive—at times hyperbolic—in the Orientalists' assessment of the Indian contribution. Indian achievements in mathematics were especially emphasized. Colebrooke drew attention to the brilliant work of the fifth-century mathematician and astronomer Aryabhata. William Robertson, who had made his reputation as a historian of Scotland, Charles V, and the colonization of America, reminded his readers that the numerals they took for granted and referred to as "Arabic" had actually been devised by the Indians. In a lengthy appendix to his study of Western knowledge of India, in which he surveyed current findings on the state of Indian scientific learning, Robertson also noted that the practice of counting by tens and many other mathematical operations had been worked out in ancient India, whence they were transmitted to the Arabs and then to the Europeans. The Orientalists, who shared the French philosophes' adulation for classical Greek and Roman civilizations (thanks to the classics-oriented public school education that most of them had received), frequently paid the Indians the high compliment of comparing their scientific work to that of the ancient Greeks. Some writers, including Colebrooke, argued that in many areas of algebra the Hindus had surpassed the Greeks and may have been the source of methods ascribed to the genius of such thinkers as Pythagoras. Robertson, who equated the Hindu religious center of Banaras with Athens, argued that the early Indians' work on logic and metaphysics (both of which he included in the sciences) was more than a match for that of the Greeks. John Playfair candidly admitted that Indian scientific thinking had developed independently of Greece and other centers of ancient civilization. He confessed to being "astonished at the magnitude of that body of science, which must have enlightened the inhabitants of India in some remote age, which, whatever it may have communicated to the western nations, appears to have received nothing from them."[90] Indian

---

[89]For examples, see the essays in Jones, *Supplemental Volumes;* John Bentley, "On the Antiquity of the Surya Siddhanta, and the Formation of the Astronomical Cycles therein Contained," *Asiatick Researches* 6 (1799), 537–58; Reuben Burrow, "A Demonstration of One of the Hindoo Rules of Arithmetic," *Asiatick Researches* 3 (1792), 145–47; and Robert Barker, *An Account of the Brahmins' Observatory at Benares* (London, 1777).

[90]John Playfair, "Remarks on the Astronomy of the Brahmins," *Transactions of the Royal Society of Edinburgh* 2/2 (1790), 192; see also pp. 174, 180, 190–91.

achievements in many areas of scientific inquiry convinced the Orientalists that the ancient Hindus had attained a high level of civilization, one that merited careful study by European authors. Robertson was the most explicit in the application of science as a gauge of the level of overall social development. He judged that India's ancient scientific discoveries alone were sufficient to warrant regarding it as one of man's greatest early civilizations.[91]

Like French writers earlier in the century, however, British scholar-officials were careful to distinguish between India's ancient accomplishments and the rather sorry state of the sciences in the subcontinent in their own era. Robertson, drawing on the earlier works of Galaisière and others, insisted that the Brahmins of his day did not understand the principles on which their astronomical tables were based.[92] Robert Bentley, who provided a detailed description of the instruments in the Banaras observatory in the 1770s, noted that they exhibited a degree of accuracy and precision quite beyond the skills of contemporary Indian craftsmen—except perhaps those under the "immediate direction of a European mechanic."[93]

Orientalists who attempted to explain the reasons for the decline of scientific inquiry in India stressed climatic factors and the stifling effects of despotism, as had French authors earlier in the century. Sir William Jones went so far as to argue that despotism, which he believed characteristic of Asian societies, had played a decisive role in the eclipse of India and other Asian civilizations by the nations of western Europe. Q. Craufurd added a reason hinted at by Pierre Poivre but not fully developed by any French author: he attributed India's supposed stagnation in the sciences, as well as its general decline, to the Muslim conquest of the subcontinent. Craufurd believed that the Islamic invasions had spread bigotry, ignorance, and barbarism throughout India—indeed, in every area that had fallen under Muslim control. He also suggested that the historic isolation of India and the secretiveness of the Brahmin castes, which sought to monopolize India's ancient learning, had discouraged critical and creative thinking and hence advances in scientific knowl-

[91]Robertson, *Historical Disquisition*, pp. 96, 295, 301–10, 335; T. E. Colebrooke, *Life of H. T. Colebrooke*, vol. 1, pp. 246, 249, 304–10; H. T. Colebrooke, *Miscellaneous Essays* (London, 1873), vol. 3, pp. 384–86, 392–404; Q. Craufurd, *Sketches Chiefly Relating to the History, Religion, Learning, and Manners of the Hindoos* (London, 1790), p. 65.

[92]Robertson, *Historical Disquisition*, p. 306. See also S. Davis, "On the Astronomical Computations of the Hindus," pp. 264–65. Davis disagreed with Robertson's assertion but conceded that most European observers shared Robertson's viewpoint. Playfair's findings ("Astronomy of the Brahmins," pp. 36, 138–39) supported Robertson's contentions.

[93]Bentley, "Antiquity of the Surya Siddhanta," p. 5.

edge. Robertson and other writers also deplored the exclusiveness of the Brahmins, noting with chagrin their indifference or open hostility to European scientific instruments and techniques.[94] In the Orientalist view, tropical lethargy, rapacious and uncaring rulers, and scholarly arrogance had overwhelmed the early impulses toward scientific investigation and speculation in Indian civilization. Over the centuries the forces of superstition and bigotry, which the philosophes so despised, had gained the upper hand, and much of India's scientific legacy had been neglected or forgotten entirely.

The corollary to the belief that the study of the sciences in India had stagnated and declined was the virtually unchallenged conviction that Indian scientific knowledge was vastly inferior to European. There was in fact a patronizing, if not condescending, undercurrent in much of the British writing on Indian sciences in this period. Though Colebrooke and Playfair believed that ancient Indian thinkers had anticipated findings of seventeenth- and eighteenth-century European scientists, particularly in mathematics,[95] the general consensus of Orientalist scholars was that India had little to contribute in terms of new understanding to what had been discovered in Europe. The study of Indian texts was recommended as an exercise in the history of science, as a means of gaining insights into Indian culture, or as a diversion for gentlemen who might, as Samuel Davis said, "amuse themselves" through the investigation of "curious information."[96] Sir William Jones conceded that Hindu writings might yield some "helpful hints" to European scientists but thought they would provide little in the way of new methods or information.[97] The Orientalists' sense that India had little to teach the West in the sciences was rooted in the conviction that Indians, ancient or modern, were indifferent to the methods and discipline that in Europe had been considered central to scientific investigation for centuries and

[94]Jones, "Tenth Anniversary Discourse," p. 114; Craufurd, *Sketches*, pp. 10, 89–90, 190, 341–43; Robertson, *Historical Disquisition*, p. 306. Craufurd's sweeping condemnation of all that is Islamic reflected a long-standing and more general hostility to Muslim civilization on the part of many European intellectuals, which Edward Said has explored in depth; see his *Orientalism* (London, 1978).

[95] T. E. Colebrooke, *Life of H. T. Colebrooke*, vol. 1, p. 251; H. T. Colebrook, *Essays*, vol. 3, p. 339; Playfair, "Astronomy of the Brahmins," p. 183.

[96]Davis ("Astronomical Computations of the Hindus," p. 264) was disagreeing with what he believed to have been prevailing views in the 1780s. See also H. T. Colebrooke, *Essays*, vol. 3, pp. 339, 375–77.

[97]Jones, *Works*, vol. 3, pp. 232–33; "Eleventh Anniversary Discourse," in *Discourses Delivered*, pp. 38–40. Jones did qualify these judgments by noting that many Indian texts had yet to be examined, and those then coming to light contained information whose "importance cannot be doubted" ("Eleventh Anniversary Discourse," p. 51; and "Third Anniversary Discourse," p. 24).

essential since the time of Newton. Though specific examples of remarkably precise Hindu measurements had been documented,[98] the Indians were generally considered careless, satisfied with approximations, and prone to error in their calculations. The British thought Indian instruments antiquated and badly flawed. They believed that most of the Indians' theories were based on speculation alone rather than derived from empirical observation and careful testing. In addition, most of the Orientalists conceded that Indian scientific thinking was adulterated by flights into fantasy, mysticism, and mythology.[99]

The annual discourses of Sir William Jones delivered to the Asiatic Society in the decade between 1784 and 1794 provide a revealing summary of the Orientalists' attitudes toward Indian achievements and abilities in the sciences and, to a lesser extent, technology. Jones chronicled the findings of various British scholar-officials and conveyed a sense of the excitement and commitment that spurred on their work. Perhaps better than any other Orientalist, he also expressed the ambivalence that was central to British writings on India in this period; he captured their admiration for ancient Indian discoveries on the one hand and their dismissal of contemporary Indian learning as debased and outdated on the other. More than colleagues like Colebrooke, Jones stressed Indian achievements in logic, grammar, ethics, and metaphysics—which he considered sciences, but most nineteenth-century writers would not. At his most laudatory, Jones exclaimed that the texts of the ancient Hindus had anticipated all the metaphysics and philosophy of Newton himself, but on another occasion he remarked that if Newton were alive and in India, the Brahmins would worship him as if he were a god.[100] Like his co-workers, Jones left no doubt of his certitude that eighteenth-century Europe was superior to India, not only in the sciences but in the "useful arts" as well. In a telling use of an image that would be readily understood by most nineteenth-century observers, Jones declared that with regard to their accomplishments in the sciences, Asiatics were "mere children" when compared with Europeans.[101]

Assessments of this sort by such renowned scholars as Jones, who were considered admirers and defenders of Indian civilization, played

[98]Playfair, "Astronomy of the Brahmins," pp. 139, 153; R. Barker, *Brahmins' Observatory*, pp. 4–5; T. E. Colebrooke, *Life of H. T. Colebrooke*, vol. 1, pp. 23, 26; Robertson, *Historical Disquisition*, pp. 304, 307.

[99]T. E. Colebrooke, *Life of H. T. Colebrook*, vol. 1, pp. 248–51, 253–54, 305–6; H. T. Colebrooke, *Essays*, vol. 3, pp. 286, 307, 327, 339, 349–50, 363.

[100]As Peter Gay (*The Enlightenment*, vol. 2, p. 556) has noted, comparisons to Newton represented the highest compliment that Enlightenment thinkers could bestow.

[101]Jones, "Second Anniversary Discourse," pp. 10–11; "Third Anniversary Discourse," p. 24; "Eleventh Anniversary Discourse," p. 49; and *Works*, p. 246.

into the hands of men such as James Mill, who were highly critical of India and advocated sweeping reforms in the subcontinent. The evidence provided by the Orientalists and earlier French scholars of both India's past capacity for scientific thinking and its contemporary deficiencies also exerted a major influence on British educational policies and patterns of recruitment to the Indian Civil Service in the nineteenth century.

African Achievement and the Debate over the Abolition of the Slave Trade

In the last half of the eighteenth century, the peoples and cultures of Africa came under close scrutiny as both defenders of the slave trade and their abolitionist opponents culled the reports of explorers, traders, and missionaries for evidence to support their increasingly irreconcilable positions. The heated debates over abolition fundamentally altered the use to which African travel literature was put by European intellectuals. Accounts of Africa were no longer just contributions to a larger process of global exploration and discovery, or fascinating but peripheral forays into exoticism. They became key factors in a series of decisions on commercial policy in England and France which vitally affected the course of economic and social development on four continents. Though descriptions of the conditions of enslaved Africans in the plantation societies of the New World also served as major sources of information for those engaged in the abolitionist debate, European impressions of Africans in their native environment carried added weight. African traits observed on the Guinea coast or in the Niger delta could not, it was thought, be explained away by appeals to the influences of the condition of enslavement itself.

Defenders of the slave trade tended to stress its economic importance to their respective national economies and the European-dominated plantation colonies, as well as issues relating to property rights and free commerce. But they supplemented these arguments with descriptions of the barbarity and immorality of African life and culture. Charges of cannibalism, sexual promiscuity, and human sacrifice usually prefaced pro-slavery disquisitions on how fortunate enslaved Africans were to be transported to Christian and civilized societies where their propensities for chaos and sloth could be curbed by supervised and steady employment.[102]

[102]For contemporary statements of this view, see Samuel Estwick, *Considerations on the Negro Cause: Commonly So Called* (London, 1788), p. 60; S. M. X. Golberry, *Travels in Africa, 1785–7* (London, 1803), vol. 2, pp. 371–72; and Christopher Meiners, "Ueber die

Though most of the abolisitionists had a rather low opinion of African culture, if not of the Africans themselves, they made great efforts to refute charges of African cruelty and savagery.[103] Again, conditions in Africa took on a special significance because the abolitionists argued that it was difficult to judge the aptitudes or potential of peoples in bondage.[104] Therefore, proponents of the anti-slavery party gave even greater emphasis than the defenders of the slave trade to the nature of African societies and the quality of life in Africa itself. At times the abolitionists drew examples from the same accounts as their opponents, but they relied mainly on reports—such as those of William Smith, Anthony Benezet, or Mungo Park—which were on the whole favorable to African culture. On the basis of these accounts the abolitionists strove not only to demonstrate the horrors of the slave trade but also to discredit the view of the pro-slavery forces that the Africans were brutal, insensitive, ignorant, and even subhuman—in short, deserving of enslavement.[105]

For the abolitionists the fact that the African was a "man and brother" did not necessarily mean that he was the equal of a European. Even the most unrelenting opponents of the slave trade conceded that African societies were wanting in the sorts of attainments that Europeans had come to associate with civilized peoples.[106] This conviction owed much to the responses of European travelers in the early centuries of expansion. Eighteenth-century accounts of Africa differed little from earlier reports in subject matter and overall conclusions about the level of

---

Natur der Afrikanischen Neger . . ." *Göttingisches Historisches Magazin* 6 (1790), 390–91. For more recent discussions of these attitudes, see Ralph Austin and Woodruff Smith, "Images of Africa and British Slave Trade Abolition," *African Historical Studies* 2/1 (1969), 77–78; David B. Davis, *The Problem of Slavery in Western Culture* (Ithaca, 1966), p. 24; and Anthony Barker, *The African Link: British Attitudes towards the Negro in the Era of the Atlantic Slave Trade, 1550–1807* (London, 1978), pp. 68–82.

[103]Philip Curtin, *The Image of Africa* (Madison, Wis., 1964), pp. 30, 48–56, 239–43; A. Barker, *African Link*, pp. 196–97; and Christine Bolt, *Victorian Attitudes towards Race* (London, 1971), pp. 211–12. As Austin and Smith have pointed out ("Images of Africa," pp. 75–76), the abolitionists generally knew less about the actual conditions in Africa than did the defenders of the slave trade.

[104]For rousing eighteenth-century affirmations of this position, see James Beattie, *Essays on the Nature and Immutability of Truth* (Edinburgh, 1776), pp. 311–12; and Anthony Benezet, *A Short Account of that Part of Africa Inhabited by the Negroes* (Philadelphia, 1762), pp. 33–34. See also the discussion of the abolitionist Thomas Clarkson's views in Eric Williams, *British Historians in the West Indies* (London, 1966), pp. 23–24.

[105]For examples, see E. Williams, *British Historians*, pp. 23–24; Curtin's discussions of the impact of the writings of Park and other travelers, in *Image of Africa*; and Wylie Sypher, *Guinea's Captive Kings: British Anti-Slavery Literature of the XVIIIth Century* (Chapel Hill, N.C., 1942), pp. 16, 32.

[106]Cf. pp. 10–14, and 37, in Arthur Lee, *An Essay in Vindication of the Continental Colonies of America* (London, 1764), with pp. 38–39.

African social development. But writers in the Enlightenment period were much more concerned than earlier observers to explain why that development had been stunted and to relate the accepted fact of African backwardness to broader questions of policy and philosophy. Whereas defenders of the slave trade blamed the African environment or, more ominously, the innate deficiencies of the Africans themselves, many abolitionists saw the ravages of the slave trade as directly responsible for the sorry state of African culture. To offset charges of the cruel and vindictive "nature" of African peoples, instances of African hospitality and kindness such as those reported by Mungo Park in perhaps the most famous travel account of this period, were combined with narratives of incidents both fictional and real in which enslaved Africans displayed great loyalty or courage.[107]

European perceptions of African religious beliefs, marriage patterns, and political institutions continued to be the main attributes by which African societies were evaluated. But in an age that celebrated reason and insisted upon the application of scientific procedures to the study of humanity, inventiveness and especially the capacity for scientific thinking grew in importance as gauges of African achievement and potential. Unlike most travelers of the early period of expansion, eighteenth-century observers provided fairly detailed descriptions of African tools, and an apparently low level of technological development became for the first time important supporting evidence for claims that Africans were inherently inferior to Europeans. Though discussions of their science remained rare and brief, African medical practices and views of such natural phenomena as eclipses received some attention. More critically, in view of the growing debate over the quality of the Africans' mental endowment, numerous eighteenth-century writers tackled the question of their capacity for scientific thinking. Scientific criteria also began to shape European attitudes in a very different way as a small minority of authors, claiming to be scientists or to be using scientific procedures, set out to demonstrate empirically that Africans were a different species from Europeans.

If David Hume's condemnation of the "Negro" is taken as typical,[108]

---

[107]Mungo Park, *Travels in the Interior of Africa* (London, 1799), vol. 2, pp. 263–64; Winterbottom, *Native Africans*, vol. 1, pp. 161–2, 201–4. On the theme of the faithful slave in Romantic literature, see Sypher, *Guinea's Captive Kings*, chap. 3; Léon-François Hoffman, *Le nègre romantique* (Paris, 1973), pp. 74–75, 85–89, 96–97; and Hoxie Fairchild, *The Noble Savage: A Study in Romantic Naturalism* (New York, 1928), pp. 198–99, 289–91.

[108]As present-day authors such as Richard Popkin ("The Philosophical Basis of Eighteenth-Century Racism," in H. E. Pagliaro, ed., *Racism in the Eighteenth Century* [Cleveland, 1973], pp. 245–46) and Eric Williams (*British Historians*, pp. 20–21) have

the standing of Africans in the eyes of eighteenth-century Europeans had clearly fallen further from the rather lowly position they had occupied in the early centuries of expansion. In what must be one of the most cited footnotes of all time, which Hume added in 1753 to a new edition of his 1748 essay "Of National Characters," he charged that the Africans had never exhibited "any symptoms of ingenuity" and that they had "no arts . . . no indigenous manufactures."[109] Though these opinions accorded with those of the French philosophes Buffon, Rousseau, and Montesquieu[110] and were frequently cited by late eighteenth-century authors,[111] they were not in fact typical of his age. They ran counter to much of the evidence contained in eighteenth-century travel accounts, whose authors were for the most part a good deal more generous in their opinions of African manufactures than travelers in the first centuries of expansion had been. James Beattie, another Scottish philosopher, who penned a spirited rebuttal of Hume's views just a decade after they appeared, came much closer to capturing the consensus of late eighteenth-century travelers. Beattie declared that both Africans and Amerindians "were known to have many indigenous manufactures and arts among them, which even Europeans would find it no easy matter to imitate."[112] Presumably Beattie was drawing upon the evidence of such travelers as John Atkins and Francis Moore, who had a high regard for the quality of African cloth and the work of African gold and iron smiths.[113] Even travelers who were highly critical of the Africans conceded their skill in handicraft manufacturing. John Barbot, for example, who characterized the Africans as dishonest, lazy, and oversexed, com-

---

suggested that it was. Williams also discusses Clarkson's arguments against Hume's opinions.

[109]In Hume, *Essays*, vol. 1, p. 252. For a discussion of the generally hostile reception to Hume's assertions in England, see A. Barker, *African Link*, pp. 115–17, 162–64. Hume's views were more favorably received overseas; see, e.g., Lamiral, *L'Afrique*, p. 50, and Meiners, "Afrikanischen Neger," p. xiii.

[110]See Montesquieu, *Spirit of the Laws*, p. 332; William B. Cohen, *The French Encounter with Africans* (Bloomington, Ind., 1980), p. 67; Hoffman, *Le nègre romantique*, pp. 70–72; and Roger Mercier, *L'Afrique noire dans la littérature française* (Paris, 1962), pp. 74–75, 94–96, 104–5.

[111]See the references in n. 109; Estwick, *Considerations*, pp. 78–79n; Gordon Turnbull, *An Apology for Negro Slavery* (London, 1786), p. 34; and the work of Edward Long, who is discussed below. For a use of Montesquieu's views on these issues, see Lee, *Essay in Vindication*, p. 14.

[112]Beattie, *Nature and Immutability of Truth*, p. 311.

[113]John Atkins, *A Voyage to Guinea, Brasil, and the West Indies* (London, 1735), p. 100; Francis Moore, *Travels into the Interior Parts of Africa* (London, 1738), pp. 72–73; A. Barker (citing Clarkson), *African Link*, p. 180; Park, *Travels in the Interior*, vol. 1, pp. 281, 283, 285; Isert, *Voyages en Guinée*, pp. 124–25; Archibald Dalzel, *The History of Dahomey* (London, 1793), p. xxiv.

pared their pottery and metal utensils favorably with those of European manufacture.[114]

However dexterous the workmen and pleasing the final product, eighteenth-century observers agreed that African tools, like those of India, were primitive and inefficient. As early as the first decade of the century, the missionary Laurent de Lucques surmised that European workers would be "astonished" that the Africans could work metal with such crude tools. He noted that they lacked such basic instruments as hammers and anvils and that their bellows were little more than pieces of wood on which skins were stretched.[115] European visitors to Africa also concluded that dexterity and patience were no substitute for effective technology when it came to production or resource extraction. Since it was in this period that Europeans first were able to travel to the gold mines in the interior, commentaries on African mining techniques provide the best expressions of an emerging European conviction that the Africans had been unable to exploit the resources of the lands they occupied because they lacked the proper machine technology.

By far the most extensive critique of African mining was contained in an account of French activities in Senegal by the Dominican friar Jean-Baptiste Labat. His five-volume survey was based on the unpublished writings of André Brué and Michel de la Courbe, two colonial administrators who had served in the area during the period preceding the publication of Labat's compilation in 1728. Labat based the policy recommendations in his discussion of mining on the assumption that the lands of the Bambuk and neighboring peoples in upper Senegal abounded in gold ore of a very high quality. Making a comparison that was sure to evoke visions of huge and easy profits, he predicted that if properly worked, the Sengalese gold mines would prove richer than the silver mines of Peru.[116] He assured his readers that the gold was quite accessible and could be easily extracted by those who had the knowhow and suitable technology. Unfortunately, he complained, the Africans who lived in the area had neither. Because they were lazy and able to

[114]John Barbot, *A Description of the Coasts of North and South Guinea* (London, 1746), pp. 34–38, 40–41. See also Winterbottom, *Native Africans*, pp. 94–96; Lamiral, *L'Afrique*, p. 49; and the general discussion in A. Barker, *African Link*, pp. 109–11.

[115]J. Cuvelier, *Relations sur le Congo du Père Laurent de Lucques, 1700–1717* (Brussels, 1953), pp. 134–40. For similar observations, see Golberry, *Travels in Africa*, p. 274; Barbot, *Description*, pp. 22, 77–78; François Le Vaillant, *Travels from the Cape of Good Hope into the Interior Parts of Africa* (London, 1791), vol. 2, pp. 229–35; A. Barker (citing John Leyden), *African Link*, p. 181; Jean-Baptiste Durand, *Voyage au Sénégal*, (Paris, 1802), pp. 376–77; and the ubiquitous Sonnerat, *Voyage*, vol. 2, pp. 35–36.

[116]A claim also made by the eighteenth-century German explorer Ludewig Römer in his *Nachrichten von der Küste Guinea* (Copenhagen, 1769), p. 4.

work the mines only sporadically, depending on the permission of local leaders, they could not begin to exploit the great resource that was theirs for the taking. Most crucially, they possessed none of the tools or machines essential for mining ore from rich veins deep in the earth. According to Labat, the Africans could only "scratch" the surface or dig shallow holes, which were abandoned at the first sign of a cave-in or with the onset of the rains that swamped them with water and mud. Consequently, a vein of ore—usually a branch rather than the main lode—would be only haphazardly worked before the Africans were forced to begin anew in another spot. Labat noted that the Africans had no instruments for determining where the richest veins were located, and that they lacked the ladders, pumps, and techniques of gallery construction needed to dig true mine shafts. He added that they knew nothing of smelting or the mercury displacement technique and therefore had to be content with a crude crushing system to separate the gold from the ore they had extracted.[117]

In a number of ways Labat's account of African mining anticipated some of the conclusions that nineteenth-century observers would draw from similar inquiries into African tools, techniques, and work habits.[118] Labat argued that because the Africans had neither the technology to extract their abundant resources effectively nor the intelligence to develop this technology, it was incumbent on the French to seize these resources and see that they were exploited.[119] Other than providing labor, which was essential to Labat's schemes, the Africans' role in his program for mine improvement was unclear. He implied that though the indolence of African workers could be overcome only through strict supervision, it was possible to teach them how to use the machines that they themselves were unable to devise. He assured his readers that the cost of his proposals would be low in terms of both men and money. He was also confident that the Africans would come to see that they could much more fully benefit from their mines if they were controlled and run by vastly more competent French engineers and managers.[120]

[117]Jean-Baptiste Labat, *Nouvelle relation de l'Afrique occidentale* (Paris, 1728), vol. 4, pp. 39–56, 73–77. See also Golberry, *Travels in Africa*, vol. 1, pp. 293–95; Lamiral, *L'Afrique*, pp. 35, 321–24; and the Abbé Demanet, *Nouvelle histoire de l'Afrique française* (Paris, 1768), pp. 167–69.

[118] See below, esp. Chapter 4.

[119]Labat, *Nouvelle relation.*, vol. 4, pp. 41, 51. For other contemporary authors who faulted the Africans for their failure to develop resources that the Europeans felt they could easily exploit, see Poivre, *Voyages d'un philosophe*, pp. 9–12; and Römer, *Nachrichten*, p. 10.

[120]Labat, *Nouvelle relation*, vol. 4, pp. 72–79. For similar suggestions by other eighteenth-century authors, see Golberry, *Travels in Africa*, vol. 1, pp. 293–95, 342–62, 356–57; and vol. 2, pp. 301–4; Mercier, *L'Afrique noire*, pp. 117–19, 127.

Not all eighteenth-century observers shared Labat's confidence that the Africans could learn European techniques and operate Western machines. François Le Vaillant, who traveled in southern Africa at the end of the century, thought the Africans incapable of even comprehending, much less using, European weapons and tools. He grew fond of amusing himself and dazzling the "natives" with demonstrations of relatively simple European contraptions. Le Vaillant delighted in the open-mouthed awe shown by some "Hottentots" for a mouth harp that he drew out of a box with the "art and mystery of a quack." He conceded that they rather quickly mastered the "ridiculous instrument," but he never seemed to tire of describing the Africans' fascination with the shining buttons on his coat or the hatchets and axes in his wagon. He boasted that his hand weapons were equal to those of the entire "horde of Hottentots," a claim that was rendered credible by the fact that his personal weapons included two double-barreled pistols and a double-barreled shotgun. Le Vaillant best illustrated the prodigious differences that he believed existed between the technology of his world and that of the "Kaffirs" (whom he considered a step above the "Hottentots") not with double-barreled wonders of metallurgy but with a makeshift bellows that he pieced together from materials scavenged from an African encampment. The "Kaffirs" were so impressed with this humble product of his ad hoc ingenuity, Le Vaillant reported, that they leaped into the air and clapped their hands, entreating him to give them "this wonderful machine."[121]

What Le Vaillant expressed indirectly through tales of actual encounters with Africans, Edward Long stated explicitly and categorically: "Negroes" were incapable of adopting European technology or inventing their own. One of the most outspoken defenders of slavery in the late eighteenth century, Long suggested rather heavy-handedly his contention that Africans or "Negroes" were less than human by belittling their capacity for making or using tools:

> It is astonishing that, although they have been acquainted with Europeans, and their manufactures, for so many hundred years, they have, in all this series of time, manifested so little taste for arts, or a genius either inventive or imitative. Among so great a number of millions of people, we have heard but of one or two insignificant tribes, who comprehend any thing of mechanic arts,

[121]LeVaillant, *Travels*, vol. 1, pp. 268–71, 316–18; vol. 2, pp. 17, 204–6, 212–17, 219, 229–34. See also Robert Norris, *Memoirs of the Reign of Bossa Ahadee, King of Dahomey* (London, 1789), p. 102; and William Smith, *A New Voyage to Guinea* (London, 1744), pp. 15–19.

or manufacture; and even these, for the most part, are said to perform their work in a very bungling and slovenly manner, perhaps not better than an oran-outang might, with little pains, be brought to do.[122]

In succeeding decades the appropriateness of the African-ape comparison would be heatedly debated by European intellectuals and politicians. But the assumption of African technological ineptness would become almost universally accepted by the first decades of the nineteenth century, as the gap between industrializing Europe and "primitive" Africa increased palpably from one year to the next. A handful of Victorian explorers echoed the plaudits of eighteenth-century travelers for the ingenuity and dexterity of African craftsmen, but their praise was drowned out by the rising chorus of European self-congratulation.

The second major assertion of David Hume's influential footnote, that the Africans had "no sciences . . . nor even any individual eminent in either action or speculation,"[123] was much more readily accepted by those of his contemporaries who deemed themselves knowledgeable about Africa. James Beattie, whose knowledge was entirely secondhand, tersely summarized the consensus of the age: "Sciences indeed they have none." The majority of Europeans who had visited Africa would have agreed. In fact, most of Beattie's contemporaries shared his view and that of writers in the early centuries of expansion that the Africans could not have sciences because they did not have "letters."[124] But as evidence from a number of travelers revealed, the Africans' lack of literacy had not prevented some peoples from making astronomical observations that European observers found surprisingly accurate or from developing healing techniques that were remarkably effective. Michael Adanson, the official botanist of the French establishment in Senegal in the 1750s, was impressed by the knowledge of the stars and constellations that the Mandingos possessed. He admitted being surprised that "such a rude and illiterate people should know so much." Other travelers commented on the efficacy of African medicines, such as those for the treatment of snakebites, and some noted their facility with arithmetical sums.[125] Praise for African attainments in the sciences, however, was exceptional. Most travelers stressed the Africans' ignorance of eclipses and other natural phenomena and their "childishly"

---

[122]Edward Long, *The History of Jamaica* (London, 1774), vol. 2, pp. 354–55.
[123]Hume, *Essays*, vol. 1, p.252.
[124]Beattie, *Nature and Immutability of Truth*, p. 311.
[125]Adanson, *Voyage to Senegal*, p. 253–54; Isert, *Voyages en Guinée*, pp. 82–83; W. Smith, *New Voyage*, p. 253.

mythical or superstition-ridden explanations for these occurrences. Either explicitly or implicitly, by omitting any reference to African medicine or astronomy, they concurred with the views of John Atkins that the Africans were "naked of Education and science."[126]

In the context of the debate over the slave trade, the generally accepted view that the Africans had developed little scientific learning was less important than the question of whether or not they had the capacity to think scientifically. Though divided on this issue, eighteenth-century authors argued in the affirmative in numbers that would be unimaginable by the early nineteenth century. In part, authors such as Anthony Benezet and the Abbé Gregoire, who insisted that Africans (or Negroes) were potentially the intellectual equals of Europeans (or whites), mirrored the philosophes' conviction that all groups of human beings were endowed with the same basic mental equipment.[127] It followed logically that if Africans possessed an intelligence "as good, and as capable of Improvement as that of the Whites,"[128] they could learn from the "lettered nations."[129] Authors who espoused popular eighteenth-century explanations for human variations (such as Arthur Lee, who blamed African despotism, and Cornelius de Pauw, who attributed the Africans' "weak intellects" to their enervating environment) also left open the possibility that Africans, both slave and free, were improvable.[130]

Opinion was far from unanimous, however, on the issue of African intelligence or potential for instruction in the sciences. At the height of the abolitionist struggle in the last decades of the century, the view advanced by Hume that Africans (or Amerindians or Asians) were in-

---

[126]Atkins, *Voyage to Guinea*, pp. 80 (quoted portion), 81; Sonnerat, *Voyage*, vol. 2, pp. 37, 41, 143; Edward Ives, *A Voyage from England to India in the Year 1754* (London, 1773), p. 14; Moore, *Travels into the Interior*, p. 143; Golberry, *Travels in Africa*, p. 87; Joseph Corry, *Observations on the Windward Coast of Africa* (London, 1807), pp. 15, 17, 60–62, 97–100; Lamiral, *L'Afrique*, p. 50; J. Cuvelier, *Documents sur une mission française au Kakongo, 1766–1776* (Brussels, 1953), pp. 52–54.

[127]Benezet, *Short Account*, pp. 12–13, 37–38; H. Gregoire, *An Enquiry Concerning the Intellectual and Moral Faculties and Literacy of Negroes* (Brooklyn, N.Y., 1810), p. 214; and Isert, *Voyages en Guinée*, p. 180. For a general discussion of the positive assessment of the African capacity for learning in this period, see A. Barker, *African Link*, pp. 116–19, 160–61, 166–67. On the Enlightenment conviction that faculties for reasoning were fundamentally the same in all human groups, see Cassirer, *Philosophy of the Enlightenment*, pp. 219–20.

[128]Benezet, *Short Account*, p. 7. See also Mercier, *L'Afrique noire*, p. 61; Ives, *Voyage from England*, p. 10; Hoffman, *Le nègre romantique*, p. 17; Golberry, *Travels in Africa*, p. 244; Cuvelier, *Documents* , p. 83; and A. Barker, *African Link*, pp. 116–17, 161.

[129]Atkins, *Voyage to Guinea*, p. 82. It is noteworthy, and reflective of the refreshing lack of rigidity of opinion on race issues at the time, that this judgment was made by an avowed polygenesist—one who believed that the Africans and Europeans had been created in separate divine acts.

[130]Lee, *Essay in Vindication*, pp. 38–39; Mercier, *L'Afrique noire*, p. 104.

nately inferior to "whites" in mental capacity was espoused by a number of influential writers. Samuel Estwick rejected the notion that human nature is universally the same and embraced Hume's opinion that "nature had . . . made an original distinction betwixt [the] breeds of men" and that Africans were "naturally inferior."[131] Though Estwick stressed moral differences, whereas Hume emphasized the ability to reason, Estwick believed that the Africans were incapable of "progress in civility or science." He noted that all attempts to educate the "natives" in the Cape of Good Hope region had proved "hopeless."[132]

Although his conclusions regarding the limits of the African intellect were similar to those of Estwick, Dominique Lamiral took a rather different approach, which amply illustrates some of the pitfalls that the increasingly popular eighteenth-century notion of the noble savage contained for the African, Amerindians and other peoples. Most writers followed Lahontan and Bougainville in locating their noble savages in the Americas or the South Seas,[133] but such Africans as Aphra Behn's famous "Oroonoko" were sometimes included among those "primitives" who had remained close to nature and thus escaped the vices of civilized societies.[134] In a confused and contradictory account of his years as a merchant on the west coast, Lamiral also took up the themes of primitive innocence and bliss in an African setting. He described in vivid detail African landscapes rampant with vegetation and teeming with exotic birds and animals, scenes worthy of the late nineteenth-century "naive" painter Henri Rousseau.[135] Lamiral recounted that in traveling in the Moorish regions one seemed to be carried back to the "cradle of the human species," to an "age of gold" in which men had fewer possessions and consequently fewer needs and desires. The Africans, "who lived under the natural laws of the innocent," had vices, he admitted (and described them in detail in other sections of his account), but before the coming of the Europeans they had known nothing of poverty, greed, theft, flattery, or murder.[136]

Elsewhere in his narrative, Lamiral described a different sort of savage, brutalized by a harsh and unforgiving environment. Both the paradisiacal existence of the "Moors," he concluded, and the degrading struggle for survival of the "Negro" peoples of the interior had sapped

[131]Hume, *Essays*, vol. I, p. 252.
[132]Estwick, *Considerations*, pp. 72–74, 78–79n, 79–80.
[133]Fairchild, *Noble Savage*, pp. 1, 69–70; Urs Bitterli, *Die Entdeckung des Schwarzen Afrikaners* (Zurich, 1970), pp. 82–84; Davis, *Problem of Slavery*, pp. 174–75.
[134]Fairchild, *Noble Savage*, pp. 34–41; Sypher, *Guinea's Captive Kings*, pp. 9–10.
[135]For a fine discussion of this aspect of Lamiral's work, see Mercier, *L'Afrique noire*, pp. 180–83.
[136]Lamiral, *L'Afrique*, pp. 143–47.

their energy and initiative, rendering them improvident, incapable of invention, and unable to break out of existing routines. Lamiral believed that as a result the Africans had done little to develop the arts and sciences. Like Hume, he explicitly invoked a scientific standard for measuring human worth when he surmised that Africans ranked much lower on the "Great Chain of Being" than Europeans because "by nature" they lacked the capacity to explore the "profound secrets of nature." Raising arguments that would be widely accepted in the nineteenth century, he judged that the "savage" Africans were guided by instinct rather than reason and that their "simple" languages left them incapable of expressing or comprehending the principles that underlay occurrences in nature or devices of human invention.[137]

Though authors more sympathetic to the Africans took up the theme of the noble savage, Lamiral's account vividly demonstrates the vulnerability of amorphous romantic notions of savage innocence in an age when scientific and technological measures of human worth were gaining wide acceptance. The sense that they were ignorant of arts, crafts, and invention had been associated with the concept of the noble savage as early as Walter Hamand's work on Madagascar in the 1660s.[138] These failings would reduce peoples regarded as savages, noble or otherwise, to objects of pity and ridicule in the industrial age, when a culture's worth was increasingly judged by the extent to which it had mastered its material environment.

Writers such as Estwick and Lamiral offered their arguments concerning the Africans' limited capacity for intellectual improvement in support of claims that slavery was beneficial to them and that the slave trade ought to be continued. Despite Lamiral's impassioned denunciation of the vices that civilized Europeans who "glorified the arts and sciences" had brought to the innocent African savages via the slave trade, in his 1789 account of Africa he defended the trade. He insisted that the Europeans were "justified" in enslaving Africans because of the latter's lower intelligence and the fact that they were not mentally able to handle living in freedom.[139] Thus, however noble, savages—at least African savages—were fit for enslavement.

A decade and a half earlier Edward Long had made many of the same arguments in defense of slavery in his *History of Jamaica*. In phrases virtually identical to those used by Estwick, who was clearly familiar

137Ibid., pp. 49–50, 180–84, 192, 196–200.
138See L. B. Wright, "The Noble Savage in Madagascar in 1640," *Journal of the History of Ideas* 4 (1943), 115–18.
139Lamiral, *L'Afrique*, pp. 158, 166–71, 204–12, 378–79 (defending the slave trade); pp. 144–47 (denouncing its effects).

with Long's work but did not actually cite it, Long pointed to the "rude state" in which he claimed the Africans had existed for thousands of years. He argued that their incapacity for improvement or "progress in civility or science" was clear evidence of their "natural inferiority."[140] Like Estwick, Long believed that Africans were a different "species" from "whites" and innately inferior to them. But Long went further to back up his racist convictions. He not only cited evidence of the Africans' utter lack of attainments or abilities in the arts and sciences; he enlisted, however feebly, the revered techniques of the scientists themselves to support his claims.[141] Though he had never been to Africa, Long asserted that his extensive observations of slaves in the plantation colony of Jamaica, where he was a resident, gave an empirical validity to his findings. He sought to satisfy the prevailing eighteenth-century sense that classification was a preeminently scientific endeavor by working out a rather crude human/African/ape scale that he managed to fit nicely into the Great Chain of Being.[142] A decade earlier Claude Le Cat, who shared Long's belief that Africans and Europeans were separate species, proposed an even cruder white/black or temperate/tropic gradation.[143]

The notion that Europeans and Africans might be separate species had, of course, been proposed centuries earlier, perhaps first by Paracelsus.[144] Schemes to group the varieties of humankind had also been attempted well before Le Cat and Long. In the 1680s François Bernier had identified five main "races," but made little attempt to rank them, in a rather brief essay, that was mainly devoted to his impressions of the attractions of women in the many lands he had visited.[145] The work of

[140]Long, *History of Jamaica*, vol. 2, pp. 351–56.

[141]The following discussion of Long's views is based primarily on ibid., vol. 2, pp. 336–37, 353–56, 373–74. A substantial literature has developed on pseudoscientific racism, some of which deals at length with its origins in the eighteenth century. See especially Curtin, *Image of Africa*, pp. 28–48; A. Barker, *African Link*, chaps. 3 and 9 on England; Cohen, *French Encounter*, pp. 84–99, 210–15; and William Stanton, *The Leopard's Spots* (Chicago, 1960), pp. 3–24.

[142]Michael Banton, *The Idea of Race* (Boulder, Colo., 1977), pp. 28–29; Jacques Barzun, *Race: A Modern Superstition* (London, 1938), p. 52; and Curtin, *Image of Africa*, pp. 37–40.

[143]Le Cat, *Traité de la couleur de la peau humaine* (Amsterdam, 1765), pp. 2–4. For a discussion of Le Cat's theories, see G. S. Rousseau, "Le Cat and the Physiology of Negroes," in Pagliaro, *Racism*, pp. 369–86.

[144]See Thomas Bendyshe, "History of Anthropology," in *Memoirs Read before the Anthropological Society of London* 1 (1863–64), pp. 353–54; and T. K. Penniman, *A Hundred Years of Anthropology* (London, 1935), p. 41.

[145]The article was published anonymously by "un fameux voyageur" but has been clearly identified as the work of François Bernier. See "Nouvelle division de la terre, par les différentes espèces ou races d'hommes qui l'habitent," *Journal des Sçavans*, 24 April 1684, pp. 85–89. For earlier classificatory efforts, see Margaret Hodgen, *Early Anthropology in the Sixteenth and Seventeenth Centuries* (Philadelphia, 1964), pp. 166–69.

the Swedish botanist Carolus Linnaeus in the first decades of the eighteenth century had given great impetus to efforts at classification, and the consequent obsession with delineating and ranking the human "races" continued into the nineteenth century. Even the German zoologist Johann Blumenbach, who argued for the fundamental unity of humankind and the Africans' capacity for improvement, was caught up in this parody of scientific procedures which invariably reinforced the growing sense of European, "white," or Caucasian superiority.[146]

Long's ventures into biology and physical anthropology in search of proof for his claim that Africans were a separate, subhuman species were as crude as his classificatory schemes. His claims that mulattoes (products of "mixed breeding" between whites and blacks) could not have offspring and that Africans had been known to have sexual intercourse with apes found little support among his contemporaries, even though nineteenth-century authors repeated them.[147] Long's attempts to demonstrate physical differences amounted to little more than exercises in racial narcissism in which he decried the Africans' "bestial fleece, instead of hair" and "bestial and fetid smell."

In the decades after the publication of Long's work, the task of establishing basic anatomical differences between "whites" and "Negroes" was taken a good deal further by such men as S. T. Soemmering and Charles White. Soemmering, a German physician who firmly believed that Africans were "true men as good as whites" and descended from a common ancestor,[148] nevertheless compared a limited sample of "white," "Negro," and ape cadavers and concluded that there were significant physical differences. In general, he said, of all the varieties of humans, the "Negroes" were anatomically closest to the apes. At century's end Charles White, an English physician, carried out his own measurements on a sample of cadavers that appears to have varied in size at different points in time. White was not above freely plagiarizing his German counterpart's findings, however, and, not surprisingly, arrived

[146]Johann Blumenbach, "De Generis Humani: Varietate Nativa (On the Natural Variety of Mankind)," (1775), in Thomas Bendysche, ed., *The Anthropological Treatises of Johann Friedrich Blumenbach* (London, 1865), pp. 264–69. See also Henry Home (Lord Kames), *Sketches of the History of Man* (Edinburgh, 1774), who disputed the environmental emphasis of Blumenbach and many other eighteenth-century authors and argued for fundamental differences between the world's "races."

[147]For discussions of contemporary scientific reactions to these claims, see A. Barker, *African Link*, pp. 52–58; and Philip A. Sloan, "The Idea of Racial Degeneracy in Buffon's 'Histoire naturelle,'" in Pagliaro, *Racism*, pp. 293–321. For samples of these arguments in the nineteenth century, see Georges Pouchet, *De la pluralité des races humaines* (Paris, 1858), pp. 24–29, 58–60, 133–34.

[148]S. T. Soemmering, *Ueber die körperliche Verschiedenheit des Negers vom Europäer* (Frankfurt, 1785), p. xx.

at similar conclusions. In virtually all ways, from the shape of their skulls to the size of their penises, Africans more closely resembled apes than Europeans. While Soemmering opined that many Africans surpassed some Europeans in understanding, White was prepared to concede only that Africans generally were equal to "thousands of Europeans" in "capacity and understanding."[149] Both physicians approvingly cited the observations of the Dutch anatomist Pieter Camper on African and European "facial angles," thus giving a boost to the phrenological investigations that would dominate "scientific" racism throughout much of the nineteenth century.[150]

As Anthony Barker has observed,[151] authors who claimed that their views on blacks were based on scientific testing were in a small minority in the eighteenth century. The argument that Africans were a separate species found little support even among defenders of the slave trade and from men who, like Soemmering and White, shared Edward Long's interest in applying scientific techniques to the study of human types. In fact, Long's views were directly contradicted by leading scientists; Johann Blumenbach declared that he could find no physical evidence for classifying the blacks as a separate species and added that "there is no so-called savage nation known under the sun which has so distinquished itself by such examples of perfectibility and original capacity for *scientific* culture, and thereby attached itself so closely to the most civilized nations of the earth, *as the Negro.*"[152] The prevailing eighteenth-century view was that Africans, Europeans, Chinese, and Indians were all subgroups of the same human species; that despite differences in physical appearance, all were the same in terms of their basic nature.[153] Most thinkers attributed differences in appearance to environmental factors, and varying levels of cultural development to institutional constraints.[154]

[149]Ibid., p. 20; and Charles White, *An Account of the Regular Gradation in Man* (London, 1799), pp. 42–43, 51, 56–61, 80–88, 137 (quoted portion).

[150]Curtin, *Image of Africa*, pp. 39–40; and Barzun, *Race*, pp. 52–58. Perhaps the first to apply phrenological techniques was the Frenchman Louis Daubenton. See A. Firmin, *De l'égalité des races humaines* (Paris, 1885), p. 7.

[151]A. Barker, *African Link*, pp. 157–77.

[152]Blumenbach, "De Generis Humani," p. 305 (original emphasis). See also Beattie, *Nature and Immutability of Truth*, p. 313.

[153]Cassirer, *Philosophy of the Enlightenment*, pp. 6, 219–20; Bryson, *Man and Society*, pp. 53, 242–43; and Gay, *The Enlightenment*, vol. 2, pp. 168–69, 338–39, 380.

[154]For an example of the former, see J. F. Blumenbach, *Über die natürlichen Verschiedenheiten im Menschengeschlechte* (Leipzig, 1793), pp. 98–99; for authors who combine the two approaches, Montesquieu, *Spirit of the Laws*, and Falconer, *Influence of Climate*. For more recent discussions of these trends, see Marshall and Williams, *Map of Mankind*, pp. 129–31, 135–37, 275–76; Cohen, *French Encounter*, pp. 73–78, and Sloan, "Racial Degeneracy in Buffon," in Pagliaro, *Racism*, pp. 293–321. Not all eighteenth-century

As Philip Curtin has shown, however, such writers as Long and Soemmering laid the groundwork for the pseudoscientific works that dominated thinking on race in the nineteenth century.[155] The Africans entered into these deliberations at a great disadvantage. Though the Europeans' esteem for China and, to a lesser extent, India declined in the eighteenth century, it was rare for an author to suggest that their stagnation and inferiority to Europe in the sciences were caused by racial constraints. The alleged backwardness of the Africans, by contrast, had long been vaguely linked to innate or biological differences. In the eighteenth century, Long and White and others gave these connections a clearer definition and a scientific gloss. They introduced a number of assumptions about Africans that would become accepted as fact in the thinking of most nineteenth-century Europeans, from naturalists and explorers to missionaries and colonial officials.

No eighteenth-century author presented these ideas with more force and conviction than did Christopher Meiners in a long essay published in 1790, "On the Nature of the African Negro." Drawing on the work of Soemmering and other physiognomists, Meiners detailed the deficiencies of the "Negro" skull and brain and their similarities to those of apes. He argued that Africans were impervious to diseases that were fatal to Europeans; that they had a much greater innate capacity to resist the heat of the tropics; and that they were insensitive to death and suffering. Without noting the apparent contradiction, he also stressed his belief that Africans had a much more developed sensory system than Europeans. This predisposed them to indolence and lasciviousness, making it difficult for them to advance in the arts and sciences. Scoffing at "tales" of Negroes who had shown the ability to master Western learning, Meiners pointed out that African peoples had improved little, despite centuries of contact with Western civilization. Their "natural stupidity," he argued, "rendered them unfit for anything, but menial labor."[156]

Scientific Gauges and the Spirit of the Times

The growing importance of scientific achievement and, to a lesser extent, inventiveness as gauges by which Europeans evaluated other

---

thinkers accepted the environmentalist position; David Hume, e.g., sought to refute it in his essay "Of National Characters" (*Essays*, pp. 244–58). See also John C. Greene, "Some Early Speculations on the Origins of Racial Degeneracy," *American Anthropologist* 56 (1954), 33–34.

[155]Curtin, *Image of Africa*, esp. pp. 43–48.

[156]Meiners, "Afrikanischen Neger," pp. 403–13, 418–21, 423–28, 430–32, 436, 444–46.

civilizations in the era of the Enlightenment was but one manifestation of the ascendancy of bourgeois priorities and values in European thinking as a whole. Most of the philosophes, especially those with a strong interest in overseas peoples and cultures, were from middle-class origins: Voltaire, for example, was the son of a notary; Colebrooke came from a family of bankers, and Jones from a family of yeomen farmers.[157] But it was not so much their social origins as their educational experiences, the common corpus of books and periodicals they read, and their frequent exchanges with scholars of varying classes at home and abroad that shaped the philosophes' views, whether on metaphysics or on the nature of government in China.[158] Thus, thinkers of noble birth, most notably Montesquieu and Leibniz, shared many of the basic assumptions that informed the writings on India and Africa of fellow philosophes from bourgeois origins.

Though Englightenment thinkers, regardless of their social origins, gave much greater attention to scientific and technological attainments than most writers had done in the early centuries of expansion, these remained subordinate indices of the worth and potential of non-Western peoples. The significance of non-Western scientific learning and mechanical devices was ultimately determined by the ways in which they indicated the presence or, more commonly, the absence of the political, economic, and intellectual freedoms that the philosophes championed above all other causes.[159] The backward state of scientific inquiry in China, for example, and the Chinese failure to develop the full potential of such key inventions as gunpowder and paper were linked to the despotic character of the Chinese government. In this view, corrupt and oppressive bureaucrats, acting on behalf of an autocratic emperor, stifled the free expression, the experimentation, and the exchange of ideas or "freedom of the pen" that most Enlightenment thinkers were convinced constituted the sine qua non of a healthy social order.[160] The consequences included an excessive veneration for order and tradition, intellectual and social stagnation, and China's loss to other civilizations of the lead it had held in earlier centuries in political, military, and technological development. Eighteenth-century writers also used the

[157]On the philosophes in general, see Gay, *The Enlightenment*, vol. 2, pp. 48–49; and Robert Anchor, *The Enlightenment Tradition* (Berkeley, Calif., 1967), p. 16.

[158]On these influences in the Scientific Revolution, see Clark, *Science and Social Welfare*, pp. 87–88.

[159]The most detailed study of the philosophes' assault on despotism in France remains Mornet, *Origines intellectuelles*. For a discussion of these themes with a broader geographical scope, see Gay, *The Enlightenment*, vol. 2, esp. chaps. 8 and 9; and Anchor, *Enlightenment Tradition*. For one of the most explicit attacks on despotism for stifling scientific inquiry and invention, see David Hume, "Of the Rise and Progress of the Arts and Sciences," in *Essays*, esp. pp. 177–87.

[160]Cassirer, *Philosophy of the Enlightenment*, pp. 251–52.

example of China to illustrate the ways in which the corruption and government restrictions that they believed were invariably associated with despotic rule impeded the growth of commerce and the spread of new technologies. Defoe and Anson, among others, railed against the policies and practices of the Qing bureaucracy which, they argued, were the very antithesis of the free-trade, laissez-faire principles that had gained increasing favor among eighteenth-century political economists. These same principles would later be powerfully championed by Adam Smith as the central pillars of the modern liberal state.[161]

Enlightenment thinkers had also held the denial of "natural liberties" by the despotic princes of India responsible for stunting the scientific inquiry that had shown such promise in early times. But for most writers the chief culprits in India were the Brahmin priests and Muslim pundits who had promoted among the Indian masses the superstitions and magic that the philosophes so despised. At the same time, these guardians of the Oriental version of *l'infame* had stifled scientific exchange by jealously monopolizing and slavishly preserving the discoveries that had been passed down from ancient civilizations. Though African rulers also came increasingly to be regarded as arbitrary despots, thinkers in this era tended to link Africa's backwardness in science and technology to environment and historical circumstance rather than to political oppression. Most writers assumed that there was little inquiry or invention to stifle or stunt in African societies. African culture and especially African potential to master European ideas and tools took on their greatest significance in the context of the debate over the abolition of the slave trade. Like the growing hostility to commercial policies in the Chinese empire, the slavery controversy centered on questions of property and free trade. The long dispute basically turned on whether or not some human groups had the right to deprive other groups of their natural liberties and reduce them to commodities in the pursuit of economic advantage.

Ironically, the image of China, the most technologically advanced of the civilizations the Europeans encountered overseas and among the most advanced scientifically, was the most diminished by the application of Enlightenment standards. China had, of course, more to lose. Early travelers and the Jesuit writers had extolled its virtues to a degree unequaled in their responses to any other civilization. But in the age of the Enlightenment, when the techniques and standards of scientific investigation became vital concerns for major European thinkers, the apparent backwardness of Chinese astronomy and medicine and the

---

[161]On these themes, see esp. Dermigny, *La Chine et l'Occident*, vol. 1, chap. 1.

Chinese failure to develop the full potential of their wonderful inventions raised questions about Chinese civilization as a whole. When doubts about Chinese science and technology were coupled—as they were in Defoe's fiction and Anson's exposé—with angry denunciations of the corruption and inefficiency of the mandarin officials of China's once vaunted bureaucracy and the poverty of its masses, they shattered the image of China as a utopian society ruled by philosopher-kings. From a model for Europe, China had been transformed by the last decades of the century into the antithesis of the commercially oriented, bourgeois-dominated, liberal state that the philosophes acclaimed as the ideal for all human societies.

With the late eighteenth-century "discovery" of India's early achievements in the arts, philosophy, and literature, as well as mathematics and the sciences, its standing in the eyes of European observers actually improved. But its time of favor was brief. In the very decades when European scholars were gaining access to its ancient learning, the collapse of the political and economic underpinnings of Indian civilization was brought home to European thinkers by the steady rise of British hegemony in the subcontinent. At the same time, India's early advantages in textile production and trade were nullified by the onset of Britain's Industrial Revolution, which was based on the mechanization of the very same sector of the British economy and increased British efforts to capture overseas markets to absorb their surplus production. Above all, the myth of the subjugation of hundreds of millions of Indians by a handful of Europeans with superior organization and (though this was less true) superior military technology fed a growing European sense of superiority. The vital roles played in the conquest by Indian sepoys, allied princes, and Hindu financiers would not be acknowledged, much less examined in any depth, until well into the era of decolonization a century and a half later. Most European observers were convinced that colonial subjects, like slaves, possessed fundamental deficiencies that accounted for their fall to such a lowly status. The fact that India's great discoveries in the sciences and philosophy were products of past ages when India's peoples ruled themselves could only bolster this conviction.

The sense of such scholars as Barker and Curtin that, in general, European attitudes toward the Africans became more favorable in the late eighteenth century than they had been in the early centuries of expansion is well founded. But this shift had little to do with new revelations concerning African attainments in science or technology. Rather, even though they did not usually explicitly condone African "fetishism" or other customs that were considered bizarre, the philosophes' relativistic approach to foreign peoples and their firm commitment to tolerance caused them to regard the aspects of African culture that had proved so offensive to earlier travelers as innocuous curiosities or as social differences that required explanation. Enlightenment authors saw nudity, for example, much less as a sign of moral depravity than as an understandable adaptation to hot and humid climates. The romantics of the era even extolled the nakedness of the Africans and Tahitians as emblems of their innocence, their primitive lack of material needs, and their freedom from the fetters of social convention that constricted and distorted social interaction in Europe.

Defenders of the slave trade, of course, sought to reinforce earlier criticisms of African peoples and cultures. But their writings concentrated on slavery as an institution and on conditions in the plantation colonies, and added little that was new about Africa itself. Opponents of slavery, who nonetheless disapproved of African customs, often blamed these on the pernicious effects of the slave trade or concentrated on favorable aspects of African culture and, more commonly, character. They emphasized African hospitality, kindness, loyalty, and courage under stress, celebrating these traits in poems, plays, and novels about noble slaves rescuing kind masters or suffering with dignity the oppression of cruel ones.

Little new was added to what had already been written about African technology or scientific knowledge. In fact, European writers continued to believe that the former was at best primitive, the latter virtually nonexistent. But a few writers, including the Scotsman James Beattie, ridiculed those who dismissed the Africans and Amerindians as barbarians simply because they had not matched the Europeans in science and technology. In a retort remarkable for its cultural relativism, Beattie charged that a few individuals, often by accident—not Englishmen or Frenchmen as a whole—were responsible for the discoveries and inventions that some claimed were proof of European superiority. An African or Amerindian visitor to Europe, Beattie asserted, was unlikely to notice the esteemed inventors or scientists but would very likely be appalled by Europe's "most fashionable duellists, gamblers and adulterers," as well as other "specimens" of its "brutish barbarity and sottish infatuation." Great Britain and France, he pointed out, were as "savage" some millennia earlier as America or Africa in his own day. He insisted that there was no reason to assume that the Africans and Amerindians possessed any less potential to develop the sciences and technology than the Europeans had demonstrated.[162]

---

[162]Beattie, *Nature and Immutability of Truth*, pp. 267, 317.

As the Industrial Revolution widened the gap in science and technology between Europeans and Africans, however, the perceived lack of inventiveness and scientific curiosity on the part of the Africans came to be interpreted as evidence of their innate lack of potential in these spheres. This trend, which culminated in the nineteenth century in the dominance of biological or racial, rather than environmental or cultural, explanations for African (or Asian or Amerindian) backwardness, exposed the vulnerability of the somewhat patronizing defense of the docile, faithful "noble Negro." Hospitality was transposed into indolence, docility into cowardice, loyalty into the necessary submission of the racially inferior.

PART TWO

THE AGE OF
INDUSTRIALIZATION

Before men begin to think much and long on the peculiarities of their own times, they must have begun to think that those times are, or are destined to be, distinguished in a very remarkable manner from the times which have preceded them. Mankind are then divided, into those who are still what they were, and those who have changed: into the men of the present age, and the men of the past. To the former, the spirit of the age is a subject of exultation; to the latter, of terror; to both, of eager and anxious interest.

John Stuart Mill, "The Spirit of the Age"

White men believed and still believe in the incontestable superiority of their brain. They believed and still believe that this superiority gives them domination over those men who are not white. The importance of scientific discoveries in Europe, the ingenuity and number of their applications, have contributed a great deal to the establishment of this new dogma.

Pierre Mille, "La race supérieure"

But the inheritance of property by itself is very far from an evil; for without the accumulation of capital the arts could not progress; and it is chiefly through their power that the civilized races have extended their range so as to take the place of the lower races.

Charles Darwin, *The Descent of Man*

Visit of the "Ambassadors" of Foutah Djallon to the printing establishment of the "Petit Journal," an engraving from the *Journal Illustré* (1882), suggests the delight of the French in the wonderment with which the African notables regard their machines. (William H. Schneider, *An Empire for the Masses: The French Popular Image of Africa, 1870–1900*, Contributions in Comparative Colonial Studies, No. 11, Greenwood Press, Inc., Westport, Conn., 1982, p. 161. Copyright © 1982 by William H. Schneider. Reprinted with permission)

The East Indian Railway, lithograph from *The Illustrated London News* (19 Sept. 1853), conveying a sense of the imperial power and presence that these great machines manifested for the subjects of the Raj.

A LITTLE TEA PARTY.

Britannia. "A LITTLE MORE GUNPOWDER, MR. CHINA?"   China. "O—NO—TAN—KE—MUM."

A Little Tea Party, a mid-nineteenth-century cartoon from the satirical magazine *Punch*, rather smugly expresses British pride in the most obvious manifestation of their technological superiority over the once mighty Chinese empire. (Reproduced by courtesy of the Trustees of the British Museum)

Global Hegemony and the Rise
of Technology as the Main Measure
# of Human Achievement

THE RISE of the industrial order in western Europe was a good deal more gradual and cumulative than the standard combination of the terms "industrial" and "revolution" would suggest. In some sectors the forces that made steam-powered factory production possible had been building for centuries, and even in England traditional hand-icraft manufacturing predominated in most industries until the second half of the nineteenth century.[1] Nonetheless, from the 1780s in Britain there was an acceleration in the pace of invention and, perhaps even more critically, an increasing application of new inventions and earlier advances in engineering to mineral extraction, manufacturing, and transportation. This shift was noted before the end of the decade by such perceptive contemporary observers as Arthur Young, who, after witnessing the spread of steam-driven machines from the cotton to the woolen textile industry, declared that a "revolution was in the making" which would transform the "appearance of the civilized world."[2] R. H.

---

[1]For recent studies stressing the gradual and incremental development of industrial technology, see A. E. Musson, *The Growth of British Industry* (London, 1978), and G. N. von Tunzelmann, *Steam Power and British Industrialization* (Oxford, 1978), for England; see Clive Trebilcock, *The Industrialization of the Continental Powers, 1780–1914* (London, 1981), for continental Europe. This view was, of course, advanced decades ago by such authors as J. H. Clapham, Paul Mantoux, T. S. Ashton, and esp. J. U. Nef. Nef's extreme arguments for pushing back the beginnings of industrialization as far as the sixteenth century have received rough treatment at the hands of subsequent scholars. On the persistence of handicraft production, see Musson, *British Industry*, and the references in Chapter 2, n 22.

[2]Quoted in H. Heaton, "Industrial Revolution," in R. M. Hartwell, ed., *The Causes of the Industrial Revolution* (London, 1967), p. 35.

Tawney's estimate that the early decades of industrialization brought about the most profound reshaping of the "material appearance" of England since the last geological upheavals rather dramatically underscores the degree to which Young's predictions were fulfilled.[3]

As industrialization spread both within England and to Belgium, the Rhineland, Saxony, and parts of France in the early decades of the nineteenth century, European travelers, missionaries, and colonial policymakers grew more and more sensitive to the fundamental differences that set their societies off from all others. They increasingly stressed Europe's uniqueness and invariably proclaimed its superiority to even the most advanced of its civilized rivals. Though by no means eschewing the sense of religious righteousness or the physical narcissism that had been the preeminent standards by which Europeans compared themselves with overseas peoples in the first centuries of expansion, European observers came to view science and especially technology as the most objective and unassailable measures of their own civilization's past achievement and present worth. In science and technology their superiority was readily demonstrable, and their advantages over other peoples grew at an ever increasing pace. This was particularly true after Europe and its North American progeny entered a new phase of industrial development based on steel, electrification, and chemical production in the last decades of the nineteenth century. Prominent social theorists and policymakers drew varying, often conflicting, conclusions from the undeniable fact of Europe's material mastery and its concomitant global hegemony, but few disputed that machines were the most reliable measure of humankind.

In the last two decades of the eighteenth century more patents were issued in Great Britain than had been granted for new inventions in all the preceding years of the century combined.[4] The successful application of Watt's rotative steam engine to the spinning of yarn on rollers in 1785 and to Crompton's spinning "mule" in 1790 greatly facilitated the concentration of cotton textile production in factories and urban areas and vastly increased British output. Between 1782 and 1800, British

[3]Quoted in ibid.

[4]This overview of the first Industrial Revolution is based primarily on T. S. Ashton, *The Industrial Revolution, 1760–1830* (Oxford, 1948), and *An Economic History of England: The 18th Century* (London, 1966); Samuel Lilley, "Technological Progress and the Industrial Revolution, 1700–1914," in Carlo Cipolla, ed., *The Fontana Economic History of Europe* (Glasgow, 1974), vol. 3, pp. 187–255; Musson, *British Industry*; D. S. L. Cardwell, *Steampower in the Eighteenth Century* (London, 1963); Peter Mathias, *The First Industrial Nation* (New York, 1969); Phyllis Deane, *The First Industrial Revolution* (Cambridge, Eng., 1965); and David Landes, *The Unbound Prometheus* (Cambridge, Mass., 1969), chap. 2. Only additional sources or specific statistics are cited below.

broad cloth production grew by three-fourths, its printed cloth by four-fifths, and its cotton piece goods by nine-tenths. Between 1785 and 1822, overall British cotton textile production increased ten times. In 1783–84 Henry Cort developed the "puddling" and rolling process of iron production that allowed the substitution of coke, which Britain possessed in abundance, for charcoal, which centuries of deforestation for shipbuilding and manufacture had rendered a scarce resource. Cort's innovations also made it possible for the British to use their own lower grade of iron ore, rather than importing higher quality ore from the Baltic, and led to a dramatic increase in British iron production from an estimated 61,000 tons in 1788 to over 227,000 tons in 1806. In the decades following Cort's breakthrough, iron steadily replaced wood, leather, and stone in virtually all fields of technical endeavor from ship- and bridgebuilding to the manufacture of machines and farm implements. These substitutions made possible great increases in the scale of building and machine design. In the 1790s, for example, the first five- and six-story buildings, supported by iron frames, were constructed both in England and on the Continent.[5]

In this same period steampower was also applied to a wide range of industries from brewing and distilling to corn milling and papermaking. The multiplication of Watt's rotative engines alone provides a striking index of the quickening pace of change. The first engine was in operation by 1784; a decade later 150 were employed in a variety of industries; and just five years after that, more than 500 Watt engines were in use.[6] In the last decades of the century in Britain and France, as well as across the Atlantic in the fledgling American republic, efforts were under way to harness the power generated by the increasingly efficient steam engine to transportation. After several failures the Marquis Jouffroy d'Abban's steam-driven vessel successfully navigated a short stretch of the river Saone near Lyons in 1783. Just over two decades later the American painter-turned-inventor Robert Fulton proved that paddle-driven steamers could be commercially viable. Steamboats would not dominate oceanic shipping until decades later, but their potential to do so was established at the outset of the nineteenth century. In the same period attempts were made in France and England to apply steam power to land transport. Wooden rails had been in use in German and British mining and iron manufacturing areas since the sixteenth century, providing a more or less regular running surface for horse-drawn wagons, but the earliest experiments with steam-driven land transport were per-

[5]Carl W. Condit, "Building and Construction," in M. Kranzberg and C. Pursell, eds., *Technology in Western Civilization* (London, 1967), vol. 1, p. 368.

[6]Mathias, *First Industrial Nation*, p. 135; Ashton, *Industrial Revolution*, p. 70.

formed on ordinary streets and roadways. As early as 1769, a steam carriage trundled—briefly—through the streets of Paris. In the first years of the nineteenth century, Richard Trevithick achieved much greater success in a series of road trials in Cornwall and London which led to the introduction of the smokestack and a steam gauge. Trevithick's subsequent experiments with locomotives on cast-iron tramways inspired George Stephenson's "Rocket," which launched the railway age with its victory in the 1829 competition on the newly constructed Liverpool to Manchester tracks.

The coal- and steam-powered revolution in production, transportation, and extraction had been compressed into a remarkably brief span of three or four decades. These transformations were so rapid and fundamental relative to change in all preceding eras that only the most myopic or isolated of George III's subjects could remain unaware or unaffected. "The people of the day were not deceived by the pristine air of much of Britain's landscape," David Landes has observed. "They knew they had passed through a revolution."[7] An 1808 article in *The Times*, for example, assumed that "the extraordinary effect of mechanical power" was "already known to the world" and that "the novelty, singularity and powerful application [of machines] against time and speed" had "created admiration in the minds of every scientific man." In 1815 Sir Richard Philips recounted how a walk through London had left him with vivid impressions of the "triumphs of mechanics" and "the precision and grandeur of action that was really sublime," a source of astonishment to every onlooker.[8]

In the mid-1830s the great essayist Thomas Babington Macaulay celebrated the uniqueness of this revolution and Britain's distinction in having initiated it. He confidently judged that the English were the "greatest and most highly civilized people that ever the world saw." In support of this sweeping assertion, Macaulay drew attention to Britain's vast empire and its powerful maritime fleet, which, he argued, could "annihilate in a quarter of an hour the [preindustrial] navies of Tyre, Athens, Carthage, Venice and Genoa together." He also stressed as proof of Britain's greatness the great advances that had been achieved in medicine, transportation, and "every mechanical art, every manufacture . . . to a perfection that our ancestors would have thought magical." In

---

[7]Landes, *Prometheus*, p. 122. For a fuller discussion of contemporary awareness of these changes, see Maxine Berg, *The Machinery Question and the Making of Political Economy, 1815–1848* (Cambridge, Eng., 1980), esp. pp. 9, 20.

[8]These reactions are recorded in the excerpts included in Humphrey Jennings, *Pandaemonium, 1660–1886: The Coming of the Machine Age as Seen by Contemporary Observers* (New York, 1985), p. 128. For workers' and farmers' reactions, see pp. 235, 238.

an earlier essay he had boasted that the British were better fed and clothed than any people who had previously existed because of the great gains in wealth that industrialization had made possible.[9] Though less exuberant than Macaulay and more sensitive to the adverse effects and potential dangers of industrialization, John Stuart Mill best expressed the sense of a large majority of the English middle class that machines had provided the means for the progressive improvement of humanity. In his *Principles of Political Economy*, first published in 1848, Mill also stressed the convergence of science and technology, which was of increasing importance as industrialization advanced and spread abroad:

> Our knowledge of the properties and laws of physical objects shows no sign of approaching its ultimate boundaries: it is advancing more rapidly, and in a greater number of directions at once, than in any previous age or generation, and affording such frequent glimpses of unexplored fields beyond, as to justify the belief that our acquaintance with nature is still almost in infancy. This increasing physical knowledge is now, too, more rapidly than at any former period, converted, by practical ingenuity, into physical power. The most marvellous of modern inventions, one which realizes the imaginary feats of the magician, not metaphorically but literally—the electro-magnetic telegraph—sprang into existence but a few years after the establishment of the scientific theory which it realizes and exemplifies. Lastly, the manual part of these scientific operations is now never wanting to the intellectual: there is no difficulty in finding or forming, in a sufficient number of the working hands of the community, the skill requisite for executing the most delicate processes of the application of science to practical uses. From this union of conditions, it is impossible not to look forward to a vast multiplication and long succession of contrivances for economizing labour and increasing its produce; and to an ever wider diffusion of the use and benefit of these contrivances.[10]

Some Britons, of course, were a good deal less enthusiastic about the effects of industrialization. Critics as diverse as Robert Owen and William Blake, Charles Dickens and Elizabeth Gaskell drew attention to the pollution of the factory towns and the miserable condition of the workers whose ranks, particularly in textile mills, included a high proportion of women and children in the early industrial era.[11] Though

[9]Quoted in Walter E. Houghton, *The Victorian Frame of Mind* (New Haven, Conn., 1957), pp. 39–40.
[10]John Stuart Mill, *Principles of Political Economy* (London, 1909), pp. 696–97.
[11]For a useful survey of literary responses to industrialization (which, however, ne-

some critics, particularly among the Tories, condemned machines themselves for blighting the English countryside and supplanting paternal ties with class hostility, many Tory intellectuals and radical reformers made a distinction between technological accomplishments and abuses caused by defects in the political and social systems into which machines were introduced. As Maxine Berg has shown, working-class spokesmen and radical intellectuals could both deplore the dislocations and suffering brought on by the "iron monster[s] with a pulse of steam" and promote machines "adapted to the needs of co-operative production" as essential elements in the workers' utopias they aspired to create. Even Owen and his disciples stressed the many ways in which mechanization could relieve workers, both male and female, from the drudgery of routine tasks, from sawing and grinding to cooking and washing clothes.[12]

Thomas Carlyle, one of the most influential Tory critics of industrialization, displayed a similar ambivalence. He decried the degradation of the industrial laboring classes and the filth of the mining and industrial towns without abandoning faith in the potential of the new machines that had produced them. He despised the factory organization that reduced workers from skilled artisans to appendages of machines, but he eschewed "criticism of industry as such." Carlyle delighted in Britain's "industrial animation," which he believed could, with proper supervision and a different form of organization, give rise to a society in which labor was vastly more productive and fulfilling than in any that had preceded it. His impulses to retreat into the Middle Ages or the seventeenth century were balanced by an "equally idealistic and often imperialistic faith in British industrialism, progress and labor."[13] Carlyle chided those who could see only "smoke and dust," "tumult and contentious squalor," in the great industrial centers such as Manchester. In his 1839 essay "Chartism" he urged his numerous readers to look beneath the surface ugliness of the industrial landscape in order to exult with him in the awesome precision and power, the triumph of man over

glects the fine novels of Elizabeth Gaskell), see Herbert Sussman, *The Victorians and the Machine: The Literary Response to Technology* (Cambridge, Mass., 1968). For early debates over the effects of industrialization, see Gertrude Himmelfarb, *The Idea of Poverty: England in the Early Industrial Age* (New York, 1984); and Berg, *Machinery Question*. For sample contemporary criticisms of industrialization, see Jennings, *Pandaemonium*, pp. 145–46, 165–66, 171–72, 233–35, 273, 319.

[12]Berg, *Machinery Question*, pp. 271–74, 279–81 (quoted portions, pp. 270, 271).

[13]Albert J. LaValley, *Carlyle and the Idea of the Modern* (New York, 1968), pp. 185, 203–8 (quoted portions pp. 203, 205). For an earlier and more typical handling of Carlyle's views on industry, which nonetheless concedes his faith in the potential of the new technology, see Frederick W. Roe, *The Social Philosophy of Carlyle and Ruskin* (New York, 1921), esp. pp. 48–70, 88–89, 107, 118–19.

nature, that the factories represented. He insisted that "a precious substance, beautiful as magic dreams, and yet no dream but a reality, lies hidden in that noisome wrappage; a wrappage struggling indeed . . . to cast itself off, and leave the beauty free and visible there."[14]

A decade later the popular novelist Charles Kingsley gave a teleological twist to the British celebration of their industrial achievements. The hearts of his middle-class readers must have swelled with pride as they read the passage in which Lancelot Smith, the spirited protagonist of Kingsley's best-selling novel *Yeast*, dismisses his Catholic cousin's "Romish Sanctity" and extolls the virtues of his own English "Civilization." Deriding the meager achievements of centuries dominated by saints and Jesuits, Smith champions "the political economist, the sanitary reformer [and] the engineer," who invented, repair, and run the machines—"spinning jenny and the railroad, Cunard's liners and the electric telegraph"—which he sees as the surest signs of Britain's power and greatness. His impromptu discourse builds to a remarkable rhetorical finale that expresses sentiments shared by Britons at all social levels in this era. Smith informs his abject cousin that the technological advances of the British are "signs that we are, on some points at least, in harmony with the universe; that there is a mighty spirit working among us, who cannot be your anarchic and destroying Devil, and therefore must be the Ordering and Creating God."[15] Kingsley's novel proved a fitting prelude to the Great Exhibition in Hyde Park two years later, which provided an occasion for the British to congratulate themselves on their truly remarkable material accomplishments and to impress their European and American rivals, as well as their colonial subjects, with the commanding lead they enjoyed in most aspects of technological development.

Although widespread industrialization in France occurred at a later date than in England, French thinkers and inventors had for centuries played leading roles in Europe's scientific and technological advance.[16] In the late seventeenth century the Académie des Sciences had commit-

---

[14]Thomas Carlyle, "Chartism," in *English and Other Critical Essays* (London, 1925), p. 219.

[15]Charles Kingsley, *Yeast* (London, 1849), pp. 77–78. Another clergyman, John Cumming, in *The Great Tribulation* (London, 1859), pp. 295–96, dubbed the scientific and industrial revolutions "a grand regenesis."

[16]Trebilcock's *Industrialization of the Continental Powers* includes an excellent synthesis of recent work on French industrialization. In numerous essays F. Crouzet has explored the formulative decades of French mechanization from a variety of perspectives. His work can be supplemented by the more general works of Rondo Cameron, C. Fohlen, and Charles Ballot.

ted itself by statute to the search for practical applications of its scientific investigations. The French led all other peoples in the systemization and institutionalization of the study of science. They were the first to establish government agencies to oversee civil engineering projects, an advanced school of engineering (l'Ecole des Ponts et Chaussées, founded in 1747), and a technical college (l'Ecole Polytechnique, founded in 1794).[17] French artisan-engineers such as J. M. Jacquard, who devised major improvements in the draw loom, and Jacques de Vaucanson, who made major contributions to machine development from lathes to looms, belie the stereotype of the French as theoreticians rather than inventors. As we have seen, the French pioneered efforts to apply steam power to land and water transport. French scientists such as Nicolas Leblanc, who founded the soda industry, and Claude Louis Berthollet, who developed chlorine for bleaching and thereby broke one of the main bottlenecks in textile manufacture, also advanced the process of industrialization.

However wide of the mark Louis Jaucourt's predictions, in his articles on manufacturing in the *Encyclopédie*, that factory production could never withstand competition from handicraft producers,[18] many French intellectuals displayed an avid interest in new inventions and a strong awareness of the changes they were setting in motion. Denis Diderot, editor of the *Encyclopédie*, prided himself on frequent visits to the "ablest craftsmen" in France.[19] And though some of the plates in the supplementary volumes to the *Encyclopédie* were dated or lifted from previous publications,[20] the production of eleven volumes of exquisitely rendered and expensive illustrations of tools and techniques is itself ample testimony to the great interest in technology among the educated classes in late eighteenth-century France.

The works of such writers as Condorcet and Julien Virey indicate an early and acute sensitivity to the roles of science and technology in long-term historical transformations. Characteristically, they continued to stress the importance of science, perhaps more than their British counterparts did, but to a much greater degree than writers earlier in the century they linked science and technology as historical forces. In the successive editions of his *Histoire naturelle de genre humaine*, Virey extolled discoveries in the sciences as the "true lever of power for man" and argued that experiment and invention were the true source of Eu-

[17]For discussions of these developments, see Rondo Cameron, *France and the Economic Development of Europe, 1800–1914* (Princeton, N.J., 1961), pp. 34–43; and G. Pinet, *Histoire de l'Ecole Polytechnique* (Paris, 1887).

[18]John Lough, *The Encyclopédie* (New York, 1971), pp. 360–62.

[19]Peter Gay, *The Enlightenment* (London, 1979), vol. 2, pp. 252–53.

[20]Lough, *Encyclopédie*, pp. 85–91.

rope's recently won dominance over nature and all other peoples. In a characteristic passage he marveled at the way in which so simple a device as a compass ("une petite aiguille aimantée, placée sur un pivot") had made possible the discovery of unknown worlds, which in turn had brought Europe vastly greater wealth than all of the "Romans' pillaging." Likewise, he argued, the elementary chemical mixture that produced gunpowder made possible Europe's emergence as a global power able to draw tribute from "the most opulent nations."[21] Virey's contemporary B. G. E. Delaville Lacépède included a lengthy paean to European discoveries in science and technology in his continuation of Buffon's monumental *Histoire naturelle de l'homme*. Lacépède saw these achievements as clear proof of his claim that the civilizations fringing the Mediterranean, and especially those that had arisen in Europe, represented the height of human accomplishment and the most nearly perfect episodes in human history.[22]

By the middle decades of the nineteenth century, the industrializing nations of western Europe and North America had greatly enhanced the advantages in the mastery of the material world which they had gained centuries earlier over all other societies. In the last decades of the century a new "cluster of innovations" launched a second wave of industrialization which increased Europe's superiority exponentially in virtually all fields of science and technology. Henry Bessemer's experiments in the 1850s with new ways of firing pig iron and the improvements made by Frederick Siemens and the Martin brothers in the following years made possible the manufacture of cheap and abundant steel. With the opening of the Suez Canal in 1869, oceangoing steamships soon eclipsed sailing vessels in global commerce and travel. The development of internal combustion engines in the last decades of the century led to the substitution of oil for coal as a source of energy, particularly for transportation. In the decades before the Great War, diesel and petrol engines enabled Europeans to travel at even greater speeds on land and sea. Electricity supplied yet another new source of power for the second Industrial Revolution. Innovations in production design and factory organization, increases in plant scale and both vertical and horizontal integration within and between industries, and remarkable advances in precision instrumentation and machine-tooling were also hallmarks of the new era. The new wave of inventions and improvements in the organization of production spawned in turn a rapid proliferation of durable consumer

[21]Julien Virey, *Histoire naturelle de genre humaine* (Paris, 1826), pp. xxii–xxiv.

[22]B. G. E. Delaville Lacépède, *Histoire naturelle de l'homme* (Paris, 1827), pp. 294–96, 304–20.

goods in the last decades of the century.[23] Perhaps even more than steel mills or power plants, machine-made draperies and carpets and vacuum cleaners (introduced in the 1850s and 1860s), telephones, cash registers, and typewriters (first marketed in the 1870s), and sewing machines and bicycles made the average citizen of western European and North American societies aware of the extent of the transformations that were under way.

The greatest gains from the second Industrial Revolution were achieved in emerging rather than established industrial centers. The entrepreneurs and engineers of Germany and the United States were less committed to obsolescent techniques and physical plants than those of Great Britain and thus better positioned to exploit new power sources and modes of production. Both follower nations had surpassed Great Britain in electrification, steel production, and machine-tooling well before the Great War. Both nations, but especially Germany, made extensive use of scientific expertise in industrial development. By the last decades of the century, large staffs of scientists and advanced laboratories were integral parts of major industrial firms. The knowledge and experimental techniques of chemists and other scientists were essential to the growth of the synthetic dye industry, which the Germans pioneered in the late nineteenth century. The Germans and French also solidified the links between science and industrial production through the establishment of a network of state-supported polytechnical schools—a pattern that the British adopted only belatedly and for the most part grudgingly.[24] The connections between science and technology—which had significantly increased but remained sporadic, individualistic, and informal during the first Industrial Revolution, were institutionalized and professionalized in the second. By century's end, science and technology had become "mirror-image twins": one oriented to experiment, theory, and knowing; the other to application, design, and doing; but both joined in a systematic endeavor to uncover the secrets and harness the energies of the natural world.[25]

[23]The fullest study of the consumer side of the Industrial Revolution remains Siegfried Giedeon, *Mechanization Takes Command* (Oxford, 1948). See also Carroll J. Pursell, Jr., "Machines and Machine Tools, 1830–1880," in Kranzberg and Pursell, *Technology*, pp. 396–404; and Stephen Kern, *The Culture of Time and Space* (Cambridge, Mass., 1983), esp. chap. 5.

[24]The most detailed study of technical education, focusing on England but making frequent comparisons with continental Europe, is D. S. L. Cardwell, *The Organisation of Science in England* (London, 1972).

[25]Lilley, "Technological Progress," pp. 235–42; Peter Mathias, *The Transformation of England* (London, 1979), pp. 79–86; Cardwell, *Organisation of Science*, pp. 13–18, 22–28; D. S. L. Cardwell, *Technology, Science, and History* (London, 1972), pp. 215–18; Edwin Layton, "Mirror-Image Twins: The Communities of Science and Technology in 19th-Century America," *Technology and Culture* 12/4 (1971), 562–80.

As industry spread, locomotives increased in speed and comfort, inventions multiplied, and European scientists probed ever more deeply into the workings of nature, Europeans translated their material superiority into global hegemony. Industrialized nations—first Great Britain and later France, Germany, and the United States—flooded African and Asian markets with cheap machine-made consumer goods, ranging from cotton textiles to kerosene lamps. The great advances in weapons design and production that accompanied the process of industrialization permitted the Europeans to subdue forcibly any overseas peoples who resisted their efforts to trade, convert, or explore. When combined with military forces which, well before the Industrial Revolution, had achieved clear superiority in discipline and organization, mass-produced weapons, railway and telegraph lines, and iron-clad steamships made it possible for the Europeans to conquer and rule directly—or defeat and control through indigenous surrogates—virtually all African and Asian peoples.[26] Each year between 1871 and 1914 the European imperialist powers added an area the size of France to their empires.

But it was not the conquests themselves that most impressed European observers (victory had become rather routine, despite temporary reverses such as those the French suffered in the 1880s in Vietnam); it was the relatively low cost and the growing ease of colonial conquest. Though only small forces and not always the most advanced weapons were committed to colonial wars, conflict with the Europeans meant that African warriors or the banner armies of China were subjected, as D. A. Low points out, to "vastly greater, more lethal demonstrations of force than any which [they] had experienced from any quarter in the past."[27] An operation such as the destruction of large flotillas of Chinese war junks by small squadrons of British men-of-war or the annihilation of thousands of determined Mahdist warriors at Omdurman served to bolster the consensus among European thinkers, politicians, and colonial administrators that they had earned the right (and duty) to be the "lords of humankind."

As the evidence of their material achievement multiplied and per-

---

[26]The military dimensions of European technological superiority have received more attention than any other. The best cross-cultural analysis can be found in William H. McNeill, *The Pursuit of Power* (Chicago, 1982). For Europe's "splendid little wars" in the nineteenth century, see William McElwee, *The Art of War: Waterloo to Mons* (Bloomington, Ind., 1974), esp. chaps. 3 and 7. On the broader technological and scientific dimensions of European global conquest in the nineteenth century, see Daniel R. Headrick, *The Tools of Empire* (New York, 1981), and *The Tentacles of Progress* (New York, 1988); and William Woodruff's *Impact of Western Man* (New York, 1967), esp. chaps. 5 and 6.

[27]Quoted in Terrence Ranger, "Connections between 'Primary Resistance' Movements and Modern Mass Nationalism in East and Central Africa," *Journal of African History* 9/4, pt. 1 (1968), 440.

vaded all aspects of life in industrializing societies, Europeans and (increasingly) Americans grew more and more conscious of the uniqueness and, they believed, the superiority of Western civilization. Those involved in the colonies and intellectuals who dealt with colonial issues came to view scientific and technological achievements not only as the key attributes that set Europe off from all other civilizations, past and present, but as the most meaningful gauges by which non-Western societies might be evaluated, classified, and ranked. Science and technology were often conflated as criteria for comparison, rather than treated as distinct endeavors, as they had tended to be in earlier centuries. In addition, generalized assertions of superiority or inferiority supplanted the detailed descriptions of individual tools or particular ideas which had been characteristic of earlier accounts of overseas cultures. This shift reflected a growing sense on the part of overseas observers that African and Asian peoples had little to offer Europe in techniques of production and extraction or in insights into the workings of the natural world. Increasingly industrialized European (and North American) cultures *as a whole* were seen to be a separate class, distinct from all others. The polarities were numerous and obvious: metal versus wood; machine versus human or animal power; science versus superstition and myth; synthetic versus organic; progressive versus stagnant. All aspects of culture could be linked to these polarities, to the fundamental dichotomy between industrial and preindustrial societies. Beggars, for example, who had once wandered in great bands throughout France and England and whose sparse numbers in China and Africa had once been noted with approbation by European observers, now scandalized visitors to the "Orient" and the "Dark Continent." As numerous travelers recounted on arriving in a non-Western port (which increasingly meant a colonized area), the very tempo of speech and motion slowed noticeably, and even the spatial arrangement of the material environment altered disconcertingly.

Most nineteenth-century observers mixed nontechnological or nonscientific gauges—systems of government, ethical codes, treatment of women, religious practices, and so on—with assessments of African and Asian material mastery. As the century passed, however, colonial administrators and missionaries, travelers and social commentators increasingly stressed technological and scientific standards as the most reliable basis for comparisons between societies and civilizations. In an age when what were held to be "scientific" proofs were increasingly demanded of those engaged in the study of natural history and social development, material achievement and anatomical measurements proved irresistible gauges of human capacity and worth. Mechanical principles and mathematical propositions could be tested; bridges, machines, and head sizes could be measured and rated for efficiency. Thus, unlike earlier gauges by which the Europeans compared societies, those favored in the nineteenth century were believed to be amenable to empirical verification and were especially suited to the late Victorian penchant for "statistical reductiveness."[28]

Improved and cheaper techniques for mass printing, advances in graphics and statistical enumeration, and (from the middle of the century) developments in photography all reinforced the preference for evaluative criteria that were tangible and testable. Empirical observation and "objective" evaluation appeared to elevate comparisons between Europeans and African and Asian peoples to the realm of "demonstrated truths," which John Seeley—one of the great champions of late nineteenth-century British imperial expansion—viewed as the main source of the greatness of Western civilization. Contrary to widely accepted contemporary assumptions, Seeley suggested that British superiority over Indians and other subject peoples did not arise from larger brains or cleverer ideas. British dominance, Seeley argued, was based on ideas that were "better tested and sounder."[29] Some decades earlier an aside by Gustave d'Eichthal, a disciple of the famed French sociologist August Comte, had provided an even more striking example of the degree to which nineteenth-century thinkers were convinced that empirical testing had advanced the reliability of the standards by which they ranked humankind. D'Eichthal observed that even if one disregarded the physical appearance of blacks—in itself enough to indicate that they were a "different sort of humanity" from whites—the division between the two races could not be contested, because it had been "scientifically demonstrated" since the end of the eighteenth century.[30] The belief that European views of non-Western peoples were grounded in scientific investigation was popularized in the last decades of the nineteenth century by such writers as Winwood Reade. In his suitably lurid *Savage Africa*, Reade admitted that he found it difficult to resist those who argued for African inferiority because it was futile to "struggle against the sacred facts of science."[31]

Because nineteenth-century Europeans believed that machines, skull size, or ideas about the configuration of the solar system were culturally

[28]On this propensity, see Richard D. Altick, *Victorian Peoples and Ideas* (New York, 1973), pp. 244–45.

[29]John Seeley, *The Expansion of England* (Chicago, 1971), p. 193.

[30]Gustave d'Eichthal and Ismail Urbain, *Lettres sur la race noire et la race blanche* (Paris, 1839), p. 2.

[31]Winwood Reade, *Savage Africa* (London, 1863), pp. 508–9.

neutral facts, evaluative criteria based on science and technology appeared to be the least tainted by subjective bias. Blatantly narcissistic gauges of the worth of non-European peoples—skin color, fashions in or lack of clothing—receded in importance; measurements of cranial capacity, estimates of railway mileage, and the capacity for work, discipline, and marking time became the decisive criteria by which Europeans judged other cultures and celebrated the superiority of their own. The fact that these criteria were as culturebound as earlier gauges and even more loaded in favor of the industrializing West was grasped by only a handful of intellectuals, who continued to value the achievements of non-Western peoples and lament the erosion or utter disappearance of African and Asian ideas, customs, and institutions in an age of European global dominance.

As Martin Weiner has demonstrated, more than a handful of British intellectuals regretted the passing of an idealized rural England.[32] Tory and radical critiques of the filthy factory districts and miserable workers' slums—familiar features of industrialization in what Lewis Mumford has termed the paleotechnic phase—grew into a broad-based anticapitalist and anti-industrial backlash in the late nineteenth and early twentieth centuries. A gentlemanly, aristocratic amateur ideal was championed in a struggle against the ascendancy of a profit- and productivity-obsessed elite of industrialists, financiers, and technicians. But this assault on William Blake's "dark Satanic Mills" and the values of the capitalist entrepreneur was for the most part a domestic campaign. Weiner's contention that "Imperialists and 'patriotic' writers [in the late nineteenth century] rarely saw industrial progress as an appropriate source of inspiration"[33] is at odds with the frequent invocation in this period of technological and scientific proofs of British superiority. These were used to explain how Britain had gained its place of world dominance and to justify the manner in which it ruled vast populations of non-European peoples around the globe. The tensions of "Janus-faced" industrial Britain longing for its village, artisan past were intensified by a clear recognition, even on the part of those of aristocratic birth who found refuge in imperial administration in the late nineteenth century,[34]

---

[32]Martin Weiner, *English Culture and the Decline of the Industrial Spirit* (Cambridge, Eng., 1981). Some of the authors and themes discussed by Weiner are also considered in Raymond Williams, *The Country and the City* (New York, 1973); and Sussman, *Victorians and the Machine.*

[33]Weiner, *English Culture*, p. 55.

[34]Francis Hutchins, *The Illusion of Permanence: British Imperialism in India* (Princeton, N.J., 1967), esp. pp. 124–36 and chap. 9.

that the advantages the British had gained as a result of the scientific and industrial revolutions were essential to the growth and maintenance of Britain's imperial order.

Perhaps no more striking illustration of the importance of technological superiority as a justification for imperial dominance can be found than in Mary Kingsley's recollection of the sense of pride and reassurance that Britain's industrial technology gave her as she explored the rainforests and savannas of west Africa. Kingsley was one of the most intrepid and independent-minded of nineteenth-century travelers, and her works are studied today mainly because of her sincere and, at the time, rather rare admiration for African beliefs and customs. Yet however strong her defense of African peoples and cultures, she made no secret of her conviction that the British were superior and that technological accomplishment was the best test of their superiority:

> All I can say is, that when I come back from a spell in Africa, the thing that makes me proud of being one of the English is not the manners or customs up here, certainly not the houses or the climate; but it is the thing embodied in a great railway engine. I once came home on a ship with an Englishman who had been in South West Africa for seven unbroken years; he was sane, and in his right mind. But no sooner did we get ashore at Liverpool, than he rushed at and threw his arms round a postman, to that official's embarrassment and surprise. Well, that is just how I feel about the first magnificent bit of machinery I come across: it is the manifestation of the superiority of my race.[35]

The apparent contradiction between Weiner's vision of a rising tide of anti-industrial criticism in the late nineteenth century and the predominance of scientific and technological standards as gauges of levels of human achievement in this era can in part be resolved by distinguishing between technology per se and industrialization in its capitalist guise. Much of the hostility to industrialism which Weiner so ably explores was aimed at excessive commercialism, unbridled competition, and change as an end in itself. Few of the critics of industrialization sought to deny that the British had displayed a remarkable capacity for invention and scientific investigation. The traditions of the gentleman scientist and inventor as part-time tinkerer were quite consistent with a "gentrified," intellectual, and entrepreneurial elite. In giving the world both the Newtonian and industrial revolutions, the British demonstrated beyond all doubt that they could excel in scientific inquiry and invention. Even

[35]Mary Kingsley, *West African Studies* (London, 1901), pp. 329–30.

the craft techniques and tools that such implacable critics of industrialization as William Morris and John Ruskin worked to preserve or restore had been used in earlier centuries as a gauge of European superiority in the "arts" over non-Western peoples. In the anticapitalist, anti-industrial utopia that Morris created in his 1891 *News from Nowhere*, all manner of machines were available, but his utopians used only those they found "handy" and ignored those that seemed unnecessary.[36] Thus, despite recurring depressions, England's eclipse as the leading industrial power by Germany and the United States, and a growing sense of the perils of unlimited mechanization, numerous late-Victorian authors exulted in Britain's scientific acumen and inventive genius.

Many of those who saw scientific and industrial achievements as the best proof of British "racial" or national superiority had little use for the aesthetic misgivings of Ruskin or for Morris's socialist critiques. As Weiner points out, few Englishmen fought the antiscientific, antitechnical trend among the public school elite more fiercely than Frederic Farrar, headmaster at Harrow and later at Marlborough.[37] But Farrar not only berated those who thought that scientific learning was not essential for the education of a gentleman; he viewed invention and scientific discovery as the best tests of the "aptitudes of races." In his 1867 essay of that title, he systematically dismissed all claims of non-European contributions in these areas. He argued that the Semites and the Aryans were the only races that had created "true" civilizations and that they had been responsible for "every noble discovery, every thought and influence that has enabled and purified the white race."[38] Farrar contrasted the vigor and progressive advance of European civilization with the savagery of Africa and the "stolid unprogressiveness of the Mongol." Though he attributed all civilized development to the Aryans and Semites, he gave particular stress to advances in science and technology as irrefutable proof of the superiority of these races over all others. Thus, though he may have had a difficult time convincing other headmasters to include more science in their curricula, Dean Farrar did much to propagate these gauges of human worth among several generations of public school boys who went on to dominate English politics and run the overseas empire.

Alfred Russel Wallace, who shared with Darwin the distinction of originating the theory of evolution by natural selection, outdid even Farrar in his unqualified praise for the blessings of industrialization. In

---

[36]William Morris, *News from Nowhere* (London, 1970), p. 146.

[37]Weiner, *English Culture*, p. 19.

[38]Frederic Farrar, "Aptitudes of the Races," *Transactions of the Ethnological Society of London* (hereafter *T.E.S.L.*) 5 (1867), 125.

his 1899 postmortem on the Victorian era, fittingly entitled *The Wonder-ful Century*, Wallace calculated that there had been only seven inventions of the "first rank" in all the centuries before 1800. The great minds of the nineteenth century he reckoned had produced thirteen, and he sought to capture his countrymen's unqualified pride in these accomplishments: "We men of the nineteenth century have not been slow to praise it. The wise and the foolish, the learned and the unlearned, the poet and the pressman, the rich and the poor, alike swell the chorus of admiration for the marvelous inventions and discoveries of our own age, and especially for those innumerable applications of science which now form part of our daily life, and which remind us every hour of our immense superiority over our comparatively ignorant forefathers."[39] In Wallace's view, the anti-industrial barbs of the Ruskins and Morrises had had little influence on most of their countrymen.

Even those who expressed doubts about the effects of industrialization and capitalism on British society tended to set Britain and Europe more generally apart from non-Western cultures on the basis of the unparalleled advances the former had made in material culture. Walter Bagehot, for example, might deplore the "rough and vulgar structure of English commerce,"[40] but in his best known work, *Physics and Politics*, he set Europe off from the rest of the globe as a sort of industrialized Middle Kingdom: "The miscellaneous races of the world [can] be justly described as being upon the various edges of industrial civilization, approaching it from various sides, and falling short of it in various particulars."[41] Bagehot contrasted Europe—where "habitual instructors," "ordinary conversation," and "inevitable and ineradicable prejudices" joined forces to make progress a "normal fact in human society"—to Oriental and "savage" societies that had no conception of progressive change. He reasoned that the inability of the Indians to appreciate fully the "blessings" conferred by British rule arose from their rigid adherence to custom and an "old feeling" of "fixity" which clashed sharply with the Englishman's "modern feeling," rooted in a commitment to social improvement and economic growth.[42]

Benjamin Kidd, whose best-selling *Social Evolution* disseminated his

---

[39]Alfred Russel Wallace, *The Wonderful Century* (New York, 1899), p. 1.

[40]Walter Bagehot, quoted in Weiner, *English Culture*, p. 137.

[41]Walter Bagehot, *Physics and Politics* (London, 1872), p. 16.

[42]Ibid., pp. 156–57; and Bagehot, "Physics and Politics," pt. 2, *Fortnightly Review*, n.s. 3 (1868), 433, 452. Similar views were shared by Henry Maine and John Crawfurd. See John Burrow, *Evolution and Society: A Study in Victorian Social Theory* (Cambridge, Eng., 1966), p. 159 (on Maine), and Crawfurd, "On the Physical and Mental Characteristics of the European and Asiatic Races of Man," *T.E.S.L.* 5 (1867), 73.

ideas even more widely than Bagehot's had spread, shared the fears for Britain of many late Victorians in a world of growing competition for markets and resources. Nonetheless, Kidd began his work with a eulogy to the scientific and mechanical advances of the industrial age. Following a fairly standard catalogue of inventions and discoveries, he proclaimed that the best was yet to come. Even the tremendous achievements of the past, he declared, were "destined to be eclipsed at no distant date" by the efforts of the "vigorous and virile" Anglo-Saxons who had been responsible for the Industrial Revolution. Even though Kidd believed that Europe's distinctiveness as a civilization was rooted in its Christian foundations, he repeatedly fell back on evidence of material accomplishment to illustrate the great differences separating Europeans from all contemporary peoples, as well as those who had built civilizations in the past. He concluded that the contrasts between Europeans and the rest of humankind were most readily apparent "in the great advances which have been made in the arts of life, in trade, in manufactures, and commerce, in the practical appliances of science, and the means of communication."[43]

Although his arguments were a good deal more defensive than Kidd's, A. H. Keane displayed the same concern to affirm the superiority of the Anglo-Saxon or English "race" in scientific exploration and technological innovation. After a series of pithy caricatures of the other peoples of the globe, he launched into an effusive tribute to England's accomplishments. Beginning with allusions to Shakespeare and other English "greats" in the literary and artistic realm, Keane moved quickly to an extensive listing of English discoveries in chemistry, physics, mechanics, engineering, and the natural sciences. He indignantly rebutted the notion (perhaps suggested by an envious Frenchman?) that the English were "dull" by again supplying a long list of scientific and inventive geniuses from Newton and Priestly to Faraday and Edison. (His inclusion of Edison reflected his view that Americans could be classified as Anglo-Saxon, though he apparently excluded the other main rival of the English, the Germans, from this select category—or at least the German peasants, whom he characterized as "heavy, dull and superstitious.") Keane concluded that the Anglo-Saxons were the best strain of the best race, the Caucasian, and vehemently rejected the contention that the main English contribution to scientific and technological advance had consisted of their improvements on French inventions. He dismissed the French as a people of inferior racial stock. "For one Frenchman," he declared, "it would be easy to produce ten or a dozen English and American inventors."[44]

[43]Benjamin Kidd, *Social Evolution* (London, 1894), pp. 6–7, 45, 120–21.
[44]A. H. Keane, *The World's Peoples* (London, 1908), esp. pp. 341, 379–80.

Even those who questioned the validity of science and technology as gauges of human worth admitted that they had dominated British attitudes toward colonized peoples. Though Rudyard Kipling retreated late in life into a rural refuge free of telephones and other modern contrivances, for decades his stories and poems had celebrated the technical skills, assertiveness, and commitment to hard work that had made the British rulers of more than a quarter of the globe. However much he might deplore the fact, C. F. G. Masterman could not deny that British imperialists had long equated physical power and sheer size with greatness and had "neglected and despised the ancient pieties of an older England."[45] Even Dean William Inge, who emerged as one of the leading critics of England's industrial order in the first decades of the twentieth century,[46] readily acknowledged that nineteenth-century Englishmen routinely equated material achievement with civilized advance. Writing in the shambles that the Great War had made of European societies, Inge mused on the Victorians' reasoning that if they could travel sixty miles per hour by train, they must be five times more civilized than non-Europeans who could do at best twelve miles per hour by nonmechanized transport.[47]

The current of hostility to industry and capitalist values was stronger in France than in England, so much so that some economic historians have seen these attitudes and the French attachment to handicraft production as major reasons for France's late industrialization relative to its cross-Channel rival.[48] France's slower and less extensive industrialization may also account for the fact that most French writers adhered more than the invention-conscious British to the eighteenth-century preference for scientific rather than technological gauges of human worth, though the two were often blended. Gustave Le Bon and Arthur de Gobineau, for example, who did much to shape nineteenth-century thinking on racism and attitudes toward non-Western cultures, found much to criticize in industrial Europe. But both viewed peoples of European stock as racially superior to all others and cited the white or Aryan race's achievements in science and, to a lesser extent, technology as clear proof of the high level of civilization it had attained. Gobineau

[45]For Kipling's and Masterman's anti-industrial sentiments, see Weiner, *English Culture*, pp. 59–60. For Kipling's celebration of capitalist and industrial values as the key to Europe's "civilizing mission," see below, Chapter 4.

[46]Weiner, *English Culture*, pp. 111–13.

[47]William Inge, *Outspoken Essays* (London, 1922), p. 162.

[48]For a general discussion of these currents in revolutionary and postrevolutionary France, see Trebilcock, *Industrialization of the Continental Powers*, pp. 136–39. For a critical appraisal of this line of argument, at least for the prerevolutionary era, see François Crouzet, "England and France in the Eighteenth Century: A Comparative Analysis," in R. M. Hartwell, ed., *The Causes of the Industrial Revolution* (London, 1967), pp. 139–74, esp. 157–70.

found the Europe of his day clearly inferior to ancient Greece and Egypt and even India in artistic production but argued that Europe had been responsible for virtually all important breakthroughs in the sciences. Though he acknowledged Europe's technological advances and advantages, he believed that the Europeans' capacity for discerning underlying principles and developing general theories was more critical than the inventions themselves. Like Farrar, he contended that Europeans had been responsible for all civilized development. Even India and China, which Gobineau included in his list of ten "true" civilizations, had advanced to that level because of the ideas and skills transmitted by Aryan invaders.[49] Le Bon also doubted that the "age of steam and electricity" was one in which the arts had peaked, but he was firmly convinced that the racial genius and superiority of Western peoples had been indisputably demonstrated by their scientific discoveries and unmatched inventiveness.[50]

By no means all late nineteenth-century writers felt the need to qualify their praise for France's industrial growth. The novelist and essayist Edmond About, for example, contrasted the slow pace of change and resistance to new technology in earlier civilizations with the recognition accorded to scientists and inventors in European society. He believed that the status and material rewards gained by achievers in these fields, as well as the competition and interaction between scientists of different nations, were the keys to the progress that he felt set Europe off from all other civilizations.[51] Decades later the journalist Pierre Mille wryly summed up the importance that generations of Frenchmen had ascribed to material advance as the standard by which to judge a society's level of development. He observed that his countrymen generally accepted the notion that the indisputable superiority of European mental capacity over that of all other races had been clearly demonstrated by European scientific discoveries and their application to production, war, and communications.[52]

In the late eighteenth century and the first decades of the nineteenth, an awareness of European material mastery was confined to small numbers of educated travelers, colonial administrators, social theorists, and missionaries. As the pace of scientific discovery and technological inno-

---

[49]Arthur de Gobineau, *Essai sur l'inégalité des races humaines* (Paris, 1853), vol. 1, pp. 172, 264–65, 360ff; vol. 2, pp. 11, 349–52.

[50]Gustave Le Bon, *Les civilisations de l'Inde* (Paris, 1887), esp. pp. 550–52. See also Robert Nye, *The Origins of Crowd Psychology: Gustave Le Bon and the Crisis of Mass Democracy in the Third Republic* (Beverly Hills, Calif., 1975), pp. 43–45.

[51]Edward About, *Le progrès* (Paris, 1864), pp. 34–38.

[52]Pierre Mille, "La race supérieure," *Revue de Paris* 1 (1905), 820–21.

vation quickened in Europe and North America, while societies in other areas appeared to stagnate or break down, growing numbers of writers sought to determine the causes of Europe's unique transformation and the meaning of what they viewed as the failure of non-European peoples to initiate their own scientific and industrial revolutions. Through romantic novels or adventure stories, and especially in the popular press, the conclusions of phrenologists, social theorists, ethnographers, and storytellers such as Rudyard Kipling and Pierre Loti were disseminated widely among the middle and literate working classes.[53] Often corrupted and vulgarized, invariably oversimplified and sensationalized, these ideas have played a major role from the nineteenth century to the present in shaping popular attitudes in Europe and North America toward African and Asian peoples and cultures. Notions of white supremacy and racial superiority, jingoistic slogans for imperialist expansion, and the vision of a dichotomous world divided between the progressive and the backward have all been rooted in the conclusions drawn by nineteenth-century thinkers from the fact that only peoples of European stock had initiated and carried through the scientific and industrial revolutions.

Africa: Primitive Tools and the Savage Mind

In contrast to the civilizations of India, China, and Islam, which most writers conceded had made significant contributions to scientific inquiry and technological development early in their histories, African cultures were considered by almost all nineteenth-century European observers to be devoid of scientific thinking and all but the most primitive technology. This assessment became more and more central to efforts to demonstrate the validity of the long-standing view of the Africans as backward and inferior peoples. Johann Blumenbach's spirited defense of "Negro" mathematical and scientific aptitudes was forgotten or ignored. Echoing, often unknowingly, an argument that Edward Long had advanced in the 1770s, numerous nineteenth-century authors asserted that the Africans had not been responsible for a single scientific discovery or any mechanical invention. Julien Virey developed this

[53]For surveys of the popular literature on Africa and "the Orient" in this era, see Léon Fanoudh-Siefer, *Le Mythe du nègre et de l'Afrique noire dans la littérature française* (Paris, 1968); E. F. Oaten, *A Sketch of Anglo-Indian Literature* (London, 1908); Susanne Howe, *Novels of Empire* (New York, 1949); and William Schwartz, *The Imaginative Interpretation of the Far East in Modern French Literature* (Paris, 1927). For a more analytical treatment of the processes by which these ideas were popularized in the late nineteenth century, see William Schneider, *An Empire for the Masses: The French Popular Image of Africa, 1870–1900* (Westport, Conn., 1982).

theme at some length. Unlike the Egyptians, he asserted, the Africans had never built great cities or monuments but were content to live in "primitive" huts. In contrast to the Indians, they were incapable of manufacturing textiles. They had produced no art worth the name and no inventions. In fact, Virey argued, the Africans had never even displayed the sort of ingenuity that the Indians had evinced in devising the game of chess or that the Arabs had revealed in their delightful stories, especially the *One Thousand and One Nights*.[54] The American writer Josiah Nott was equally sweeping in his assertion that in their entire history the Africans had produced no cities, no monumental architecture, no "relic" of science or literature, and not even a "rude" alphabet. Similarly, James Hunt, the first president of the Anthropological Society of London, challenged those who argued for African equality with whites to name one "Negro" who had distinguished himself in any field. In listing careers in which evidence of African achievement might be sought (but never found), Hunt ranked "man of science" above all others.[55]

In the view of most nineteenth-century authors of works on Africa, these categorical dismissals of black material achievement appeared to be confirmed by the accounts of European explorers, missionaries, and colonial officials. Though some authors made exceptions for particular peoples, Richard Burton, one of the most wide-ranging and frequently quoted of British explorers, captured the sentiments of the great majority of the on-the-spot observers when he declared that technology in Africa was limited to weaving, cutting canoes, making "rude" weapons, and "practising a rough metallurgy."[56] Less hostile and better-informed observers than Burton made distinctions between different African peoples in terms of their level of technological development. In general, there was a strong correspondence between praise for a particular people's tools and weapons and favorable judgments about its society as a whole. The German explorer Karl Peters, for example, had little good to say about most of the peoples whom he encountered during his travels in East Africa. His stereotypes—from the "thieving" and "impudent" Kikuyu to the "proud but savage" Masai—were standard fare for late nineteenth-century readers of explorers' accounts. Relative to his undisguised disdain for most African peoples, however, Peters had high

[54]Virey, *Histoire naturelle*, pp. 48–49.

[55]Josiah Nott, "The Negro Race," *Popular Magazine of Anthropology* 3 (1866), 107–9; James Hunt, *On The Negro's Place in Nature* (London, 1863), p. 37. See also Charles Smith, *The Natural History of the Human Species* (Edinburgh, 1848), p. 196.

[56]Richard Burton, *A Mission to Galele, King of Dahomey* (London, 1864), vol. 2, p. 202. For a similar summary, see Edmond Ferry, *La France en Afrique* (Paris, 1905), pp. 221–22, 229.

praise for the powerful Ganda, whom he considered extremely skillful builders. Yet he qualified even this rare praise by his supposition that the Ganda's engineering feats had been stimulated by Egyptian and European influences.[57]

As H. A. C. Cairns has shown, the Ganda's skill at roadbuilding and their capacity for metalworking were vital in winning very favorable assessments of their level of social development from a number of British travelers and missionaries in this period.[58] The British were particularly impressed by the complexity of Ganda political organization, engineering and architectural skills, and proclivity for experiment and innovation. These were seen as evidence that they shared many of the traits the British considered central to their own superiority. The fact that the Ganda were eager to learn and imitate British techniques, specifically those related to metalworking, only served to reinforce the impression that they were a cut above the rest of the Africans in social development and potential for improvement.

Another of the great experts on colonial affairs of the day, Sir Harry Johnston, distinguished the "Bayansi" from the other "savages" of the Congo as a people who had developed a "decided indigenous civilization of their own." He based the distinction solely on the Bayansi's superior engineering skills, which were evinced in their large houses with swinging doors, sophisticated handicraft industries, and skill in the working of metals.[59] In response to a lecture on Africa delivered some years later by Johnston, a missionary named Milum vehemently rejected Johnston's sweeping assertion that before colonization all Africans were savages. But in his defense of the Yoruba and claims for the high level of development they had achieved, Milum relied on the standards invoked earlier by Johnston, standards that were overwhelmingly technological: "What were they [the Yoruba] before they came into contact with Europeans? Certainly not savages. They had smelting furnaces, and they made iron and excellent steel. They were dyers, and to this day their dyes are the envy of European countries. They made their own cloth. They were by no means naked savages, for they dressed in the most decent manner, and I should like to commend the native dress of the

---

[57]Karl Peters, *New Light on Dark Africa* (London, 1891), pp. 417–18.

[58]H. A. C. Cairns, *Prelude to Imperialism: British Reactions to Central African Society, 1840–1880* (London, 1965), pp. 109, 112.

[59]Harry Johnston, "On the Races of the Congo and the Portuguese Colonies in West Africa," *Journal of the Anthropological Institute* 13 (1884), 475. For other examples of this pattern, see David and Charles Livingstone, *Narrative of an Expedition to the Zambesi and Its Tributaries, 1858–1864* (New York, 1866), pp. 122–25; William Allen and T. R. H. Thompson, *A Narrative of the [Trotter] Expedition to the River Niger in 1841* (London, 1848), vol. 1, p. 324; and Louis Figuier, *The Human Race* (London, 1872), p. 495.

Yorubas as the most fit and proper costume, and as being superior to the European style."[60]

Yet, however much travelers or colonial officials may have been impressed with the technical skills of particular peoples, their praise was always qualified by the underlying assumption that a vast gap existed between the capacities of Africans as a whole and those of the Europeans. The French-born explorer Paul Du Chaillu captured this ambivalence in noting that the "cannibal" Fon of the Congo region refused to barter for European or American iron for their knives and arrowheads, preferring metal forged on their own "ingenious bellows." Nevertheless, Du Chaillu saw Fon material culture as woefully primitive: they had not invented, he observed, "a thing so simple as a handle for a hammer."[61]

Perhaps the most striking manifestation of the assumption that the African peoples lacked either technological development or the capacity for invention can be found in works on the riddle of Zimbabwe, which preoccupied so many travelers to southern Africa in this period. The massive and skillfully constructed stone edifices found in what was to become the British colony of Southern Rhodesia appeared to contradict the widespread belief that Africans were utterly lacking in engineering skills and had produced little or no monumental architecture. The ruins of Zimbabwe had to be explained or these assumptions reassessed. The former approach was much more congenial to nineteenth-century authors. The mystery of Zimbabwe's origins was conceded by all who wrote about the impressive ruins, whose impact was considerably enhanced by the paucity of comparable stone buildings in the rest of sub-Saharan Africa. The absence of written records relating to the ruins and the infant state of archeological and anthropological work in the Zimbabwe region left much room for speculation. But one thing was clear to most observers: black Africans were incapable of the architectual, engineering, and stoneworking feats that those who built the great walls and towers of Zimbabwe had obviously possessed.[62]

[60]Recorded in *Proceedings of the Royal Colonial Institute* 20 (1888–89), p. 118. Similarly, Richard Freeman ranked the Muslim peoples of the Sudan above the pagan peoples of the forest almost wholly on the former's superiority as toolmakers and textile producers; see *Travels and Life in Ashanti and Jaman* (Westminster, Eng., 1898), pp. 201, 417–18.

[61]Paul Du Chaillu, *Explorations and Adventures in Equatorial Africa* (London, 1861), pp. 90–92. See also the Livingstones' *Expedition to the Zambesi*, p. 562.

[62]The following discussion of the Zimbabwe riddle is based on Lionel Decle, *Three Years in Savage Africa* (London, 1898), vol. 1, p. 191; Robert M. Swan, "Some Notes on Ruined Temples in Mashonaland," *Journal of the Anthropological Society of Great Britain and Ireland* 27 (1897), 2–13 (quoted portions, pp. 9–10); Frederick C. Selous, *Travel and Adventure in Southeast Africa* (London, 1893), pp. 331–41; J. T. Bent, "Origins of the Mashonaland Ruins," *Nineteenth Century* 34 (1893), 991–97 (quoted portion, p. 996). For variations on these arguments, see James Bryce, *The Relations of the Advanced and Back-*

The British explorer Lionel Decle summarized the general consensus of nineteenth-century writers when he concluded that the construction of buildings as grand as those suggested by the "magnificent ruins" of Zimbabwe was simply beyond the capacity of *any* African people; the structures in "their extent, their gigantic proportions, and their general plan indicate a loftiness of conception very far superior to the present ability of the Negro race." Decle's contemporary Robert Swan agreed. He cautioned readers of the prestigious journal of the Royal Anthropological Society against assuming that the "Kaffirs" had constructed the edifices simply because Africans then living in the vicinity often built their homes on top of the ruins. Swan declared emphatically that none of the tribes "now living anywhere near Mashonaland" could ever have had "even the small knowledge of geometry and astrology [*sic*] that was necessary in planning these temples." Frederick Selous, a big-game hunter in the 1890s with strong opinions on all manner of African affairs, was among the tiny minority who believed that the Bantu peoples might have played a role in the construction of Zimbabwe. But his conviction arose from a low opinion of the ruins rather than a maverick respect for African abilities. He called Zimbabwe a "rude structure" and surmised that its architects could have achieved only the low state of development which he associated with Africans.

Though most writers dismissed the Africans as candidates, there was some disagreement over who had in fact built Zimbabwe. The dominant view was advanced by J. T. Bent, who wrote a detailed essay on the origins of the structures in 1893. Drawing on the work of German and Austrian archeologists in addition to the standard British sources, Bent speculated that the great edifices had been constructed centuries earlier by Arabs who had migrated inland from the Swahili coast. He surmised that if Africans had played any role in the construction, it had been as slave laborers. He found this conclusion inescapable, given "the well accepted fact that the negroid brain never could be capable of taking the initiative in work of such intricate nature." Until recent years, similar views were espoused by the white settlers who dominated Rhodesia. In official publications and textbooks the African role in the construction of Zimbabwe was vigorously denied.[63] That the name Zimbabwe

---

ward Races of Mankind (Oxford, 1902), pp. 93–97; and R. N. Hall, *Great Zimbabwe* (London, 1905), pp. xxvi–xxviii, 18, 80.

[63]Perhaps the earliest expression of this view at the popular level was included in John Buchan's adventure for schoolboys, *Prester John*, originally published in 1910. In the novel an English schoolteacher named Wardlaw summarily dismisses the notion that Zimbabwe was erected by the "natives," insisting that "the men who could erect piles like that . . . were something more than petty chiefs" (New York, n.d.), p. 65.

was chosen by the new nation which emerged from the struggle for black majority rule in Rhodesia indicates the importance that the discovery of the Bantu origins of the great complex has played in the efforts of Africans to rebut centuries of misinformation about their historic achievements.

Attempts to deny the African contribution to the technological or architectural accomplishments of ancient civilizations had been made, of course, long before European explorers prowled the ruins of Zimbabwe in the late nineteenth century. In the 1850s the American authors Josiah Nott and George Gliddon, who were widely considered authorities on the "Negro question," had declared that the civilization of Meroe, long noted for the high level of ironworking it had attained, was the product of Egyptian and not "Negro" genius. They also rejected the suggestion that black Africans had played a role in the development of Egyptian civilization and its wondrous architectural feats, other than perhaps the provision of brute labor. Confronted with evidence that the Mandingos and Fulani had historically displayed considerable technical skill and achieved a high level of civilization, Nott and Gliddon countered that these peoples were "less black" ("mahogany") than other Africans and had been highly influenced by Arab migrants and Islamic civilization.[64] A decade later Frederic Farrar conceded that the Africans had made some contributions to Egyptian civilization but asserted that these had been highly inflated by defenders of the "Negro race." Farrar concluded that the achievements of Egypt were largely the product of Aryan and Asiatic efforts and that black Africans had supplied mainly "strains of cruelty and Fetishism."[65] William Clark, who in the 1850s wrote a highly favorable account of the Yoruba of present-day Nigeria, inexplicably traced the origins of what he regarded as an exceptional people to the Canaanites of ancient Palestine.[66] These claims do much to explain why authors of African descent, from Antenor Firmin in the late nineteenth century to Cheikh Anta Diop in the twentieth, have been so concerned to demonstrate the importance of black African contributions to ancient Egyptian civilization.[67]

[64]Josiah Nott and George Giddon, Types of Mankind (London, 1854), pp. 52, 186–88.

[65]Farrar, "Aptitudes of Races," pp. 119–20.

[66]William Clark, Travels and Explorations in Yorubaland, 1854–8 (London, 1872), pp. 287–92.

[67]Antenor Firmin, De l'égalité des races humaines (Paris, 1885), esp. pp. 333–77; Cheikh Anta Diop, The African Origins of Civilization (New York, 1974); Anthony Noguera, How African Was Egypt? (New York, 1976). A surprising number of late nineteenth-century European writers conceded important African influences on Egyptian civilization; see, e.g., Marius-Ary LeBlond, "La race inférieure," Revue de Paris 4 (1906), 109–113.

Nowhere was the technological gap that grew ever wider between nineteenth-century Europeans and Africans more graphically depicted than in the hundreds of incidents in which travelers, settlers, and missionaries reported the awestruck responses of Africans to even the simplest mechanical devices. Paul Du Chaillu recalled how the "natives" regarded his clock as an "object of wonder" and believed it to be his guardian spirit. On a later journey, Du Chaillu delighted in the Africans' awe (and fear) of his "galvanic battery," magnet, and photographic equipment; even his black-tinted beer bottles were "held in very high estimation by the chiefs."[68] Anna Hinderer, the wife of a missionary in Yoruba country, told of crowds of Africans with "eyes and mouths wide open" gathering around her bungalow to listen to her play the harmonium. A visiting African leader insisted on seeing and attempting to play the instrument. When he could not, he remarked, according to Hinderer, that "only white people can do anything great like make wood and ivory 'speak.'"[69] Lovett Cameron, a British explorer, recorded an even more exuberant response. Having inspected the watches, guns, compasses, and other instruments that Cameron's party carried, the uncle of another African "chief" exclaimed: "Oh, these white men! They make all these wonderful things and know how to use them! Surely men who know so much ought never to die; they must be clever enough to make a medicine to keep them always young and strong, so that they will never die."[70] The old man, Cameron claimed, believed that the Europeans were thousands of years old and able to conjure up these devices from their inner consciousness.

The English traveler Richard Freeman "astonished and delighted" his African bearers by allowing them to view the moon through his telescope. Freeman derived even greater pleasure from African wonderment at the flame that engulfed his Christmas pudding (he reported that the "native" onlookers burst into applause) and his use of a fork.[71] The Scottish explorer Henry Drummond, reminding his readers that he was without books or newspapers, reported that he amused himself and was able to "entertain the savages" by lighting matches, buttoning his coat, "snapping" his revolver, or using his mirror to set fire to their clothing. He added that he found such pleasure in the Africans' awestruck response to these diversions that he sometimes indulged in them three or

---

[68]Du Chaillu, *Explorations and Adventures*, pp. 412, 417.
[69]Richard B. Hone, ed., *Seventeen Years in the Yoruba Country: Memorials of Anna Hinderer* (London, 1872), pp. 34–38, 44. See also William Junker, *Travels in Africa during the Years 1875–1878* (London, 1890–92), vol. 2, p. 173.
[70]Lovett Cameron, *Across Africa* (London, 1876), p. 100.
[71]Freeman, *Travels and Life*, pp. 109, 159.

four times a day.[72] As late as the decade before the Great War, the young Elspeth Huxley described similar African reactions to such commonplace objects as matches, wheels, and oil lamps. She puzzled over the fact that these devices made a much greater impression on the "ignorant" and "backward" peoples of Kenya than did such modern marvels as airplanes and radios.[73]

Perhaps no inventions elicited as much astonishment and respect from Africans as European firearms. As early as the 1820s, G. Mollien, a French traveler with considerable sensitivity to the Africans, reported that when he fired his double-barreled gun, the Bahene people cried out in astonishment, "We are only beasts."[74] Some decades and great advances in firearms manufacture later, Du Chaillu told of the Ashira's praise for his revolver. Even though they had muskets, he wrote, they could not even begin to comprehend the workings of a gun that "fired time after time without stopping."[75] Du Chaillu's contemporary, the French explorer M. E. Mage, noted a similar "avid interest" in his Colt revolver on the part of the ruler Ahmadou. Mage recalled with amusement the obvious amazement of the proud leader, who sought to maintain an air of indifference in all situations. The gun's performance, however, prompted Ahmadou to lose his composure and openly marvel at the "small copper cartridges" that could carry so far.[76] Though the superiority of European firearms to the weapons wielded by Africans was readily apparent, some travelers were not above resorting to ruses to amuse themselves and dazzle the "natives" further. William Devereux recounted how he and his companions told a group of "fine strapping natives" who had boarded his steamship to look it over that the funnel was a large cannon. The visitors were understandably "awe-stricken" by what they believed to be a monstrous gun, but they were even more impressed—Devereux confided—by the white women on board.[77] Ironically, it was the great missionary-explorer David Livingstone who most accurately and succinctly summarized the advantages for Europeans in Africa of vastly superior firearms: "Without any bullying, fire-

[72]Henry Drummond, *Tropical Africa* (London, 1888), pp. 103–4.

[73]Elspeth Huxley, *The Flame Trees of Thika* (Harmondsworth, Eng., 1983), p. 31. Decades earlier, David Livingstone had recorded similar responses (*Expedition to the Zambesi*, p. 242).

[74]G. Mollien, *Travels in the Interior of Africa to the Sources of the Senegal and Gambia* (London, 1820), p. 64.

[75]Du Chaillu, *Explorations and Adventures*, p. 412.

[76]M. E. Mage, *Voyage dans le Soudan Occidental (Sénégambie-Niger)* (Paris, 1868), p. 229.

[77]William Devereux, *A Cruise in the "Gorgon"* (London, 1869), p. 193. For a similar account of African confusion about naval artillery, see Paul Du Chaillu, *A Journey to Ashango-Land* (New York, 1867), pp. 61–62.

arms command respect, and lead [African] men to be reasonable who might otherwise feel disposed to be troublesome."[78]

The apparently overwhelming impression that firearms and mechanical devices made on peoples throughout sub-Saharan Africa led many explorers and envoys to the courts of African rulers to recommend guns and machines as gifts for African "chiefs" and kings. In the account of his 1817 mission to the ruler of Ashanti, Edward Bowdich described at great length the king's great interest in British medicines and botanical books and his obvious delight in the many mechanical devices that Bowdich had brought along as presents. Bowdich advised that "ingenious novelties"—including, in this instance, telescopes, pistols, kaleidoscopes, watches, a microscope, a pocket compass, matches, and a camera obscura—had played a critical role in winning the favor of the powerful African ruler.[79] Perhaps acting on Bowdich's instructions, Joseph Dupuis, who led a similar mission to the Ashanti capital at Kumasi three years later, selected "mainly mechanical contrivances" as gifts for the African ruler. Unfortunately, some of the items he had chosen met with a decidedly less favorable reception than Bowdich's presents had enjoyed. The Ashantehene, Dupuis reported, was frightened by a small organ, which had been hauled through the rain forest with great effort, and he was disappointed by a lathe, which Dupuis conceded was "too mechanical for a royal present." Even a watch and a music box failed to win favor. Rather than admit his own failings as a purveyor of royal presents, Dupuis blamed the vast distance separating European and African in technological mastery: "The task of winding up the watch or the musical box required a degree of care foreign to the comprehension of the king; it was requisite, therefore, to put the mechanism in motion each time."[80]

Despite occasional setbacks, mechanical devices were favored as gifts by most travelers and explorers throughout the nineteenth century. In the 1870s Wilhelm Junker assured his readers that African leaders took "childish delight in the merest mechanical trifles." He recommended knife blades, mirrors, photographs, and even empty candy boxes as presents that were sure to please the "natives."[81] John Hanning Speke,

[78]Livingstone and Livingstone, *Expedition to the Zambesi*, pp. 213, 310. See also Decle, *Savage Africa*, p. 111; Cameron, *Across Africa*, pp. 125, 161; Francis Galton, *The Narrative of an Explorer in Tropical South Africa* (London, 1853), p. 201; and Cairns, *Prelude to Imperialism*, pp. 43–47, 76, 119, 162.

[79]Edward Bowdich, *Mission from Cape Castle to Ashanti* (London, 1819), 44, 97–98, 455–57.

[80]Joseph Dupuis, *Journal of a Residence in Ashanti* (London, 1820), pp. 93–94, 100.

[81]Junker, *Travels in Africa*, vol. 3, p. 2. See also Allen and Thompson, *Expedition to the River Niger*, vol. 1, pp. 315–18.

one of the most lionized explorers of the late nineteenth century, related numerous incidents in which mechanical presents or demonstrations of technological prowess won his expedition the protection or at least grudging acceptance of local "chiefs." A box of "lucifers," the accuracy and rapid fire of his revolvers, and even the opening and shutting of his umbrella (which Speke apparently carried throughout his travels) proved sufficient to overawe hostile warriors or mollify the suspicions of African leaders, from the powerful ruler of the Ganda to a local headman who, Speke reported, literally begged him for a box of matches.[82]

As Speke's experience suggests, innocuous, nonmilitary contraptions could sometimes be as effective for compelling cooperation or dispelling hostility as the advanced weapons that have all but monopolized the attention of historians of colonialism. Henry Drummond observed that even a piece of glass used to set fire to some dry grass could become a source of wonder and terror.[83] An equally striking example of reliance on the simplest of devices is provided in Hermann von Wissman's 1891 account of his travels in equatorial Africa. Von Wissman told of how the captain of the steamship on which he was traveling sounded the whistle to frighten off a crowd of African merchants who were quarreling with members of Von Wissman's party. He noted that this tactic had also been used on an earlier journey with much the same impact. On the second occasion, "the impression was again so overpowering that all of the natives took to their heels in wild fear, disappearing in the thickets and rushing towards the village. Only one old white-haired Herculean chief, who was standing close to the river, felt ashamed to run, but was terrified to such a degree that he staggered backward."[84]

Though as far as I am aware no direct connection has been established, this incident could have been the inspiration for two encounters that occur in Joseph Conrad's novella *Heart of Darkness*. In Conrad's story the protagonist Marlowe also uses the steam whistle of his ramshackle riverboat to scare away hostile "natives" on two occasions. In the first instance the sound of the whistle brings an abrupt end to an

---

[82]John Hanning Speke, *Journal of the Discovery of the Source of the Nile* (London, 1868), pp. 131, 186–87, 228, 291, 295, 298. For another account of the use of an umbrella to scare away "threatening natives," see Joseph Thompson, *To the Central Lakes and Back* (London, 1881), vol. 1, pp. 53–54. Speke's onetime traveling companion Richard Burton ignored Speke's advice on gifts and came to grief when the king of Dahomey spurned his "fancy clothes" and chided Burton for not having brought him a fine carriage; see Burton, *Mission to Galele*, vol. 1, pp. 321–22.

[83]Drummond, *Tropical Africa*, p. 104. David Livingstone (*Expedition to Zambesi*, p. 242) also noted the "magical aura" that the colonizers' vast technological superiority gave them in the eyes of the "natives."

[84]Hermann von Wissman, *My Second Journey through Equatorial Africa* (London, 1891), pp. 19–20.

attack by the forest dwellers who have killed Marlowe's helmsman. Marlowe recalls the awesome effect that the shriek of the whistle has on the attackers: "The tumult of angry and warlike yells was checked instantly, and then from the depths of the woods went out such a tremulous and prolonged wail of mournful fear and utter despair as may be imagined to follow the flight of the last hope from the earth."[85] On the second occasion Marlowe uses the whistle to drive off the African followers of Kurtz (the trading company's agent) and prevent them from being slain as a "jolly lark" by the "pilgrims" aboard the steamship. Again the impact of the simple device is overwhelming:

> At the sudden screech there was a movement of abject terror through the wedged mass of bodies. "Don't! don't you frighten them away," cried some one on deck disconsolately. I pulled the string time after time. They broke and ran, they leaped, they crouched, they swerved, they dodged the flying terror of the sound. The three red [clay-caked] chaps had fallen flat, face down on the shore, as though they had been shot dead. Only the barbarous and superb woman did not so much as flinch, and stretched tragically her bare arms after us over the sombre and glittering river."[86]

A comparison of Von Wissman's and Conrad's accounts not only enhances Conrad's reputation as a skillful writer but also reveals his remarkable empathy, rare for a European in this period, with the African and Asian peoples who were being beaten into submission by the European imperial juggernaut.

In addition to the vital role they accorded technological superiority in subduing the peoples of Africa, European writers viewed it as a major determinant of the nature of social and economic interaction between Europeans and colonized Africans. As the builders of railways, seaports, and mine shafts and the suppliers of finance and machine capital, the Europeans thought it fitting that the Africans, whom they regarded as technologically backward, be relegated to the position of laborers. As the following incident related by H. L. Duff demonstrates, most Europeans were also convinced that all but the most truculent Africans accepted this allocation of roles in deference to the Europeans' manifest scientific and technical mastery. Duff, who was Chief Secretary in the British administration in Nyasaland, had decided to take his "boy" back to England with him. As the two were approaching the steamship on

[85]Joseph Conrad, *Heart of Darkness* (New York, 1950) p. 118.
[86]Ibid., p. 145.

which Duff had booked passage to Europe, the servant was dazzled by the lights gleaming in the darkness from the portholes and asked Duff why the flames did not consume the vessel. After patiently explaining the workings of electric lamps and pointing out other technolocial marvels on the ship, Duff teasingly asked his servant why the Africans had not invented such "wonders." The African protested that his people lacked the white man's magic and that the Europeans were favored by God, who helped them to understand many things that were kept hidden from other others. When asked whether God had not made the Africans as well as the Europeans, the African, according to Duff, replied: "Perhaps, but I think he made us to be your tenga-tenga (bearers)."[87]

For many Europeans the differences between their own highly developed, technologically and scientifically oriented societies and what they perceived to be backward and superstitious African cultures were merely manifestations of the vast gap in evolutionary development that separated "civilized" Europe from "savage" Africa. Richard Freeman's association of the "rude and primitive" tools of the Africans he encountered in his travels with the peoples of the Stone Age was rooted in the widely accepted idea of recapitulation which Stephen Gould has argued was one of "the most influential ideas of late nineteenth century science."[88] Ethnologists and natural scientists who adhered to the idea of recapitulation believed that primitive cultures represented an ancestral stage in the evolutionary development of more advanced cultures. It followed that contemporary African societies, which were widely held to be primitive or savage, provided living examples of a level of material culture, organization, and thought which European peoples had passed through millennia earlier. Those who lived and worked among African peoples, Harry Johnston observed, felt as if they were going "thousands of years into the past" to ages of savagery and brutishness.[89] That such a journey back in time offered opportunities for the study of the earliest stages of human development was, according to Henry Drummond, one of the most important reasons for African exploration. "Ignorant eyes," he

---

[87]H. L. Duff, *British Administration in Central Africa* (London, 1903), pp. 265–66. For a similar exchange, see Ferry, *France en Afrique*, p. 243. Some observers averred that the Africans were indifferent to technological marvels because the ingenuity that went into them was simply beyond their comprehension—"ça c'est affaire pour Blanc" (A. Cureau, *Les sociétés primitives de l'Afrique equatoriale* [Paris, 1912], pp. 80–81).

[88]Freeman, *Travels and Life*, pp. 417–18; Stephen Gould, *The Mismeasure of Man* (New York, 1981), p. 114. As Peter Marshall and Glyndwr Williams point out, the idea had originally been suggested by a number of eighteenth-century authors; see *The Great Map of Mankind* (London, 1982), pp. 215, 274.

[89]Quoted in Keane, *The World's Peoples*, p. 73.

noted, saw only savages, but the informed observer realized that "they are what we were once; possibly they may become what we are now." Drummond confessed that however edifying, the journey back to the dawn of civilization was a profoundly disturbing one: the juxtaposition of civilization as represented by "a steel ship; London built, steaming six knots ahead" and African savagery in the form of "grass huts, nude natives, and a hippopatamus" was so unsettling that "the ideas refused to assort themselves."[90]

No writer captured better than Joseph Conrad the sense of adventure and unease that late nineteenth-century European explorers and missionaries felt as they traveled into the African interior—and back in time:

> We penetrated deeper and deeper into the heart of darkness. It was very quiet there. At night sometimes the roll of the drums behind the curtain of trees would run up the river and remain sustained faintly, as if hovering in the air high over our heads, till the first break of day. Whether it means war, peace, or prayer we could not tell. The dawns were heralded by a chill stillness; the wood-cutters slept, their fires burned low; the snapping of a twig would make you start. We were wanderers on a prehistoric planet. We could have fancied ourselves the first men taking possession of an accursed inheritance, to be subdued at the cost of profound anguish and of excessive toil. But suddenly, as we struggled round a bend, there would be a glimpse of rush walls, of peaked grass-roof, a burst of yells, a whirl of black limbs, a mass of hands clapping, of feet stamping, of bodies swaying, of eyes rolling, under the droops of heavy and motionless foliage. The steamer toiled along slowly on the edge of a black and incomprehensible frenzy. The prehistoric man was cursing us, praying to us, welcoming us—who could tell? We were cut off from the comprehension of our surroundings; we glided past like phantoms, wondering and secretly appalled, as sane men would be before an enthusiastic outbreak in a madhouse. We could not understand because we were too far and could not remember because we were travelling in the night of first ages, of those ages that are gone, leaving hardly a sign—and no memories.[91]

[90]Drummond, *Tropical Africa*, pp. 3, 40, 60. For other contemporary examples, see Livingstone and Livingstone, *Journey to the Zambesi*, pp. 492–93; Decle, *Savage Africa*, vol. 1, pp. 509–10; and Duff, *British Administration*, pp. 187–88. For more general discussions of these themes in the colonial literature, see Cairns, *Prelude to Imperialism*, pp. 88–91; and Fanoudh-Siefer, *Le mythe du Nègre*, pp. 63, 104–5. Some writers saw the Africans as examples of a prehuman or pre-Adamite stage of development; see Samuel Baker, *The Albert N'yanza, Great Basin of the Nile* (London, 1866), vol. 2, pp. 316–17.

[91]Conrad, *Heart of Darkness*, p. 105.

India: The Retreat of Orientalism

In contrast to the generally accepted European views of Africa, which were shaped and then fixed in stereotypes through the cumulative impact of numerous writers and observers—from natural scientists and explorers to novelists and missionaries—a single author and work exerted a decisive influence on nineteenth-century European attitudes toward India. Begun in 1806 but not finally published until the end of 1817, James Mill's *History of British India* captured many of the philosophical currents and much of the common prejudice and ethnocentrism of his day in its relentless critique of Indian tradition and culture. In his blanket dismissal of Indian achievements and in the general standards he enunciated for judging the worth of any society past or present, Mill provides a striking illustration of just how important scientific thinking and technological skills had become as gauges of human worth.

Although Mill had no firsthand experience of India—a fact he noted proudly because he believed that it heightened his capacity for objectivity—as preparation for the writing of his six-volume magnum opus he had spent many years reading whatever works on the subcontinent were available.[92] Mill, the self-made son of a shoemaker, did not come to these materials as a neophyte seeking knowledge. He was a mature thinker who had long before completed his university education in Greek history and philosophy and who had already formed very definite opinions about a wide range of questions from ethics to the proper role of government in "advanced societies." Working from a rather rigidly held set of philosophical assumptions rooted in Benthamite Utilitarianism and ideas about progress set forth by the thinkers of the Scottish Enlightenment,[93] Mill sifted through massive amounts of information. He selectively gathered evidence to support his critiques of Indian beliefs and customs and to demonstrate Utilitarian positions on various aspects of political economy. He vehemently rejected the generally favorable view of Indian civilization advanced by Sir William Jones and other Orientalists. He drew heavily on parliamentary papers and such critical accounts of India as those by Robert Orme and the Abbé Dubois. Mill

---

[92]The biographical background for James Mill's *History* has been taken primarily from Duncan Forbes, "James Mill and India," *Cambridge Journal* 5 (1951–52), 19–33; Alexander Bain, *James Mill: A Biography* (London, 1882); Leslie Stephen, "James Mill," in *The English Utilitarians*, vol. 2 (London, 1900); William Thomas's useful introduction to the 1975 edition of Mill's *History of British India*; and Eric Stokes, *The English Utilitarians and India* (Oxford, 1959).

[93]See esp. Gladys Bryson, *Man and Society: The Scottish Inquiry of the Eighteenth Century* (Princeton, N.J., 1945), pp. 36, 42–43, 76; and Sidney Pollard, *The Idea of Progress: History and Society* (Harmondsworth, Eng., 1971), pp. 59–78.

sought to propound definitive judgments about the worth of Indian ideas and institutions in particular and of "Oriental" societies more generally.

The views of his fellow Scotsman Charles Grant set the tone for Mill's massive *History*.[94] As a member of the Board of Trade in India, Grant had worked closely with Lord Cornwallis in the latter's efforts to reform the notoriously corrupt East India Company administration in the 1780s and early 1790s. Grant suffered the loss of two children and severe financial setbacks during his early years an an aspiring financier in Bengal. These experiences shaped his appraisal of Indians and Indian civilization, which differed substantially from that shared by prominent officials like Warren Hastings and the members of the Orientalist clique. Driven by a religious zeal that was first manifested during the period of his successive personal crises, Grant drafted in 1792 his "Observations on the State of Society among the Asiatic subjects of Great Britain," which anticipated many of the main arguments James Mill would make in far greater detail and more vehemently two and a half decades later.

Grant found the Indians dishonest, servile, indolent, and morally stunted. Though he focused his criticisms on the Hindu religion, which he pictured as an ancient repository of degrading social customs, rituals, and supersitions, he implicitly questioned the worth of Indian society and civilization as a whole. He commented explicitly on deficiencies in Indian science and technology, which he linked to overall defects in Hindu civilization. He conceded that the Indians had made advances in a variety of forms of handicraft manufacture, but like Sonnerat he argued that these had been achieved in an age long past and that there had been little improvement in Indian techniques in thousands of years. Grant contrasted the Hindus' superstition and lack of curiosity with the "great use" that the British and other Europeans had made of reason "in all subjects" and the manner in which European scientific discoveries complemented, rather than competed with, Christian beliefs.[95] He argued that the Indians were ignorant of the natural sciences and that "invention seems wholly torpid among them." He boasted that the superiority of

[94]The best biography of Grant remains Ainslee Embree, *Charles Grant and British Rule in India* (London, 1962). See also Stokes, *English Utilitarians*, esp. chap. 1; and George D. Bearce, *British Attitudes towards India, 1784–1858* (Oxford, 1961), esp. pp. 51–64. The following discussion of Grant's views is based upon his "Observations on the State of Society among the Asiatic Subjects of Great Britain, Particularly with Respect to Morals; and on the Means of Improving It—Written Chiefly in the Year 1792," *Parliamentary Papers*, 1832, East India Company (8) "Report from Committee," vol. 1, app. 1, pp. 3–92.

[95]This argument was made by numerous nineteenth-century authors. See, e.g., J. M. Mitchell, *Hinduism Past and Present* (London, 1897), p. 271.

Europeans' understanding of the natural world could readily be demonstrated by the "sight of their machines."

Grant's "observations" were published in the *Parliamentary Papers* of 1813 and reprinted in 1832, in each instance at a point when important policy decisions regarding India were being debated. The considerable influence of Grant's opinions is perhaps best attested by the admission of Warren Hastings that the insistence of men like Grant that Indian civilization was decadent and mired in superstition, and thus much in the need of conversion to Christianity, had prevailed over his own advocacy of tolerance toward Indian beliefs and customs.[96]

Although Mill rarely cited his sources, many of his arguments also correspond to those advanced by two writers who published works highly critical of India in the years just before Mill's own study appeared. In 1811, William Ward, the head of the Serampore Mission College in Bengal, produced a formidable four-volume *Account of the Writings, Religion, and Manners of the Hindus*. Ward's main purpose was to expose the deficiencies of Indian religious beliefs and practices, but he managed to summarize most of the major criticisms of Indian science that had been raised by earlier writers. Ward declared categorically that the Hindus knew "nothing" of anatomy, surgery, chemistry, pharmacy, physics, and botany; that their geography was "wholly false"; and that they were "very imperfectly acquainted" with mathematics. He conceded that the Hindus had once made discoveries in astronomy and had some medical knowledge, but even in these fields, he added, their knowledge had long since stagnated, and they were centuries behind the Europeans. In general, he concluded, the sacred books of the Hindus contained "little if any real knowledge" and were full of the "grossest absurdities, the greatest exaggerations, and the most puerile conceits."[97] Ward's dismissal of the worth of most of Indian science was complemented by the less comprehensive but nonetheless telling critique of Indian technology published three years later by Benjamin Heyne, a surgeon in the service of the East India Company and a self-styled naturalist and amateur geologist. Heyne's criticisms were focused on Indian mining techniques, which he found "desultory and destructive, wasteful and negligent," and ironworking, which he characterized as "rude, imperfect and small-scale."[98] Taken together, these accounts by men with years of service in India belittled Indian achievements in sci-

[96]Penderel Moon, *Warren Hastings and British India* (New York, 1949), pp. 348–50.

[97]William Ward, *Account of the Writings, Religion, and Manners of the Hindus* (Serampore, 1811), vol. 1, pp. v, x, 191–93; vol. 2, pp. 333–37.

[98]Benjamin Heyne, *Tracts Historical and Statistical on India* (London, 1814), pp. 104–10, 219, 222–24.

ence and technology as a whole and provided abundant ammunition for Mill's all-out assault on the Orientalists' image of India as a great civilization.

James Mill's critique of virtually all things Indian was more comprehensive than Grant's or Ward's and largely devoid of the qualifications with which these earlier writers had hedged many of their pronouncements. Driven by the Utilitarian belief that general principles could be found on which all human polities and societies could be effectively organized and run, Mill employed what he had learned about India to illustrate how societies should *not* be structured and administered. Often using Indian beliefs and institutions as surrogates for those in England itself which he found offensive and wished to attack, he criticized Indian civilization with a lack of restraint made possible by the paucity of its defenders and the utter ignorance of the British reading public beyond a small circle of Orientalists and retired East India Company officials.

Mill found virtually nothing to praise in Indian society, past or present. Indian religion was "gross and disgusting" and Indian law "impossibly backward"; all manner of Indian creations from art and architecture to historical records were summarily dismissed as "rude."[99] Though he had met few and knew none well, Mill confidently characterized the Indians as dirty, dishonest, lacking in muscular development, and (befitting a people of the tropics) highly sensuous. He exhibited a remarkable penchant for finding fault with anything Indian. Even the game of chess, which earlier writers had cited as proof of the Indians' ingenuity, was mentioned by the dour and workaholic Mill only to demonstrate their indolent nature. According to his vision of India, Brahmins sat on rocks chanting meaningless mantras, and Indian princes played chess and hunted tigers while cities fell into ruin, bandits marauded in the countryside, and the peasants starved. Much of this vision was little more than extravagent fiction. Unfortunately, some of it was grounded in the reality of the chaos and suffering that had spread across the subcontinent in the eighteenth century following the collapse of the Mughal empire.

In Mill's view, the Indians had advanced beyond the first and lowest stage of social development but had never been "truly" civilized—a level of refinement that he believed had been attained only by the ancient Greeks and the Europeans since the Renaissance. He thought India

[99]"Rude" was apparently a favorite epithet of writers of the Scottish Enlightenment, such as Adam Ferguson. See Bryson, *Man and Society*, p. 42. This summary of Mill's general views is based on the relevant portions of the *History of British India*, vol. 1 (London, 1826).

roughly comparable to medieval Europe, though inferior to the latter in agriculture, the arts and intellectual attainments. Central to his low estimate of both Indian and medieval European achievements was his conviction that they had contributed little to scientific thought and technological advancement. Though Mill was a scholar and administrator, like many Utilitarians he showed a keen interest in scientific discoveries, particularly those that might be applied to inventions that would improve the condition of humankind. Along with other luminaries from Edinburgh, in the early 1820s Mill became actively involved in the campaign to establish mechanics' institutes to spread scientific knowledge among the artisan classes.[100] These interests were reflected in the considerable attention that he gave to science and technology in the *History of British India* and his underlying contention that material backwardness was a clear sign of a low level of social development. Mill could not deny that the Indians had long excelled in the production of cotton textiles, but he pointed out that Indian looms and tools in general were "rude," thus demonstrating that Indian "ingenuity" was only in its infancy. He conceded that India had monumental buildings in abundance but averred that they were not necessarily signs of advanced civilization. He felt, on the other hand, that good roads and lead pipes were clearly associated with civilized development, but India, he declared, had neither. He asserted that the Indians "knew nothing" about the use of glass for windows or improved sight and that their furnaces for working both glass and iron were of poor quality. The fact that "all Europeans" who had visited India had been struck by the "rudeness" of Indian tools, he concluded, was one of the surest proofs of his contention that Indians had never been truly civilized.[101]

James Mill also argued that the irrational and superstitious nature of Indian thought and the backward state of the sciences in India provided further evidence of the "low state" of the subcontinent's development. Like William Ward, he opposed the view of many earlier writers that India had made great contributions to scientific thinking in ancient times. Selectively citing the works of H. T. Colebrooke and the French scholar Pierre Laplace, Mill sought to show that discoveries attributed to the Indians in astronomy, algebra, and numerical notation were in fact the work of "bolder and more inventive peoples." Even if the Indians had invented "Arabic" numerals, he declared, that would not prove that they were civilized. If, of course, they themselves had devel-

[100]Cardwell, *Organization of Science*, pp. 36–43.
[101]For Mill's criticisms of Indian technology, see his *History*, vol. 2, pp. 1, 3, 9, 14–19, 21–22, 30–31, 42–43.

oped the algebra they possessed when the Europeans first arrived, that would "indicate a high level of civilization"; needless to say, Mill thought they had not. He considered Indian medicine backward, Indian botany "superficial," and Indian trigonometry nonexistent. Even in areas where the Indians had attained some knowledge, Mill found the ends to which they directed their inquiry unworthy of a civilized people. The Indians' cultivation of mathemathics and astronomy exclusively for "wasteful and mischievous" and "irrational" pursuits such as astrology, he argued, rather than ends that would serve utility, "infallibly denotes a barbarous nation."[102]

It is difficult to overstate the impact of Mill's polemic. His *History of British India* helped him earn a position with the East India Company, where he eventually rose to the post of Examiner of India Correspondence and where his son John Stuart was also gainfully employed. The *History* firmly established James Mill as Jeremy Bentham's most prominent disciple. It was widely regarded as a major attempt to apply to specific societies a broad range of Utilitarian tenets from those stressing education and legal reform to those advocating free trade. Mill's views on Indian society and history strongly influenced the pivotal policy decisions of reformist British administrators in India such as Lord William Bentinck and Thomas B. Macaulay. The latter declared Mill's *History* to be the "greatest historical work in English" since Gibbon's *Decline and Fall of the Roman Empire*.[103] The *History* became required reading for British youths preparing to rule, conduct trade, or win Christians in India. Numerous authors have pictured Mill's six volumes piled on the nightstands of the portside cabins of Company servants on the voyage out to India—though how many youths were dutiful enough actually to read Mill's turgid tomes is uncertain. Presumably most future administrators had some acquaintance with Mill's views, because the *History* was long used as a textbook for training candidates for the Indian Civil Service. Eric Stokes has seen the work as the chief source of the British sense of superiority over the Indians.[104] Even though more knowledgeable and balanced accounts of India appeared soon afterward, Mill's extreme opinions and blatant biases came in time to be regarded as facts and reasoned judgments. Perhaps more than any other writer, he was responsible for the image of India that prevailed in the middle and late nineteenth century in British circles, from the Viceroy's council chamber to the infamous clubs where Anglo-Indians gath-

[102]Ibid., pp. 67–70, 86–101, 125–34.
[103]Quoted in Forbes, "Mill and India," p. 33.
[104]Stokes, *English Utilitarians*, p. 313. John Burrow notes the highly detrimental effect that Mill's polemic had on Oriental studies in England; see *Evolution and Society*, p. 52.

ered to drink, gossip, and above all shut themselves away from the decadent, irrational, and bewildering peoples and customs that Mill had conjured up in his *History of British India*.

Contrary to the impression left by many of the authors who have discussed Mill's ideas and their impact, his specific assertions and even his overall assessment of Indian culture were often challenged by better-informed observers. Writers as varied in background and general approach as the East India Company official John Howiston and the famed French sociologist Gustave Le Bon readily acknowledged India's claims to a "vast and indisputable" antiquity as a civilized nation.[105] Many writers concurred with the view of the distinguished jurist Henry Maine that earlier in its history India had achieved a level of civilization roughly comparable to that of medieval Europe. Like his contemporaries who wrote on Africa, Maine believed that the study of Indian institutions and ideas was valuable not because these posed viable alternatives for Europe but because they provided valuable insights into Europe's own past. For Maine, however, the stage of development that India exemplified was well above that which most observers accorded "savage" Africa.[106]

Well-researched and carefully argued works like Maine's could not compete with Mill's. The vision propounded in the *History of British India* of a barbarous and superstition-encrusted India that carried one back into "the deepest recesses of antiquity," rather than Maine's medieval *gemeinschaft*, was the one that prevailed in nineteenth-century popular literature and ideas. It pervaded Anglo-Indian fiction in hackneyed descriptions of moonlit temples crowded with grotesque statuary; filthy, half-naked fakirs both repelling and fascinating innocent English maidens; and lewd and boisterous festival celebrations.[107] In scenes reminiscent of Conrad and other authors who pictured their travels in Africa as journeys back in time, William Arnold captured the nineteenth-century British colonizer's sense of the gulf that had opened between Europe and India. In his 1850 novel *Oakfield; or, Fellowship in the East*, Arnold's protagonist Edward Oakfield, a young East India Company cadet, remarks on the contrast between the "mud hovels of the swarm-

---

[105]John Howiston, *European Colonies in Various Parts of the World* (London, 1834), vol. 2, p. 210; and Le Bon, *Les civilisations de l'Inde*. Mill's assertions about Indian science and technology were often at odds with contemporary assessments by on-the-spot observers. See, e.g., Edward Strachey, "On the Early History of Algebra," *Asiatick Researches* 12 (1816), 158–59, 161–64, 183–84; the essays by Alexander Walker and Thomas Halcott included in Dharampal, *Indian Science and Technology in the Eighteenth Century* (Delhi, 1971); and the discussion of the views of Lancelot Wilkinson in Chapter 5, below.

[106]Burrow, *Evolution and Society*, pp. 77, 87.

[107]For a detailed study of these themes across a wide selection of Anglo-Indian literature, see Benita Parry, *Delusions and Discoveries: Studies on India in the British Imagination, 1880–1930* (Berkeley, Calif., 1972), esp. chaps. 2 and 3.

ing natives" and the swiftly moving steamship on which he is traveling up the Ganges to Calcutta. Oakfield reflects on the differences between the dynamic British, engaged in trading and empire-building on a global scale, and the languid, unchanging Indians passively spending their pre-destined lives beneath "the same scorching sky, the same rich vegetation,—the same funeral river; while the primeval Brahmins, sitting in primeval groves, asked 'Where shall wisdom be found?'" On a later journey, a fellow voyager named Middleton comments on the incongruity between the Europeans on the swiftly moving steamer and the Indian "multitudes engaged in their harsh-sounding, unpleasing, but animated devotion." Both men agree that the juxtaposition provides clear confirmation of the "inconceivable separation there apparently and actually is between us few English, silently making a servant of the Ganges with our steam-engine and paddle-boats and those Asiatics, with shouts and screams worshipping the same river."[108]

Like the eighteenth-century Orientalists, those writers who disagreed with Mill's contention that India had failed to rise above barbarism to civilization invariably stressed India's past achievements and contrasted them with its current backwardness relative to advanced European states. In this view, India had produced flourishing civilizations in the distant past, but, for reasons that varied by author—including climate, "racial inertia," and despotic governments—these once great cultures had stagnated and fallen into the decadent condition in which the Europeans found them in the eighteenth and nineteenth centuries.[109] Also like the Orientalists, many writers in the early nineteenth century singled out the stagnation of Indian scientific learning and the backwardness of its technical instruments as indicative of the experience of the civilization as a whole. From Bishop Heber's often quoted dismissal of Indian astronomy as "rubbish" to the French traveler Victor Jacquemont's characterization of Indian science as "triple nonsense for the makers and consumers," the fallen state of Indian learning was impressed upon European readers.[110] The decline in Indian mechanical and

[108]William Arnold, *Oakfield; or, Fellowship in the East* (London, 1854), vol. 1, pp. 13–14, 128. For a discussion of additional meanings in the Oakfield-Middleton exchange, see Hutchins, *Illusion of Permanence*, p. 122.

[109]For examples of these arguments, see Figuier, *The Human Race*, p. 336; Kidd, *Social Evolution*, pp. 142–43; J. S. Mill, *Political Economy*, pp. 12–14, 113–14; and George Campbell, *Modern India* (London, 1852), p. 6.

[110]Reginald Heber, *Narrative of a Journey through the Upper Provinces of India* (London, 1828), vol. 1, pp. 291, 295–97; Victor Jacquemont, *Letters from India* (London, 1835), p. 343. See also Howiston, *European Colonies*, vol. 2, pp. 49, 71; James T. Wheeler, *Adventures of a Tourist from Calcutta to Delhi* (n.p., 1868), p. 26; and the examples in R. C. Majumdar, "Social Relations between Englishmen and Indians," in Majumdar, ed., *British Paramountcy and Indian Renaissance*, vol. 10, pt. 2, of *History and Culture of the Indian People* (Bombay, 1965), pp. 337–38.

engineering skills was also cited by numerous authors. Robert Knox drew attention to the fact that the Indians of his day looked with "awe and wonder" at the "splendid structures" found throughout the subcontinent which they no longer possessed the capacity to build or even the energy to repair. John Crawfurd, citing the want of mechanical and mathematical skills among the Indians, doubted that such a people could have made the ancient discoveries and edifices that earlier writers had attributed to them. He speculated that they might have imported their learning and skills from classical Greece.[111]

As in Africa, superiority in military technology and organization was one of the more obvious manifestations of the differences that Europeans believed distinguished their level of social development from that of the Indians. In contrast to their behavior in Africa, however, the Europeans made no attempt to employ rifles and revolvers or music boxes and cameras to dazzle or entertain the "natives." Extensive involvement in the internal affairs and political struggles of the subcontinent, extending back to the early decades of the eighteenth century, had left the British with no illusions about their ability to overawe the Marattas, Sikhs, or Rajputs with mere displays of the accuracy and firepower of their latest weapons. In land weaponry, the Europeans had begun their rise to power in India in a state of rough parity with their Indian rivals. Through much of the eighteenth century the advantages the British and French enjoyed in warfare were primarily due to superb leadership, exemplified in such men as Robert Clive and Joseph Dupleix, plus superior discipline and organization. In applying the latter to struggles in the subcontinent, first the French and soon after them the British recruited Indian troops, called sepoys, into the European-led armies that vied for control with the larger but more unruly and poorly trained armies of Indian princes. In addition, Indian leaders—most successfully those of the Sikhs in the northwest—adopted European modes of training and organization and purchased large quantities of European arms in what eventually proved to be futile efforts to block the rise of British hegemony.[112] Thus, the Indians had been exposed to British military technology long before the Industrial Revolution. They felt little or none of the shock of African peoples who were drawn—often quite abruptly—into conflict with European forces in a later era when Western military advantages had been greatly enhanced by the process

[111]Robert Knox, *The Races of Men* (London, 1862), pp. 451–52; Crawfurd, "Physical and Mental Characteristics," p. 68.

[112]The best discussions of these patterns can be found in Philip Mason, *A Matter of Honour: An Account of the Indian Army, Its Officers and Men* (Harmondsworth, Eng., 1976); and Stephen P. Cohen, *The Indian Army* (Berkeley, Calif., 1971), esp. chaps. 1–3.

of industrialization. In addition to large-scale sepoy recruitment, the extensive employment of Indians in the imperial bureaucracy, British mercantile and manufacturing concerns, and later in the railway, telegraph, and postal services meant that they had a greater familiarity with the latest advances in European applied science and European technology than any other non-Western people.

Military technology more often brought together than distanced the British and Indians on a day-to-day basis, but most nineteenth-century writers were quick to point out the ways in which the superiority of the Europeans was demonstrated by the fact that they were the inventors and manufacturers of modern weaponry. From James Mill early in the century to Fitzjames Stephen in the 1880s, British observers tended to subsume military technology within a broader category of military prowess that was seen as the key to the conquest and rule of the vast Indian subcontinent by a handful of intrepid Europeans.[113] This myth of the stalwart few dominating the many hundreds of millions, which pointedly ignored the essential Indian military, administrative, and economic roles in the rise of British dominance, buttressed the claim of numerous writers that Britain had earned the right to rule India by virtue of conquest and martial excellence. In the 1860s John Crawfurd explored the links between better weapons and overall British superiority in some detail. Noting the Asians' failure to fully develop their early innovations in military organization and firearms, Crawfurd asserted that "the art of war is that which proclaims the loudest the incomparable superiority, both physical and intellectual, of the European over the Asiatic races." He contrasted the stagnation of Indian military technology since the early eighteenth century with the Europeans' development in the same period of artillery that made Clive's cannons at the battle of Plassey in 1857 seem like "popguns." He noted that the sepoys were useless without British officers to train and command them. Echoing the sentiments of numerous nineteenth-century authors, Crawfurd declared that the technology and military skills that had allowed the British to conquer and rule the hundreds of millions of Indians provided "the most signal example of the superiority of the European races over the Asiatic."[114]

Perhaps no writer better illustrates the great impact of scientific and

[113]For examples, see James Mill, "East Indian Monopoly," *Edinburgh Review* 20 (1812), 485; Fitzjames Stephen, "Foundations of Government in India," *Nineteenth Century* 80 (1883), 545; Thomas B. Macaulay, "Minute on the Black Act" (1836), in C. D. Dharker, *Lord Macaulay's Legislative Minutes* (London, 1846), p. 196; and Bearce, *British Attitudes*, pp. 47, 181.

[114]Crawfurd, "Physical and Mental Characteristics," pp. 69–70.

technological standards on the nineteenth-century decline in European esteem for India than the French sociologist Gustave Le Bon. A physician by profession, Le Bon is now remembered chiefly for his works on crowd psychology and revolution, but he was also an avid traveler who visited both the Middle East and India in the last decades of the nineteenth century. His personal impressions and inquiries were published in two large volumes, one on the Arabs in 1884 and a second, *Les civilisations de l'Inde*, in 1887. Though these works are now largely forgotten (Edward Said, for example, does not mention Le Bon's account of the Arabs in his indictment against Western Orientalists), both are in fact substantial works that capture in revealing detail many of the intellectual preoccupations and assumptions of bourgeois Europe in *la belle époque*. With regard to India in particular, Le Bon offered fresh impressions at a time when British assumptions about the Indians had reached a high degree of consensus.[115] His views were also a good deal more generous and often more sensitive to Indian values and aesthetics than those of Mill and many earlier writers. Le Bon left little doubt that civilizations had developed in India and that these were mainly the creations of the Indians themselves. He had high praise for Indian art and architecture, religion, philosophy, and some artisan skills, such as weaving, fine metalworking, and woodcarving. Moreover, he expressed strong misgivings about the level of artistic accomplishment in industrial Europe; he even had the audacity to assert that in the fine arts and philosophy, Indian civilization had once reached higher levels of achievement than Europe exhibited in the age of "steam and electricity."

Nevertheless, Le Bon's praise for Indian civilization was fundamentally qualified by his conviction that the Indians were greatly inferior to the Europeans in the fields of science and technology. Judging that Indian scientific thinking had not advanced beyond "vulgar mediocrity," he found it far inferior to the pioneering work of the Arabs and too crude to be compared with that of the Europeans. Unlike most of his contemporaries, Le Bon recognized the high level of craftsmanship that Indian artisans had attained in numerous fields, but he agreed with those who argued that such technology and scientific understanding as the Indians possessed had been borrowed long before from the Arabs and especially the Greeks. He also concurred with the widely held view that the Indians had progressed to a level of development on a par with medieval Europe and then stagnated and fallen far behind progressive Western societies. Thus, even though Le Bon's estimates of Indian cul-

[115]This overview of Le Bon's arguments is based primarily on *Les civilisations de l'Inde* (Paris, 1887). For his views on the sciences and handicrafts in India, see esp. pp. 189–94, 244–46, 547–52, 562–64.

ture and history displayed a much greater appreciation for Indian contributions than most nineteenth-century European observers allowed, his evaluations of Indian science and capacity for invention were as disparaging and categorical as those of Mill and earlier writers and virtually all his contemporaries. These perceived flaws in turn strongly detracted, in Le Bon's eyes, from the overall accomplishments of the Indian peoples. He concluded that an inherent incapacity for scientific inquiry and original invention had stranded them at a far lower level of social development than that of the western Europeans and led to their conquest and domination by peoples more proficient in these critical areas of human endeavor.

China: Despotism and Decline

As the French essayist and novelist Pierre Mille observed in the early 1900s, when merchants, scientists, and technicians replaced missionaries and philosophers as arbiters of European opinion, beginning in the late eighteenth century, European awe of and desire to emulate China shifted to hostility, contempt, and an urge to remake the country in accord with Western designs. Mille mused, not without sarcasm, that because China lacked railroads, spinning jennies, and leaders like the German General Helmuth von Moltke, it could no longer hide its scientific and technological shortcomings and general backwardness from the progressive and aggressive Western powers.[116] The Qing dynasty found its ability to resist the demands of the Western powers that they be able to trade and proselytize at will steadily diminished as overpopulation, natural calamities, and spiraling official corruption gave rise to widespread social unrest and a series of major rebellions, which sapped China's strength from within. Thus, long before the Opium War of 1839–42, which so brutally revealed the great military and technological advantages the Europeans had gained as a result of the Industrial Revolution, European assessments of Chinese civilization had turned decidedly negative. As we have seen, eighteenth-century merchants and naval commanders such as Lange and Anson were the first to broach many of the criticisms that would be directed against China in the era of industrialization. From the early decades of the nineteenth century these critiques were elaborated and intensified by travelers, missionaries, and members of the embassies that periodically visited the court at Beijing. No single writer played as pivotal a role in articulating shifting European attitudes toward China as James Mill did with regard to India, but

[116]Mille, "La race supérieure," p. 821.

John Barrow, a member of the Macartney embassy in 1793, published in 1804 an account of his experiences in China that set the tone for most later writers.

Formulated in the midst of a mission that is largely remembered for the ominous differences it underscored between British and Chinese approaches to foreign trade and diplomatic relations, Barrow's *Travels in China*, reflects the indignation and growing frustration felt by the members of Lord Macartney's retinue, who numbered nearly a hundred. In earlier centuries China's wealth, size, and unified power had permitted its rulers to display a good deal of arrogance in their dealings with European traders and missionaries. By the last decades of the eighteenth century, this posture was anachronistic, but the Chinese, oblivious to the transformations that were occurring in the West, persisted in regarding British and Dutch presents to the emperor as tribute from barbarian supplicants. They insisted that Macartney and other plenipotentiaries "kowtow" to the emperor. His refusal precipitated a major crisis for the protocol-conscious courtiers. In addition, the Qian-long emperor issued an imperious edict informing George III that Britain had nothing of importance to offer China in trade. The emperor also threatened that if the British continued to be troublesome, he would forbid the export of rhubarb and thereby cause overseas peoples the untold suffering of constipation.[117] Thus, at one level Barrow's account, as well as a similar reminiscence published by George Staunton and Macartney's own journal, can been seen as responses to the supercilious behavior of the Chinese. Given China's much diminished position relative to the industrializing nation-states of Europe, the English and other Europeans increasingly found this behavior intolerable.

Barrow's reactions were also shaped by his own experiences and personal prejudices. Like Mill, Barrow was a self-made man. Reared in a modest Lancashire household and lacking the financial support and encouragement Mill had received from his mother, Barrow was forced to end his formal schooling after completing grammar school. But, as he proudly informs the reader of his *Auto-Biographical Memoir*, he strove with some success to continue to educate himself after becoming a clerk in a Liverpool foundry at the age of fourteen.[118] Though William Proudfoot, who in the 1860s published a somewhat petty critique of *Travels in China*, heaped scorn on Barrow's pretensions to scientific

[117]An excellent introduction to the background of the mission is included in J. L. Cranmer-Bing, ed., *An Embassy to China* (London, 1962). For a discussion of the imperial edict and a translation of the original version, see Arthur Waley, *The Opium War through Chinese Eyes* (Stanford, Calif., 1958), pp. 28–31.

[118]John Barrow, *Auto-Biographical Memoir* (London, 1847), pp. 6–7, 14–16. Biographical details on Barrow are based on this memoir.

learning,[119] Barrow had in fact received considerable practical instruction in applied science in the years during which he rose from clerk to overseer of the iron foundry. He also claims that at various points he had received instruction in or taught himself additional mathematics and astronomy. He left the foundry for a chance to travel to Greenland on a whaling ship and, after returning to England, won a position as a teacher of mathematics at a boys' academy in Greenwich. Barrow's part-time work as a tutor for the son of Sir George Staunton, who served as secretary to the Macartney mission, led to his invitation to join the ambassador's entourage. Barrow's official position as Comptroller of the Household afforded him numerous opportunities for interchange with Chinese at all social levels from porters to mandarin officials. Contacts with the latter opened the way for him to travel quite extensively in the Qing empire. He claimed to have some familiarity with the Chinese language, presumably also self-taught, and was confident that his personal observations and experiences had given him the qualifications to write an authoritative insider's account of Chinese civilization.

Despite its title, Barrow's *Travels in China* was intended to be much more than a travelogue. He undertook the same sort of detailed appraisal of Chinese society and history that Mill later attempted for India. Barrow interspersed descriptions of his travels and personal experiences with lengthy discussions of varying aspects of Chinese culture, from the grand themes of legal codes and the treatment of women to such minutiae as the appalling lack of water closets or other "decent places of retirement."[120] He began with the assumption that China had once achieved a high level of civilization—higher than that of Europe until the fifteenth century—"if not in the sciences, at least in the arts and manufactures, conveniences and luxuries of life."[121] His judgments on the quality of Chinese life and material culture tended to be favorable at the beginning of his residence in China but grew more and more disparaging as time passed. On first seeing Beijing, for example, he hailed it as the greatest city on earth, with "streets that are much cleaner than those of its European counterparts." He initially commended the Chinese for their hospitality, honesty, industry, and skill. By the later stages of his travels, however, Barrow had come to regard Chinese cities as filthy and overcrowded breeding grounds for disease; he found China's laws "barbarous," its women degraded, and its lower classes oppressed. In the course of a discussion on the "gross and vile" amuse-

[119]William Proudfoot, *Barrow's "Travels in China": An Investigation* (London, 1861), pp. 15–17.
[120]John Barrow, *Travels in China* (London, 1804), p. 333.
[121]Ibid., p. 289.

ments of the general populace, he concluded that Chinese civilization "exists more in state maxims than in the minds of the people."[122]

Although similar criticisms can be found in the works of earlier writers,[123] Barrow went much further than even China's severest critics in challenging what he characterized as Jesuit fabrications that China was a major center of scholarship, wonderful inventions, and advanced material culture. Developing a theme that would become a major presupposition for China "experts" in the nineteenth century, he argued that the Chinese had nurtured in ancient times a sophisticated civilization that had advanced steadily until about the fifteenth century, when it stagnated and went into long-term decline. Unlike the philosophes and many later writers, he had little to say about the causes of China's arrested development, though he did allude to the stultifying effects of China's despotic regime and its arrogant refusal to acknowledge the discoveries or skills of foreign peoples. It is not likely that Barrow had read the eighteenth-century writings of Hume and Condorcet, who had done much to advance the notion of the progressive development of human societies, but his account as a whole did provide an implicit contrast between static, past-minded, backward China and the continually improving, forward-looking, industrializing states of Europe. Architecture, social customs, and legal codes figure in his judgment of the relative worth of the two civilizations, but much of his commentary is devoted to an extended comparison of the woeful state of scientific learning and technological innovation in China with the splendid accomplishments of the Europeans in these fields.

Barrow acknowledged in a general way that ancient China had been a major source of invention and discovery, but his narrative includes little that is favorable about specific aspects of Chinese technology or science. He found Chinese ships "clumsy" and their navigation techniques antiquated and inept. He credited the Chinese with the invention of the compass but deemed those in use at the time of his visit far inferior to those employed by European navigators—which, he added, the Chinese were unable to read. He also doubted that they had ever made great sea voyages. Barrow found little to admire about Chinese tools or machines, manufacturing techniques, or modes of transportation. He judged Chinese agricultural implements to be on a par with those of Ireland—that is, very backward indeed. He contrasted the "simple machines" of China with the size and complexity of those he had worked

[122]Ibid., pp. 90, 98–99, 138–39, 208, 222, 349.
[123]See esp. George Staunton, *An Authentic Account of an Embassy from the King of Great Britain to the Emperor of China* (Philadelphia, 1799), vol. I, pp. 9, 12–16, 33–34, 82, 106–12, 226–29.

with in the English Midlands. He pointed out that China's carts lacked springs and inside seats, which had been in use in Europe since the late seventeenth century.[124] Even in papermaking and silk textile weaving, where Barrow conceded Chinese excellence, he qualified his praise by pointing out that the techniques of manufacturing these products had not changed for centuries. Though he extolled the splendor of the Great Wall and the Grand Canal, he refused to see these structures as typical or as evidence of a general aptitude for either engineering or architecture; he found Chinese buildings for the most part monotonous and awkward in design.[125]

Barrow's views on the state of the sciences in China differed little from those of China's harshest eighteenth-century critics. He thought Chinese learning even further behind that of Europe than Chinese technology and averred that it had advanced little in two thousand years. He claimed that algebra, geometry, and chemistry were "totally unknown" to the Chinese and that their arithmetic was "mechanical" and wholly based on the abacus, a device which he thought useful for calculating sums but hardly suitable for conceptual breakthroughs. He suggested that the study of anatomy would "shock the weak nerves" of the timid Chinese and sought to illustrate the backward state of their medicine in general with the observations that they never resorted to bloodletting and that their medical texts were "little better than herbals." To impress upon his readers the extent of Chinese deficiencies in the sciences, he quoted one of the delegation's physicians, a Dr. Gregory, who reckoned that the best medical care to be found in China would be roughly equivalent to that provided by a sixteen-year-old apprentice to an Edinburgh physician.[126]

Barrow ridiculed the Chinese for their unbearably arrogant supposition that foreigners had little to offer the Middle Kingdom. He pointed out that they had borrowed heavily in both the sciences and technology in recent centuries. He stressed the role of the Jesuits in transmitting knowledge and inventions from the West and noted that the emperor was dependent on them for accurate calendars and clocks and effective artillery pieces. Barrow saw this dependence as a symptom of China's general backwardness and evidence of the degree to which it had fallen behind the West: "They can be said to be great in trifles, whilst they are really trifling in everything that is great."[127]

[124]T. K. Derry and T. I. Williams, *A Short History of Technology* (Oxford, 1960), p. 212.

[125]Barrow, *Travels*, pp. 25–26, 38–41, 61, 90, 215, 298–300, 301–5, 307–10, 312, 334–38, 564–66.

[126]Ibid., pp. 274, 295, 297, 344–54.

[127]Ibid., pp. 189, 284, 300–301, 342, 355 (quoted portion).

As Mill had done in regard to India, Barrow concluded that China had been considerably overrated by earlier writers, particularly the Jesuits and the philosophes. He deemed the Chinese "totally incapable" of excellence in machine technology and scientific theory, or even of appreciating recent European accomplishments in these fields. For this reason, and despite his early comments on the enthusiasm of the "Tartars" for European swords and carriages, Barrow did not recommend mechanical presents for the emperor or other notables. In fact, these were the very sorts of gifts that had been brought by the Macartney mission, as evidenced by the many contrivances presented to Qian-long and the presence in the British retinue of a Dr. Dinwiddie, whose chief task was to oversee the running and repair of the machines and to introduce Chinese officials to the wonders of European science.[128] In contrast to George Staunton, who reported intense Chinese interest in European inventions,[129] Barrow was dismayed by the lack of curiosity and understanding evinced by the emperor and his highest officials for the ingenious devices that the mission had carried to Beijing. He wrote that the Chinese confused planetariums with musical instruments—an error that revealed much, he felt, about the state of astronomy and mathematics in China. He further observed that because the head eunuch was unable to explain the purpose of air pumps and assorted electrical devices to the emperor, he simply informed the ruler that they were intended as "playthings" for the imperial grandchildren. So low was Barrow's estimate of the capacities of the people who had contributed such a large share of humankind's basic inventions that he strongly urged that future presents consist mainly of articles of gold and silver, childrens' toys, and other trinkets.[130]

Although *Travels in China* has received far less attention than James Mill's *History of British India*, John Barrow's doubts and criticisms anticipated most of those that were raised throughout the nineteenth century. As his harshest critic William Proudfoot conceded, *Travels in China* was widely quoted and very favorably received by reviewers, who considered it one of the best travel accounts of the period.[131] Even before it was published, Barrow's fellow ambassadors cited his opinions extensively and referred to him as an authority on Chinese affairs.[132] By 1805 the work had been translated into German and French and published in an American edition. As Proudfoot admitted, Barrow became some-

---

[128]Proudfoot, Barrow's "Travels," p. 18.
[129]Staunton, *Authentic Account*, pp. 25, 82, 226–27.
[130]Barrow, *Travels*, pp. 112–13, 311–12, 343.
[131]Proudfoot, Barrow's "Travels," p. iii.
[132]Staunton, *Authentic Account*, pp. 2, 82, 112–13, 117–18.

thing of a celebrity in the late Georgian era, and his work was long regarded as one of the fullest and most reliable accounts of China. Later travelers and missionaries did not often footnote their works, but opinions and observations almost identical to those of Barrow abound in nineteenth-century writings, and he is explicitly cited by James Mill and numerous other authors.[133] His importance has been obscured, however, by the emphasis placed by twentieth-century historians on the Qing dynasty's humiliation in the Opium War of 1839–42 as the turningpoint in European thinking about China and the Chinese. Barrow's impact is also difficult to measure because several of his contemporaries, especially J. C. L. de Guignes and John Davis,[134] published descriptions of China expressing similar criticisms, though none provided the same detail or claimed the expertise that Barrow assured his readers he possessed.

The observations and opinions of John Barrow and his contemporaries formed the nucleus of a cluster of ideas about China that informed virtually all nineteenth-century accounts of the Qing empire. These ideas also shaped the assumptions and prejudices that were a vital part of the cultural baggage carried by missionaries, diplomats, and traders when they went out to the "Far East." As time passed, successive Chinese defeats at the hands of the Western powers, and the spread of economic distress and social disturbances within China as the Qing dynasty entered a period of irreversible decline, prompted harsher judgments and more extreme condemnations of Chinese customs and institutions. Diverse aspects of Chinese culture came under scrutiny. But China's failure to develop scientific thought or to advance technologically was seen by many of its most prominent European and American critics as the central cause of its backwardness and its greatly diminished standing in the "family of nations." Through the middle and late nineteenth century, European and American writers never seemed to tire of chronicling Chinese deficiencies. Barrow had anticipated most of their criticisms, denigrating Chinese accomplishments in such convincing detail that most later writers contented themselves with sweeping generalities. The veteran missionary Evariste Huc, for example,

---

[133]Mill, *History*, vol. 2, pp. 311, 512; Jacob Abbott, *China and the English* (New York, 1835), pp. 149–51; Clarke Abel, *Narrative of a Journey in the Interior of China* (London, 1819), pp. 189, 202; F. S. Feuillet de Conches, "Les peintres européens en Chine et les peintres chinois," *Revue Contemporaine* 25 (1856), esp. 234; and William Ellis's introduction to Charles Gutzlaff, *China Opened* (London, 1838), pp. vii–viii, x–xi, xxx–xxxi.

[134]For sample criticisms, see respectively J. C. L. de Guignes, *Voyages à Peking, Manille, et l'Ile de France* (Paris, 1808), vol. 2, pp. 161, 167–68; and John Davis, *The Chinese: A General Description of the Empire of China and Its Inhabitants* (London, 1836), vol. 1, pp. 6–7, 221–23, 240–41, 275–77, 301, 309–10.

lamented the decadence that had taken hold of China, where "old vices" (not specified) were on the increase. He compared China with its chaos and decreptitude to progressive Europe where "almost every passing day marked some new discovery."[135] Equally typical was the sort of summary dismissal written by Captain John McLeod, who visited China several decades after Barrow: "With people who still imagine the earth to be a plain, and China in the middle, with all her tributory kingdoms around her; who are equally uninformed with regard to astronomy; who in the prohibition of the study of the human frame preclude the attainment of the very basis of medical knowledge; and who, in fact, are equally ignorant; and determined to continue so; it is evidently impossible to connect the term science in any shape or manner."[136]

The French economist Michel Chevalier put the case for Chinese inferiority to Europe even more succinctly. Tiny England, he boasted— in a rare instance of a Frenchman in this era lauding the attainments of his nation's old and bitter rival—contained more machines, roads, and canals and produced and consumed more iron than all of the vast Qing empire.[137] Robert Knox made the comparison both simpler and more sweeping: he wrote in the 1860s that a single English engineer possessed more practical knowledge than all the savants of China.[138] His glib dismissal lacked the eloquence of Macaulay's earlier quip that a single shelf of books from the library of an educated Englishman was worth all the learning of Asia, but Knox's remark more effectively captured the sentiments of his age. Most nineteenth-century writers came to see achievements in the applied sciences—which engineering perhaps best exemplified[139]—rather than in poetry, law, or philosophy as those most responsible for Europe's global ascendancy, as well as those that best measured the extent to which Chinese civilization as a whole had fallen behind that of the West.

From the earliest centuries of European overseas contact with China, the military vulnerability of the scholar-gentry of the Middle Kingdom had drawn sweeping censures from adventurers and sea captains such as Galeate Pereira and the feisty George Anson. From the first decades of

---

[135]Evariste R. Huc, *The Chinese Empire* (London, 1855), p. xxiii. For similar views see also W. H. Medhurst, *China: Its State and Prospects* (London, 1838), pp. 97–98.

[136]John McLeod, *Voyage of His Majesty's Ship Alceste to Cathay, Corea, and the Island of Lew Chew* (London, 1819), p. 184.

[137]Michel Chevalier, "L'Europe et la Chine," *Revue des Deux Mondes* 23 (July 1840), p. 210.

[138]Knox, *Races of Men*, p. 284. For further examples of this key theme, see James Wilson, *China* (New York, 1887), pp. 811–13, 285; and J. Dyer Ball, *Things Chinese* (London, 1892), pp. 1, 7–8, 29, 48, 229.

[139]Layton, "Mirror-Image Twins," esp. pp. 576, 580.

the Industrial Revolution, China's military backwardness was elevated from a specific but significant fault to one of the most glaring manifestations of its technological inferiority to Europe. An incident related by George Staunton nicely illustrates the deeper significance that members of the Macartney mission assigned to what they viewed as the grave military shortcomings of the Qing empire. Staunton recounted that among the presents intended for the Qian-long emperor were six "elegant" brass cannon. When the Chinese legate assigned to the Macartney party discovered the cannon among the presents about to be sent to the imperial palace, he insisted that they be left behind. Staunton observed that the "whole conduct [of the Chinese official] seemed to indicate a mind agitated with apprehension," and he concluded that the artillery pieces were refused because their obvious superiority to anything the Chinese could devise might well have led the latter "to entertain a higher idea of the prowess of the English nation than [their] own."[140]

In 1819 John McLeod told of a minor armed clash between a British frigate and a fleet of Chinese war junks. After a quarrel with the port officials at Canton, McLeod's warship fought its way past the junks and the forts in the Bogue or delta area that guarded the approaches to the city. Noting how easily the defenses of the great port had been breached, McLeod observed that "almost *any* European gunners with the same advantages would have blown the frigate out of the water."[141] In the years that followed, which were filled with recurring Anglo-Chinese tensions, numerous writers commented on the military weaknesses of the Chinese empire. Perhaps the fullest discussion appeared in an anonymous essay published at Canton in the *Chinese Repository* of 1836. The author argued that the Chinese with their antiquated cannon and unruly armies were "powerless on land" and dismissed the imperial navy as nothing more than a "monstrous burlesque." He viewed these weaknesses as mere symptoms of much more fundamental deficiencies in Chinese society as a whole. In fact, he prefaced his detailed critique with the claim that "there is, probably, at the present no more infallible a criterion of the civilisation and advancement of societies than the proficiency which each has attained in 'the murderous art,' the perfection and variety of their implements for mutual destruction, and the skill with which they have learned to use them."[142] Though most European observers would have balked at making military prowess so central a gauge of the level of social development, its growing importance in

[140]Staunton, *Authentic Account*, pp. 33–34.
[141]McLeod, *Voyage of His Majesty's Ship*, pp. 163–65.
[142]*Chinese Repository* 5/1 (1836), 165–78 (quoted portion, p. 165).

shaping European assessments of the overall merit of non-Western peoples boded ill for the Chinese, who had fallen far behind the aggressive "barbarians" at their southern gates.

With the Opium War of 1839–42, the full meaning of China's military backwardness was brutally revealed. In a series of engagements on land and sea—rather modest confrontations by European standards—British ships and British-led Indian infantry routed the numerically superior Chinese forces. Britain's decisive military advantage was perhaps most dramatically demonstrated by clashes between the iron-clad paddle steamer *Nemesis* and the Chinese war junks. In addition to the most advanced naval artillery pieces, including two pivot-mounted thirty-two pounders, the newly built *Nemesis* was armed with a rocket launcher. In what proved to be the most memorable clash of the war, it singlehandedly engaged a fleet of fifteen Chinese war junks. The British ship took the initiative by reducing the lead junk to a roaring ball of smoke and fire with a Congreve missile. As the remaining junks fled or were hastily abandoned by their demoralized crews, the *Nemesis* continued up the coast, forced the panic-stricken inhabitants of a small town to evacuate their homes, sank a second war junk and captured another.[143]

The contrast between the *Nemesis* and the Chinese junks—which with their mat sails and painted eyes struck one British officer as "apparitions from the Middle Ages"[144]—cast further doubt on the already much-contested image of China as a powerful and advanced civilization. These and later military setbacks convinced virtually all European observers that China was no match for Europe[145] and reduced the Chinese in the eyes of the European public to the pitiful creatures ridiculed in an 1859 *Punch* jingle:

> With their little pig-eyes and their large pig-tails
> And their diet of rats, dogs, slugs, and snails,
> All seems to be game in the frying-pan
> Of that nasty feeder, *John Chinaman*.
> Sing lie-tea, my sly *John Chinaman*
> No fightee, my coward *John Chinaman*

[143]The best narrative of the political and military aspects of the war is provided in Peter W. Fay, *The Opium War, 1840–1842* (Chapel Hill, N.C., 1975). The historical background is skillfully surveyed in Hsin-pao Chang, *Commissioner Lin and the Opium War* (Cambridge, Mass., 1964). Frederic Wakeman, Jr., *Strangers at the Gate* (Berkeley, Calif., 1966) treats the broader social issues in depth and with great insight.

[144]Fay, *Opium War*, p. 222.

[145]See, for examples, Huc, *Chinese Empire*, pp. 402ff.; Knox, *Races of Man*, p. 282; and John L. Nevius, *China and the Chinese* (New York, 1869), p. 278.

John Bull has a chance—let him, if he can
Somewhat open the eyes of *John Chinaman.*

In order to imagine that China was Europe's equal, John Crawfurd
mused nearly a decade later, "we must fancy a Chinese fleet and army
capturing Paris and London, and dictating peace to the French and
English."[146]

Chinese ineptness at using up-to-date military technology provided
the material for most of the anecdotes of bumbling "natives" which
European commanders and travelers, like their counterparts in Africa,
included in their memoirs to illustrate the great distance that separated
the scientifically minded, industrializing Western peoples from all others. During the Opium War, J. Elliot Bingham, a lieutenant aboard the
British corvette *Modeste*, mocked the ignorance and credulity of the
Chinese, who had attempted to block the passage of the English fleet up
the Pearl River to Canton by lining up large earthen jars in the hope that
the Europeans would believe them to be batteries of cannon. Bingham
assured his readers that such "childlish" ruses were commonly employed by the Chinese against the British invaders.[147] Later authors
cited similar attempts at deception—lighting fires in huge iron tubes to
frighten the English with smoke; wearing huge and hideous masks to
make the Europeans think they were "fighting monsters"—to demonstrate the "tricks worthy of children" to which the Chinese were forced
to resort because of their technological inferiority. Frederic Farar
sneered at the "asinine ignorance of the Chinese gunners who held lights
near their cannon to allow them to fire at night." The English journalist
Henry Norman noted that the Chinese were so foolish that they dried
percussion caps and dynamite on steam boilers. Lord Curzon claimed
that when Chinese soldiers were given modern weapons, they tended to
jam them because they regarded all rifle cartridges as identical and thus
did not bother to make sure they had the right gauge for the weapon
they were using. Curzon went on to aver that superstition was widespread among the Chinese (adding, in a characteristic burst of snobbery
rather than relativism, that the same might be said of the European
masses). He claimed that it was widely believed in China that Christian
missionaries used parts of the human body to concoct medicines and to
mix chemicals such as those used in photography.[148]

---

[146]John Crawfurd, "On the Conditions Which Favor, Retard, or Obstruct the Early
Civilisation of Man," *T.E.S.L.* (1861), 165.

[147]J. Elliot Bingham, *Narrative of the Expedition to China* (London, 1842), vol. 1, pp.
303–4, 345.

[148]Knox, *Races of Man*, p. 285; Farrar, "Aptitudes of Races," p. 124; Henry Norman,

As if to compensate for the awe with which European travelers in earlier centuries had responded to China, nineteenth-century writers delighted in stories showing Chinese amazement at and admiration for simple Western devices. Bingham and McLeod boasted of their ability to "dazzle" crowds of Chinese onlookers with crude telescopes and noted that even the lowly bilge pumps of British battleships astonished Chinese visitors.[149] The American missionary Arthur Smith reported that the Chinese were utterly bewildered by Western science and technology. Their inability to comprehend the "miracles" of Western invention, he argued, forced them to feign indifference or to seek fantastic explanations. Above all, Smith believed, the ignorance of the Chinese rendered them extremely vulnerable to deception on the part of their European rivals. To illustrate the point, Smith told the story of a carriage master at a European legation who convinced a group of curious Chinese that the springs of the vehicles in his care were able to propel them without horse or donkey; he insisted that donkeys were used only to show respect for Chinese custom.[150]

Two basic positions developed in the nineteenth century regarding China's failure to match European achievements in technology and the sciences. Elements of the first approach, which stressed early advance and then stagnation, had been suggested by eighteenth-century writers. Their stress on stagnation was supplemented in the nineteenth century by a sense of overall decline, which appeared to be confirmed by the breakdown of the Qing empire. In 1855 an anonymous reviewer of E. R. Huc's account acknowledged that China had been responsible for the three inventions—the compass, printing, and gunpowder—which had been the "principal material agents of the progress of the modern world." In singling out these inventions, the writer continued a tradition that extended as far back as the sixteenth-century works of Jean Bodin and Louis Le Roy, which had been pivotal in the gestation of the idea of progress in the Western intellectual tradition.[151] But the reviewer, writing when England's industrial supremacy was at its height,

---

The Peoples and Politics of the Far East (London, 1895), p. 286; and G. N. Curzon, *Problems of the Far East* (London, 1894), pp. 329, 352. In fact, as Frederic Wakeman has argued (*Strangers at the Gate*, pp. 27–28), the resort to trained monkeys and talismans with firecrackers to ward off the Western invaders was adopted only in desperation by Confucian scholar-officials, who were normally disdainful of the necromantic currents in Chinese culture.

[149]Bingham, *Expedition to China*, pp. 274–75; McLeod, *Voyage of His Majesty's Ship*, pp. 71–72, 165–66, 173.

[150]Arthur Smith, *Chinese Characteristics* (Shanghai, 1890), p. 42.

[151]J. B. Bury, *The Idea of Progress* (New York, 1960), pp. 40–41, 44–45, 52–62, 212, 215, 220, 292, 332; Morris Ginsburg, *The Idea of Progress* (Westwood, Conn., 1972), pp. 1, 53–54; Pollard, *Idea of Progress*, pp. 10–11, 20–22, 26–30, and chap. 2 passim.

merely used this praise for China's past accomplishments to set up an extended critique of its present backwardness and its failure to develop the full potential of these important inventions. He noted that the Chinese had not used the compass to explore the globe or expand trade, that they had not advanced beyond firecrackers after their initial discovery of gunpowder, and that their invention of printing had not led to a distinguished literary tradition. The reviewer went on to argue that though the Chinese knew about the circulation of the blood, their general knowledge of human anatomy was on a par with that of the rudest savages. He asserted that even in areas of manufacture in which they had displayed the greatest skill, such as the production of silk textiles and porcelain, Chinese tools and techniques had not improved for centuries.[152] This assumption was so routinely accepted that Paul Champion could insist on the continuing relevance of mid-seventeenth-century texts on Chinese manufacturing techniques which he edited in the 1860s, because little or no change had occurred in the intervening centuries.[153]

Numerous authors commented on the stultification of Chinese creativity and a related reluctance to import innovations from other civilizations,[154] but none matched Frederick Farrar's utter disdain for things Chinese. Having established that the "Aryan" race had excelled over all others in all fields, Farrar singled out the "semi-civilized" Chinese as a perfect example of "arrested development" and "mummified intelligence." He contended that their early inventions had "stopped short" at the "lowest point" in contrast to the constantly modified and improved devices of the Europeans. Farrar declared that the Chinese looked upon the compass as a plaything, that their ships were little more than "painted tubs" and their gunpowder "mere pyrotechny." He found all Chinese endeavors from religion to politics lacking in "progressiveness," "enthusiasm," "warmth," and "vigour." Reiterating the arguments of Captain Anson a century earlier, Farrar asserted that because the Chinese reduced everything to the "dead level of practical advantage," their learning, arts, and inventions had been marked by a

[152]*Edinburgh Review* 101 (1855), 423–44. The arguments made by John Crawfurd in an essay written half a decade later are so similar that it is reasonable to conclude that he was the author of the Huc review. See "Conditions," p. 161.

[153]See Champion's introduction to Julien M. Stanislas, *Industries anciennes et modernes de l'empire chinois* (Paris, 1869), p. 2.

[154]See, e.g., Jean Bory de Saint-Vincent, *L'homme: Essai zoologique* (Paris, 1827), pp. 259–61; Hugh Murray, J. Crawfurd, et. al., *An Historical and Descriptive Account of China* (Edinburgh, 1836), vol. 2, pp. 15–16, 301–4; William Lawrence, *Lectures on Physiology, Zoology, and the History of Man* (London, 1819), p. 483; and Le Marquis de Courcy, *L'empire du milieu* (Paris, 1867), pp. 184–85, 444–45.

"plague-spot" of "utilitarian mediocrity." Even their language, which, he said, had not developed beyond "hieroglyphics," exuded a sense of rigidity and stagnation; it was, he concluded, little more than a "petrified fragment of primeval periods."[155]

The explanations for Chinese inertia offered by the proponents of the early invention–long term stagnation approach varied considerably, from Farrar's insistence on innate racial deficiencies to a focus (similar to that of such eighteenth-century writers as Parennin and Sonnerat) on Chinese despotism and veneration for tradition.[156] John Barrow touched on the latter but emphasized the absence of theory in Chinese thought as a whole. He expressed his displeasure at several mandarins' lack of curiosity regarding the reasons why alum caused mud particles to sink to the bottom of a vessel filled with water. He commented that though they were adept at dyeing and tinting "all manner of objects," the Chinese had no theory of colors. He pronounced them "totally ignorant" of the "basic principles" of astronomy, which, he snidely added, they professed to "value so much" but in fact "understood so little." It was this lack of interest in underlying principles, Barrow concluded, that had prevented the cumulative increase of scientific knowledge in China: "The practical application of some of the most obvious effects produced by natural causes could not escape the observation of a people who had, at an early period, attained such a high degree of civilization, but satisfied with the practical part, they pushed their enquiries no farther."[157]

Barrow believed that the poverty of Chinese scientific theory had stunted technological development throughout their history. He noted that the Chinese were well aware that the heat of steam is much greater than that of boiling water and that they had long enclosed steam in ceramic vessels to soften animal horn for lanterns. Barrow was convinced that if this applied knowledge had been placed in a broader theoretical framework and improved through observation and experimentation, the Chinese would eventually have discovered the principles of the steam engine. This had not happened because the Chinese "seem not to have discovered its [steam's] extraordinary force when pent up; at

---

[155]Farrar, "Aptitudes of Races," pp. 123–24.

[156]The positions of Farrar and other racist thinkers on these issues are discussed in the third section of Chapter 5, below. For discussions stressing despotic rigidity and intense conservatism, see the anonymous review of De Guignes, *Voyage*, in *Edinburgh Review* 14(1809), 415, 422–26; Virey, *Histoire naturelle*, p. xxxiii; McLeod, *Voyage of His Majesty's Ship*, p. 191; and Samuel W. Williams, *The Middle Kingdom* (New York, 1879), vol. 2, p. 143.

[157]Barrow, *Travels in China*, pp. 284, 297–98, 340–41 (quoted portion).

least they have never thought of applying that power to purposes which animal strength has not been adequate to effect."[158]

Barrow's contemporary John Davis concurred with these pronouncements on Chinese deficiencies. Though his overall assessment of Chinese civilization was a good deal more favorable than Barrow's, Davis too was convinced that the Chinese had fallen far behind the West in many areas but especially in science and technology. Davis averred that much of the blame for China's stagnation could be traced to an obsession with custom and intense conservatism. These traits, he argued, had prevented the Chinese from fully exploiting the potential of their own inventions or those they had imported from the West and elsewhere. But like Barrow, Davis believed that the Chinese indifference to abstract knowledge and failure to apply the inductive method were critical liabilities. He noted an incident reported by another traveler, Clarke Abel, in which a Chinese scholar-official had shown no interest in potassium when Abel could not tell him of any practical use for it. Davis asserted that because the Chinese lacked curiosity, an interest in underlying principles, and a commitment to developing a cumulative body of knowledge, they must have "stumbled by mere chance upon useful inventions, without the previous possession of any scientific clue." Though Davis may have overestimated the direct impact of scientific knowledge on invention in Europe prior to the eighteenth century, he displayed an acute awareness of the increasing links between the two in his own era.[159]

The judgments of Barrow and Davis were repeated by numerous later writers. Even though E. R. Huc, for example, had high praise for Chinese bridges and metalworking and mathematical skill, he faulted Chinese invention and science for their lack of systematic investigation and "fixed general principles." Huc, who wrote one of the most widely read accounts of China in the mid-nineteenth century, concluded that the absence of these qualities rendered Chinese efforts at innovation and discovery "scattered and desultory." He claimed that the natural sciences had no place in formal education in China, that the preservation of scientific knowledge and technical knowhow was left to "ignorant workmen." Consequently, the Chinese had actually lost scientific ideas and had forgotten techniques they had once mastered. Later observers pinpointed similar failings; Charles Gutzlaff, for example, concluded

[158]Ibid., pp. 297–98.
[159]Davis, *The Chinese*, vol. I, pp. 6–7, 221–23, 240–41, 273 (quoted portion), 275–76, 301, 309–10.

that unless remedied by influences from the West, China could never produce "Bells or Pasteurs."[160]

The second and minority position regarding China's technological and scientific backwardness, despite its promising beginnings, was even less flattering to the Chinese than the early creativity-stagnation thesis. As early as 1819 William Lawrence suggested that Caucasians from Persia had been the source of the artistic and scientific accomplishments of "the East" (by which he appears to have meant China, Japan, and India). Similar theories were proposed by several authors later in the century. The most influential of these, the Count de Gobineau, substituted Aryan invaders for Caucasian migrants. As late as the first decade of the twentieth century Robert Douglass, professor of Chinese at King's College (London) and Keeper of the British Museum, argued for the Middle Eastern origins of China's discoveries. He claimed that the Chinese had originally migrated from the Fertile Crescent, taking with them the scientific learning and philosophy of the earliest civilizations.[161] Other writers were even more fanciful. Georges Pouchet, suggested that Greece had been the source of the "flowering of science in the East." The anonymous author of an 1865 essay in the *Anthropological Review* was more vague about the origins of Chinese science and technology. He felt, however, that the stagnation of Chinese inventiveness could best be explained by the fact that the ingredients for civilized development had come as a "gift from without." Thus, though the Chinese were able to advance until the impetus from these imports was spent, they lacked the "inherent intellectual vitality" to sustain their early development.[162]

In the writings of such extreme Anglo-Saxon supremacists as Robert Knox, China's backwardness and stagnation were seen as proof that its past achievements had been highly overrated. Knox insisted that the Chinese had neither "invented nor discovered" anything, and that what they had borrowed they had not understood and thus could not improve upon. Unable to generate their own scientific and technological breakthroughs, the hapless Chinese had waited passively since the time of Alexander the Great for their destruction at the hands of more creative

[160]Huc, *Chinese Empire*, pp. 301–3; Gutzlaff, *China Opened*, vol. 1, p. 507, and vol. 2, p. 159; S. W. Williams, *Middle Kingdom*, vol. 1, pp. 143–45, and vol. 2, pp. 178–79; George W. Cooke, *China* (London, 1858), p. 416.

[161]Lawrence, *Lectures*, pp. 483–84; Gobineau, *L'inégalité des races humaine*, vol. 1, pp. 353, 363–64; Robert Douglass, *China* (Akron, Ohio, 1903), pp. 504–5. See also A. Smith, *Chinese Characteristics*, p. 273.

[162]Georges Pouchet, *De la pluralité des races humaines* (Paris, 1858), p. 2; "Race in History," *Anthropological Review* 3/11 (1865), 246.

and dynamic peoples.[163] Though total denials of Chinese achievement like Knox's were more rare than contemporary dismissals of African invention and science-mindedness, or of the African origins of Zimbabwe and Meroe, the pattern and its implications were similar. The refusal to acknowledge China's great contributions to technological innovation and scientific discovery facilitated the efforts of the more extreme advocates of white supremacy to denigrate the one civilization that had clearly rivaled and, in many categories of material achievement, surpassed Europe in the preindustrial era.

In view of what they regarded as the obvious ignorance and inferiority of the Chinese, nineteenth-century observers repeatedly registered their surprise and frustration at the Chinese refusal to admit European superiority and China's need to borrow ideas and machines from the West. As early as the 1790s John Barrow had unfavorably compared the Chinese, who remained contemptuous of Western imports, with the Russians, who had admitted their backwardness and sought to remake their society to conform with those of western Europe.[164] A century later Arthur Smith compared the mind of a Chinese scholar to a "rusty, old smooth-bore cannon" on a "decrepit carriage": it required a good deal of hauling about before it could be aimed and even then was likely to miss its target.[165] But the consequences of Chinese backwardness, vulnerability, and resistance to outside influences were less amusing than Smith's analogy. In the late 1860s John Nevius, an American missionary who had resided in China for sixteen years and who had great respect for Chinese culture, regretted that "the Chinaman has almost become a synonym for stupidity and his habits and peculiarities afford abundant occasion for pleasure and ridicule. This impression has become so fixed and so genuine that correspondents and news editors to stir up interest search for the most preposterous and grotesque anecdotes they can find."[166] Writing in the same period but without Nevius's empathy, Frederick Farrar approvingly quoted Tennyson's famous preference for fifty years of Europe over a "cycle of Cathay." The schoolmaster then sought to outdo the poet in imagery and denigration of things Chinese by proclaiming the age of Pericles worth a hundred centuries of "that frightful torpor, that slumber of death, that immoral congealment which characterizes the so-called wisdom of China."[167]

---

[163]Knox, *Races of Men*, pp. 282–83, 451, 599.
[164]Barrow, *Travels*, p. 284.
[165]A. Smith, *Chinese Characteristics*, p. 133.
[166]Nevius, *China and the Chinese*, p. 275.
[167]Farrar, "Aptitudes of Races," p. 124.

Material Mastery as a Prerequisite of Civilized Life

James Mill was not only one of the first authors to stress the centrality of scientific and technological accomplishments as proof of the superiority of Europeans over non-Western peoples; he was one of the earliest to link material achievement to a new sense of what it meant to be civilized. Mill had little to say about the elaborate etiquette and genteel manners that were the essence of civilization for members of the French court cliques, who distilled the concept from earlier ideas of *civilité* in the middle of the eighteenth century.[168] He focused instead on detailed comparisons of laws, political institutions and material culture, asserting that though toolmaking abilities might not be a "proof of civilization . . . a great want of ingenuity and completeness in instruments and machinery is a strong indication of the reverse." Like his contemporary Henri Saint-Simon, Mill was convinced that there was no better "index of the degree in which the benefits of civilization are anywhere enjoyed than in the state of the tools and machinery" of a given people.[169]

These sentiments reflected the fundamental shifts that were occurring in European perceptions of themselves and their relationship to non-Western peoples. French and British writers increasingly depicted the European historical experience as unique. Though religion, physical appearance, and modes of political organization remained important emblems of this distinctiveness, the Europeans' scientific outlook and capacity for invention were more and more frequently cited as the basic attributes that set them off from all other peoples. They were responsible for the vast distance in time and level of development that Sir William Lawrence, writing shortly after Mill's *History of British India* was published, saw as separating the "highly civilized nations of Europe, so conspicuous in the arts, science, literature" and "a troop of naked, shivering, and starved New Hollanders, the hordes of filthy Hottentots," and the rest of the "more or less barbarous tribes of Africa." As Lawrence's caricature suggests, there was little tolerance for visions of noble savages in this intellectual climate. Now, savages were just that—poorly clothed and sheltered, chronically hungry, incessantly engaged in warfare and at the mercy of nature's cruel whims.[170]

168These shifts are elegantly traced by Norbert Elias in *The Civilizing Process: The Development of Manners* (New York, 1978), esp. pp. 35–40, 102–4.

169Mill, *History*, vol. 2, pp. 30–33. For Saint-Simon, see Frank E. Manuel, *The New World of Henri Saint-Simon* (Cambridge, Mass., 1956), pp. 227–28.

170Lawrence, *Lectures*, p. 244. On the repudiation of the Noble Savage ideal in the nineteenth century, see Christine Bolt, *Victorian Attitudes to Race* (London, 1971), pp. 144–45; Robert Berkhofer, Jr., *The White Man's Indian* (New York, 1978), pp. 75–76; and Léon-François Hoffman, *Le nègre romantique* (Paris, 1973), pp. 115, 138–40.

By the middle of the century most writers were lumping Africans and Melanesians together indiscriminately as savages or primitives. The term "barbarian" was reserved for peoples who, like the Chinese and Indians, had advanced somewhat and then stagnated and declined. Few thinkers would have dissented from John Crawfurd's view that "skillful industry" was "ever a proof of superior civilization,"[171] though many might have added that it was a precondition for civilization itself. They would also have concurred with his sense that China ought to be ranked above India in level of development because the Chinese had shown a much greater aptitude for the "useful arts" than the Indians. Some might have challenged his view that the Africans were superior to the American Indians because the latter could not work iron, but they were likely to support their case for the Amerindians with references to Mayan astronomy and mathematics or the genius of Inca engineering and architecture.

The new sense of what it meant to be civilized and the conviction that only peoples of European descent measured up to standards appropriate to the industrial age owed much to the growing influence of "self-made" individuals in shaping European perceptions of non-Western peoples. James Mill, John Barrow, and Gustave Le Bon, for example, who played pivotal roles in shaping attitudes toward India and China in their respective societies, were from families of modest means and little status: Mill's father was a shoemaker, Barrow's a small landholder and agricultural laborer, and Le Bon's a provincial functionary. Each of these writers spent his early years far from the centers of political and intellectual life but was driven from an early age to rise above his modest origins and make his mark in society. Each aspired in particular to acceptance by the intellectual establishment of his respective society. Each sought acclaim as an authority: Mill on India and issues of political economy more generally; Barrow on China; Le Bon on a multitude of subjects from phrenology and mass psychology to India and Arabia.

Their schooling and professional careers made these men acutely aware of the sweeping scientific and technological transformations that were occurring in their respective societies. Though his studies were concentrated in the humanities, Mill's years at the University of Edinburgh—then one of the premier scientific and medical schools in Europe—and his extensive involvement in the Utilitarian movement served to underscore in his thinking the contrast between Europe's ma-

---

[171]John Crawfurd, "On the Effects of the Commixture, Locality, Climate, and Food on the Races of Man," *T.E.S.L.* I (1861), 77–78.

terial advances and India's stagnation. Barrow had firsthand experience of the new industry during his years as a foundry supervisor; he scaled the lower rungs of the social ladder because he was able to convince others that he was competent to teach mathematics and astronomy. His knowledge of these matters goes far to explain the confidence, even arrogance, with which he criticized the techniques and tools of the Chinese and the state of their knowledge in diverse areas of scientific endeavor. Le Bon's training as a physician initiated him into the sciences in a period when the French were making great advances in medicine and other areas.[172] His penchant for invention, interest in physiology and phrenology, and success as a popularizer of contemporary scientific thought and discoveries ensured that he would devote considerable attention to the techniques of production and the modes of thinking of the peoples he encountered in his overseas travels.

Common origins and ambitions, of course, did not necessarily produce common views, even on fundamental issues. Le Bon clearly thought both the Indians and the Arabs civilized and gifted in a number of areas, whereas Mill and Barrow continually referred to the Indians and Chinese as rude and barbaric. Le Bon also had much greater reservations than his British counterparts about the disruptive effects of the process of industrialization, as distinct from the benefits of invention and technological change more broadly. But whether or not there was consensus on which peoples measured up, or on the advantages and drawbacks of the particular course that technological change had taken in Europe, there was widespread agreement in this era on the criteria by which one could distinguish civilized from barbarian and savage cultures. This was particularly true among the parvenus who were the most explicit in applying the new standards and thus instrumental in promoting their growing acceptance. Evidence of scientific and technological accomplishment was no longer peripheral or regarded as symptomatic of more fundamental values and institutional arrangements. Machines and equations, or their absence, were themselves indicators of the level of development a given society had attained. Civilization was not a state; it was a process. Individuals who through education and hard work had risen above their modest family origins placed a high premium on improvement, a term that is ubiquitous in nineteenth-century writings on colonial areas. Change was not only good; it was essential for the civilized. Stagnation and decadence were associated with barbarians; "primitive" and poorly developed material

[172]For an insightful account of Le Bon's life and thought, see Nye, *Origins of Crowd Psychology.*

culture with savages.[173] The history of civilized peoples was a tale of progress, of continuous advance; that of barbarians, a dreary chronology of endless cycles of decline and recovery.

Nineteenth-century authors vied with each other to coin the most compelling labels to capture this contrast: J. R. Seeley offered "future" versus "past" societies; Charles Caldwell, "improving" versus "stationary"; Carl Carus, "day," "twilight," and "night" peoples; and there were many variations on the dichotomies of active and passive, male and female.[174] As the century passed, these broad contrasts were refined. Specific attributes, associated with the scientific and industrial revolutions, came to be seen as characteristic of those who had achieved civilization. These ranged from ways of perceiving time and space to patterns of work and discipline. Unlike Le Bon, most European observers deemed peoples who lacked these qualities beyond the pale of cvilization. If there was controversy, it focused on how to get non-Europeans to adopt civilized ways and the extent to which different peoples were capable of benefiting from the colonizers' civilizing mission.

Though this parvenu syndrome can be traced in the lives of many of the colonial officials, explorers, missionaries, and social theorists who did so much to shape European attitudes of Africans and Asians in this era,[175] its importance ought not be overstated. Individuals from an aristocratic background (Bentinck, Dalhousie, the Marquis de Courcy) and *haute bourgeois* families (Gobineau—despite his pretensions to aristocratic origins—Bagehot, D'Eichthal) continued to play important roles in

[173]See McLeod, *Voyage of His Majesty's Ship*, p. 15; W. Cooke Taylor, *The Natural History of Society in the Barbarous and Civilized State* (London, 1840), pp. 1–5; and J. Denniker, *The Races of Man* (London, 1900), p. 127.

[174]For a general discussion of several of these classificatory schemes, see Philip Curtin, *Image of Africa* (Madison, Wis., 1964), pp. 369–70. For contemporary examples, see Seeley, *Expansion of England*, p. 176; Charles Caldwell, *Thoughts on the Original Unity of the Human Race* (New York, 1830), pp. 134–58; and d'Eichthal and Urbain, *Lettres*, pp. 15, 22.

[175]An informal survey of the lives of some fifty individuals from these groups, for whom the necessary biographical information is available, reveals that a large majority had lower-middle-class or middle-class origins, and that well over 50 percent were seriously involved in science or technology in their youth. As George Stocking has shown ("What's in a Name? The Origins of the Royal Anthropological Institute," *Man* 6/3 [1971], 380–81), the English naturalists and anthropologists who were so influential in shaping nineteenth-century European attitudes toward non-Western peoples were split between Darwinists and "ethnologicals" on one hand and "anthropologicals" on the other. The former were from well-to-do, middle-class families and well ensconced in the scientific establishment of the day. Though some members of the rival anthropologicals' organizations were drawn from respectable families, their deliberations and writings were dominated by men of lower-middle-class origins such as Richard Burton and James Hunt, whose scientific credentials were suspect.

colonization and exploration, or their writings provided key sources of information for readers interested in overseas areas. Nor, as the careers of Pierre Sonnerat and Dominique Parennin illustrate, were parvenus absent from the ranks of the authors concerned with non-Western cultures in the preindustrial era. Thus, the shift that occurred in the nineteenth century was in degree more than in kind: self-made individuals of middle-class and even working-class origins assumed an ever greater responsibility for European enterprises in Africa and Asia and became the main source of information about and opinions on these areas. As was the case in the eighteenth century, the standards of judgment applied to non-Western peoples in the industrial era and the conclusions derived from them were widely shared by literate individuals at all class levels in England and France. All had witnessed the profound transformations that were occurring in western Europe and contrasted these with what they perceived to be stagnation and decadence in Africa and Asia. Aristocrat and parvenu alike retained the eighteenth-century conviction that laissez-faire, capitalist enterprise and human liberties had been essential for Europe's development. But now they were more acutely aware of the unprecedented nature of the power and material wealth that the Europeans' unique scientific and technological advances had generated. Now they more often saw these advances as the source of Europe's global dominion, as standards by which to measure the accomplishments of others against those of Europe, and as prerequisites for civilized life as they had come to define it.

Attributes of the Dominant:
Scientific and Technological Foundations
of the Civilizing Mission

HISTORIANS HAVE rarely treated the civilizing mission as a serious ideology. Few books have been devoted to the ideas associated with the civilizing mission, even though they were pervasive in late nineteenth-century European writings on imperialist expansion in Africa and Asia.[1] The arguments of those who sought to provide an ethical justification for European global dominance have usually been allotted only a few lines or paragraphs in discussions of the causes for their unprecedented indulgence in territorial aggrandizement. In the era of decolonization the suggestion that nineteenth-century Europeans might have sincerely believed in the moral dimension of their overseas conquest and rule has met with skepticism, if not outright ridicule. At best, the notion has been seen as an expression of the seemingly limitless capacity of the Victorians for self-righteous rationalization and naiveté. More commonly, as William Langer pointed out decades ago,[2] the civilizing mission has been considered little more than a hypocritical attempt to elevate base motives with high-sounding clichés about the European destiny to better the condition of humanity.

The lack of serious attention to the civilizing mission as an ideology can be attributed in part to the ways in which its tenets were propagated by nineteenth-century spokesmen for imperialism. The ideas of the

---

[1] Some of the more detailed explorations of these ideas can be found in Martine Loutfi, *Littérature et colonialisme* (Paris, 1971), which focuses on French fiction before World War I; and Gérard Leclerc, *Anthropologie et colonialisme* (Paris, 1972), esp. pt. 1. Perhaps the fullest examination to date is literary scholar T. Walter Herbert's *Marquesan Encounters: Melville and the Meaning of Civilization* (Cambridge, Mass., 1980), which deals with American, not European, thinking on the subject.

[2] William Langer, *The Diplomacy of Imperialism* (New York, 1938), p. 91.

civilizing mission were enunciated piecemeal in parliamentary speeches defending punitive expeditions in the Sahel or Southeast Asia; in the apologias of such European leaders as Jules Ferry, driven from power by the failure of their colonial schemes; in lectures before geographical societies and colonial associations, which were major centers of imperialist sentiment in this era; and in newspaper articles, particularly those in illustrated weeklies—like the *Petit Journal* and the *Journal Illustré*—aimed at mass audiences. Perhaps the most influential expressions of the civilizing-mission ideology can be found in the poems and stories of such popular writers as Rudyard Kipling and Paul d'Ivoi (Charles Deleutre). There was no authoritative, comprehensive, and systematic exposition of the beliefs that made up the credo of nineteenth-century policymakers, missionaries, and other advocates of overseas colonization. Therefore, by the strictest of definitions,[3] the civilizing mission was not an ideology at all. But if we dismiss it because its ideas strike us as simplistic, because no thinker combined them in a coherent whole, or because in the postcolonial era they seem little more than duplicitous platitudes fabricated to cover colonial crimes, we risk distorting and neglecting assumptions and aims that affected nineteenth-century European colonial policies and activities in major ways.

Undoubtedly, claims that colonial conquests had been undertaken in order to uplift African or Asian peoples could be little more than cynical camouflage for brutal exploitation, as the Belgian King Leopold II and his rapacious agents demonstrated in the Congo in the late nineteenth century. But many of those who justified imperial expansion or colonial policies in the name of higher purposes linked to the civilizing mission were firmly convinced that they were acting in the long-term interests of the peoples brought under European rule. The civilizing mission gave a moral dimension to arguments for imperialist expansion that were otherwise limited to economic self-interest, strategic considerations, and national pride. Like most ideologies, it enabled its adherents to defend violence and suffering as necessary but temporary evils that would prepare the way for lasting improvements in the condition of the subject peoples. It lent a "humanitarian mystique" to the nasty business of conquest and domination.[4] It gave credence to the belief that the interests of all peoples could be equated with those of Britain and France. Because of it, nineteenth-century European colonizers could speak of conquest as "liberation" or "deliverance" and of repression as

[3]Such as that proposed by Edward Shils, s.v. "Ideology," in the *International Encyclopedia of the Social Sciences* (New York, 1968), vol. 7, pp. 68–75.
[4]Raoul Girardet, *L'idée coloniale en France de 1871 à 1962* (Paris, 1972), p. 86.

"pacification."[5] Politicians and writers representing all positions on the political spectrum, including at times those on the left, routinely used or accepted terminology that strikes us today as Orwellian doublethink.[6]

Though the tenets of the civilizing mission are usually associated with late nineteenth-century justifications for European empire-building in Africa and Asia, many of them had been propounded long before by apologists for colonial expansion. As early as the sixteenth century, Spanish jurists and churchmen such as Bartolomé de Las Casas had sought to provide a moral rationale for the subjugation of the Amerindian populations of the New World. Las Casas and his allies focused on refuting the arguments of such men as the formidable scholar Juan Gines de Sepúlveda that conquest through just war gave the Spanish unlimited control over the peoples of the Americas, including the right to enslave them. But Las Casas's defense of the Amerindians and objections to their enslavement were fashioned in part on an insistence that conquest imposed heavy responsibilites on the Spaniards to spread Christianity, suppress such abuses as human sacrifice, and improve the material condition of the peoples of the New World.[7] Early French policies toward colonized peoples shared the Spanish emphasis on the obligation to foster conversion to Roman Catholicism. But after the Revolution of 1789 the doctrine of assimilation—which made extensive education and acculturation of at least the elite among colonized peoples imperative—played a major role in French colonial thinking. Though the assimilationist approach was repudiated in the First and Second Empire periods, it was advocated through much of the nineteenth century for Algeria and other areas where French rule was established for the first time as well as such long-standing possessions as those on the coast of Senegal.[8]

Even more than in assimilationist doctrine, the intellectual roots of the civilizing-mission ideology can be found in the writings and policies

[5]Ibid., pp. 87–91. See also Georges Hardy, *Faidherbe* (Paris, 1947), pp. 74–75, 98, 105; and Hubert Deschamps, *Les méthodes et les doctrines coloniales de la France du XVIe siècle à nos jours* (Paris, 1953), p. 156.

[6]Though the pacification campaigns of the Vietnam War demonstrate the staying power of this sort of sloganeering and the beliefs that give rise to it. See Loutfi's discussion of the views of Jean Jaures in *Littérature et colonialisme*, p. 119; and Girardet, *L'idée coloniale*, pp. 96–98, 104–11.

[7]The fullest discussion of these disputes available in English can be found in the works of Lewis Hanke; see esp. *Aristotle and the American Indians* (Chicago, 1959), and *The Spanish Struggle for Justice in the Conquest of America* (Boston, 1949). See also Andrée M. Collard's translation of Las Casas's *History of the Indies* (New York, 1971).

[8]On early French policy, see Hubert Deschamps, *Les méthodes et les doctrines*, pp. 15–16, 20–32, 71–76, 106ff., 121–24. On the emergence of assimilationism, see esp. Raymond Betts, *Assimilation and Association in French Colonial Theory, 1890–1914* (New York, 1961), chap. 2.

of a succession of talented British administrators who oversaw the building of Britain's Indian empire in the late eighteenth and early nineteenth centuries. Personal profit and military glory were the dominant motives of such adventurers as Robert Clive, who laid the foundations of the empire and rather ruthlessly exploited the Indian populace in the middle decades of the eighteenth century. But their successors increasingly stressed the responsibilities and moral obligations that territorial control imposed upon the British conquerors. Peace, honest government, and efficient revenue systems were emphasized by early reformers such as Lord Cornwallis and Sir John Shore. The next generation of administrators included men of greater vision and a better understanding of the complexities of Indian society. Though they differed in the extent to which they wished to see India tranformed on the basis of Western precedents, they shared John Malcolm's delight in being able to effect real improvements in the lives of the steadily growing population in British-ruled areas. In a letter to his wife, Malcolm expressed both his sense of mission and the confidence that he personally was playing a decisive role in India's renewal: "I often wish that you were here to enjoy the blessings I obtain from the poor inhabitants who continue to refer their happiness to me; and it joys my heart to find myself . . . restoring great provinces to a prosperity they have not known for years."[9]

A second member of the "quartet" of distinguished British administrators who dominated British Indian policymaking in this period, Mountstuart Elphinstone, went so far as to argue that the ultimate goal of the British in India should be "the improvement of the natives reaching such a pitch as would render it impossible for a foreign nation to retain the government."[10] Though Elphinstone added that the time when such a transition could occur seemed an "immeasurable distance" away, his assertion that the British should press for improvements, even though these might eventually undermine their rule, suggested the belief that the obligation to bring good government and reform to India took precedence (at least for Elphinstone and his circle) over questions of power and political control. Admittedly, it was easier for Elphinstone, who had no Indian nationalists demanding his job, to advocate an altruistic approach to the British imperium than it would be for late nineteenth-century administrators. But his successors greatly expanded the range of the improvements that they believed it was their duty to introduce into India. They also implemented educational policies that

---

[9]Quoted in Philip Woodruff, *The Men Who Ruled India* (New York, 1964), vol. 1, p. 211.

[10]Quoted in ibid., p. 221.

resulted in the emergence of the English–educated Indian classes that played decisive roles in the nationalist movement, which fulfilled his prophecy by ousting the British from the subcontinent.

Although it must be pieced together from disparate speeches and writings, the rather amorphous configuration of ideas that made up the civilizing mission contained a number of common themes which, as Raoul Girardet has noted, were "tirelessly" repeated.[11] Invariably included among these was the conviction that European colonization was bringing peace and order to societies plagued by constant warfare and internal strife. Colonial conquests meant the overthrow of corrupt despotisms and the establishment of honest and efficient governments based on the consistent application of well-defined laws.[12] As stories such as Kipling's "Head of the District" and "Enlightenments of Pagett, M.P." and reminiscences of former officials such as the well-publicized letters of Hubert Lyautey make clear, the advocates of the civilizing mission ranked improvements in the condition of the peasantry of Africa and Asia among the central aims of colonization.[13] In their view, colonial armies would put an end to the sufferings caused by the depredations of bandits and marauders; fair taxes would replace the excessive demands of local potentates; and public works and technical training would allow cultivators to increase their productivity and income.

Underlying all these aspirations was the assumption that Europeans were the best rulers and reformers of African and Asian societies because they represented the most progressive and advanced civilization ever known. The ultimate proof for this assumption lay in Europe's unrivaled scientific and technological achievements. One could debate the merits of monarchical as opposed to parliamentary government, or the advantages of monotheism over animism. But there could be no doubt that Europeans were able to manufacture more and better firearms and machines and that they had an understanding of the material world vastly superior to that attained by any other culture. In literature, the fine arts, and philosophy, Jules Harmand pointed out in the early 1900s,

[11]Girardet, *L'idée coloniale*, p. 81.

[12]For a discussion of the importance of these ideas in British colonial thinking, see G. C. Eldridge, *England's Mission: The Imperial Idea in the Age of Gladstone and Disraeli, 1868–1880* (Chapel Hill, N.C., 1973), pp. 238–42. For the French, see Hardy, *Faidherbe*, pp. 76–78; or Léopold de Saussure, *Psychologie de la colonisation française* (Paris, 1899), pp. 103–5. The maintenance of order and prevention of communal strife among indigenous groups became increasingly important aspects of the civilizing mission as the challenges of African and Asian nationalists intensified in the twentieth century; see, e.g., Albert Sarraut, *La mise en valeur des colonies françaises* (Paris, 1922), pp. 120ff.

[13]Kipling's "Head of the District" has recently been reprinted in *Stories from the Raj*, ed. S. Cowaslee (London, 1983). For a discussion of Lyautey's writings and their considerable impact, see Girardet, *L'idée coloniale*, pp. 79–80.

every people believed that it had excelled all others, but in the sciences and technology all peoples were convinced with an "irresistible certitude" of the Europeans' superiority. Writing at a time when Asian nationalism and anticolonialism were on the rise, Harmand, one of the most influential colonial theorists of the period, asserted that science and technology alone accounted for whatever admiration non-Western peoples felt for European civilization. He believed that of all the things Europeans had to offer to colonized peoples, they were interested only in the Westerners' scientific knowledge and technical knowhow. All else was "disdained" or "detested."[14]

Many nineteenth-century writers equated the advance of European colonization with the triumph of science and reason over the forces of superstition and ignorance which they perceived to be rampant in the nonindustrialized world.[15] Spokesmen for imperialist expansion argued that without Western science and technology there was no hope of improving the condition of the impoverished masses of China and India or of civilizing the "savages" of Africa. The more extreme advocates of European global hegemony, who adhered to a survival-of-the-fittest world view, went so far as to argue with Frederic W. Farrar that scientific and technological backwardness explained and justified the decimation or (in the case of the Tasmanians) the utter extermination of "primitive peoples" who had not "added one iota to the knowledge, the arts, the sciences, the manufactures, the morals of the world."[16] But most imperialist spokesmen were content to rationalize the more unsavory aspects of colonial administration—such as demands for forced labor and the unregulated extraction of natural resources—as the inevitable consequences of the economic imbalance created by Europe's industrialization and the difficulty of diffusing Western technology among backward and suspicious peoples. The small numbers of Europeans who actually governed the colonized peoples relied on their superior technology not only for the communications and military clout that made the ongoing administration of vast areas possible but also for the assurance that they had the "right," even the "duty," to police, arbitrate disputes, demand tribute, and insist upon deference. The extent to which African and Asian peoples acquiesced to European domination out of respect for the colonizers' self-proclaimed technological superiority is hard to determine.[17] But it is clear that the confidence—or arrogance—with

[14]Jules Harmand, *Domination et colonisation* (Paris, 1910), p. 268.

[15]For discussions of this theme at various points in times, see Girardet, *L'idée coloniale*, pp. 28, 47–48, 88–89.

[16]Frederic Farrar, "Aptitudes of the Races," *T.E.S.L.* 5 (1867), 120.

[17]A number of studies have explored the ways in which the technological mystique of the Europeans contributed to their domination over colonized peoples. Among the best

which European administrators and missionaries set about the task of ruling and remaking the societies of Africa and Asia owed much to their sense of mechanical and scientific mastery.

The European colonizers' sense of their preeminence in inventiveness and organization and their vastly superior understanding of the workings of nature, not merely the conqueror's prerogative, justified their monopolization of leadership and managerial roles in colonized societies. Within the colonies, European administrators and supervisors saw to it that indolent "natives" were put to work and that hitherto untapped resources were developed. European colonizers assumed that it was both natural and mutually beneficial for advanced European societies to provide machines and manufactured goods in return for the primary products—foodstuffs, plant fibers, and minerals—exported from colonized areas and those such as China that had been "informally" divided into "spheres of influence" by the European powers. A revolution had been effected in global commercial exchanges. Europe and its North American offshoots, which had once been peripheral areas exporting mainly raw materials, had become the global centers for manufactured and finished goods, investment capital, and entrepreneurial and technical skills. India and China, which for millennia had been major sources of handicraft manufactures avidly sought in long-distance trade, were reduced to supplying cotton, tea, and raw silk to Euroamerican industries and consumers. Having set these changes in motion, the Europeans turned to transforming the very cultures of Africans and Asians to bring them into accord with modes of thought and behavior the colonizers deemed rational, efficient, and thus civilized.

There is perhaps no more striking illustration of the central position that science and technology assumed in the nineteenth-century ideologies that shaped European interaction with non-Western peoples than the missionaries' frequent invocation of European superiority in these areas in their efforts to win converts among African and Asian peoples. The missionaries, of course, continued to be motivated primarily by the conviction that Christianity was superior to all other religions and vital to the salvation of pagan peoples. But even in earlier centuries, as we have seen, the Jesuits had made use of their scientific knowledge and technical skills to gain access to and win converts. In the nineteenth century, when European military dominance secured almost

---

are G. Jahoda, *White Man* (Oxford, 1961); and Octave Mannoni, *Prospero and Caliban: The Psychology of Colonization* (London, 1956). In the 1880s the British traveler Henry Drummond, put it most succinctly when he commented that superior technology was "the whole secret of a white man's influence and power"; see *Tropical Africa* (London, 1888), p. 105.

unlimited possibilities for proselytization, missionaries increasingly stressed not just discrete advances but the essential base that Christian civilization had provided for the emergence of the scientific and technological revolutions. Though religion remained the main influence shaping missionary attitudes toward non-Western peoples, beyond the missionary community it assumed an increasingly subordinate role. At the same time, science and technology, which earlier had only peripherally affected most Europeans' views of non-Western peoples, now became key gauges by which even the missionaries compared other civilizations to their own and vital sources of their sense of righteousness and purpose. David Livingstone, for example, who saw himself as a "cog in God's machinery," regarded railroads and telegraphs as important instruments for breaking down barriers to Christian conversion. He went so far as to speculate that the "stagnation of mind" which had checked technological development in non-Christian societies was part of God's larger purpose. Their failure to advance, he reasoned, had ensured that "the greatest power derivable from science and art might be associated with the religion that proclaims peace and good-will to man."[18]

As early as the 1830s, the Baptist missionary and Sanskritist William Carey bluntly dismissed the suggestion of some American donors that the funds they had recently sent in support of his school be spent only for religious and educational purposes: "I have never heard anything more illiberal. Pray can youth be trained for the Christian ministry without science? Do you in America train up youths for it without any knowledge of science?"[19] George Smith, Carey's biographer, viewed his response as typical of all "great missionaries," who like David Livingstone believed that scientific learning was the "handmaiden" of religion.[20] In midcentury Abbé Boilat confirmed the equal enthusiasm with which Catholic missionaries sought to "yoke science to the chariot of Christian truth,"[21] remarking that teaching African youths astronomy was a most effective way of demonstrating the "power and majesty of God."[22] Boilat's contemporary the Reverend John Cumming saw science not only as a means of exalting the Christian God but also of weakening adherence to "heathen" beliefs, in this case those held by the

---

[18]David and Charles Livingstone, *Narrative of an Expedition to the Zambesi and Its Tributaries* (New York, 1866), pp. 627–29; and Leclerc, *Anthropologie et colonialisme*, pp. 22–23. For similar views, see J. F. A. Ajayi, *Christian Missions in Nigeria, 1841–1891* (Evanston, Ill., 1965), pp. 15–16; and *Chinese Repository* 7/1 (1838), 42.

[19]Quoted in George Smith, *The Life of William Carey* (London, 1885), pp. 327–28.

[20]Ibid., p. 295.

[21]Ibid., p. 296.

[22]Abbé Boilat, *Esquisses sénégalaises* (Paris, 1853), p. 13.

Indians. "We can," he wrote, "upset the whole theology of the Hindoo by predicting an eclipse."[23] At the end of the century W. A. P. Martin, an American missionary, underscored the importance of the scientific connection to efforts to spread Christianity in China. Martin conceded that the literati of China had very little interest in "the spiritual elements of our holy faith" but did see Christianity as "a powerful agency, co-operating with the diffusion of science, to emancipate [their] country from the bondage of superstition."[24]

In the nineteenth century, missionaries increasingly equated material backwardness with heathenism in its many varieties. They argued that Christian nations were more likely to be progressive, scientifically minded, and technologically proficient than non-Christian ones.[25] Mission stations came to be viewed as centers for the dissemination of technical skills and, depending upon the Europeans' estimate of the people in question, scientific learning. Mission hospitals and clinics were seen as enclaves of civilization in degenerate societies, and points from which superior Western ideas regarding bodily cleanliness, community hygiene, and social discipline could be transmitted. In agriculture and craft skills at the very least, the technological advance of backward or unprogressive peoples became an integral part of the missionary enterprise.[26]

The continuing importance of Christian proselytization among the "benighted" peoples of Africa, Asia, and Oceania serves as a reminder not only of earlier civilizing offensives overseas but also of those that had been going on for centuries within Europe itself. Much of the millennium after the fall of Rome had been occupied by a long struggle to bring the "barbarian" peoples of northern and eastern Europe into the Christian fold. In the seventeenth century new campaigns were undertaken by both church and state in France and elsewhere to capture the minds, bodies, and souls of the great mass of the people, who until then had succeeded in maintaining a high degree of autonomy.[27] The kings and bishops who oversaw this campaign and the village curés who

---

[23]John Cumming, *The Great Tribulation* (London, 1859), p. 281.

[24]W. A. P. Martin, *The Chinese* (New York, 1898), p. 248. For a similar view, see B. C. Henry, *The Cross and the Dragon* (New York, 1885), p. 432.

[25]Ronald Hyam, *Britain's Imperial Century, 1815–1914* (New York, 1976), pp. 56–57.

[26]Peter Buck, *American Science and Modern China, 1876–1936* (Cambridge, Eng., 1980), pp. 16, 25, 31–33; and H. A. C. Cairns, *Prelude to Imperialism: British Reactions to Central African Society, 1840–1890* (London, 1965), pp. 79, 114, 149–56. See also the sections on missionary education in Africa, India, and China in Chapter 5, below.

[27]This process has been of increasing interest to historians. The following overview is based largely on Robert Muchembled, *Popular Culture and Elite Culture in France, 1400–1750* (Baton Rouge, La., 1985), pt. 2.

carried it to the rural masses sought to supplant regional distinctiveness and provincialism with national uniformity and at least a semblance of cosmopolitanism. They sought to root out folk "superstitions" and vestiges of pagan worship and replace them with sacramental rites of passage, standardized holy images, and tales of the lives of saints and virtuous kings. Through harsh punishments and the persecution of witchcraft and other practices that were seen as antithetical to the emerging nation and true Christianity, peasants and urban artisans were taught deference to their superiors and acceptance of their place, however modest, in the transformed hierarchy of power and status generated by the process of state centralization. The uncoordinated and often unconscious agents of this internal civilizing offensive sought to impress the virtues of sobriety, honest labor, and emotional and bodily control upon groups who were perceived as bawdy, crude, unruly, and potentially dangerous to those who dominated both church and state.

Campaigns directed against the peasantry were paralleled by civilizing offensives within the elites themselves as courtiers and the *haute bourgeoisie* sought to propagate their sense of refinement and culture among the "middle levels" of emerging national societies. As we have seen, this process led in eighteenth-century France to an ever more stringent definition of what it meant to be civilized and the genesis of the concept of civilization itself. Though the nobility and bourgeoisie of northwestern Europe were confident that they were fully civilized by the first decades of the nineteenth century,[28] they were equally convinced that the process was far from complete among the "lower orders"—especially with respect to such attributes as foresight and discipline and the new attitudes toward time, work, and nature which had come to be seen as essential to civilized life. Well into the twentieth century, laborers continued to balk at routinized work rhythms, and peasants over much of Europe lived in village worlds where biological, natural, and ritual rhythms, much more than clock time, determined the pace of daily life.[29]

The recognition that Europe's civilizing mission among non-Western peoples was in some ways an extension of centuries-old campaigns to civilize the peasant and working classes of Europe itself is suggested by rather frequent comparisons in nineteenth-century writings between

---

[28]Norbert Elias, *The Civilizing Process* (New York, 1978), p. 50.

[29]For studies that deal with the persistence of these mentalities, see Eugen Weber, *Peasants into Frenchmen* (Stanford, Calif., 1976), esp. 19–22, 482–84; Rudolph M. Bell, *Fate and Honor, Family and Village* (Chicago, 1979), esp. chaps. 4, 6, 7; and Michael Seidman, "The Birth of the Weekend and the Revolts against Work: The Workers of the Paris Region during the Popular Front (1936–38)," *French Historical Studies* 12/2 (1981), 249–76.

these groups and the colonized. Ethnologists—James Frazer and E. B. Tylor, for example—recommended the study of European peasant beliefs and "survivals" as the starting point for the investigation of "primitive" cultures overseas.[30] Francis Galton attributed unemployment among Europe's working classes to the same lack of endurance and unsettled disposition that he had observed in "savage" Africa. Alfred Marshall blamed the runaway birth rates of both the European poor and peoples "in an early state of civilization" on their improvidence.[31] Some colonial officials found it useful to compare the regularity of school attendance among Algerian children with that of French peasants; others equated France's obligation to provide schooling for its African subjects with that of educating its own peasants.[32] The strictures that European travelers and missionaries leveled at Africans and Asians for their indolence, improvidence, and disregard for punctuality were applied as readily by middle-class authors to the working classes, peasants, and entire "racial" groups within Europe itself.[33]

The civilizing mission, then, was more than just an ideology of colonization beyond Europe. It was the product of a radically new way of looking at the world and organizing human societies. Though aristocrats had contributed much to the scientific and technological breakthroughs that underlay this new approach to the material world, it was the European bourgeoisie who acted on its premises to revolutionize production and social organization and to transform individual behavior and consciousness. And though aristocrats participated in the civilizing offensives both within Europe and overseas, by the nineteenth century it

[30]See Fred A. Voget, "Progress, Science, History, and Evolution in Eighteenth- and Nineteenth-Century Anthropology," *Journal of the History of the Behavioral Sciences* 3 (1967), 147. As Gay Weber has noted, a number of nineteenth-century authors compared the urban poor to "savage" peoples; see "Science and Society in Nineteenth Century Anthropology," *History of Science* 12 (1974), 276–77.

[31]See Greta Jones, *Social Darwinism and English Thought* (Brighton, Eng., 1980), pp. 145–46, 158.

[32]Martin D. Lewis, "One Hundred Million Frenchmen: The 'Assimilation' Theory in French Colonial Policy," *Comparative Studies in Society and History* 4/2 (1962), 140; Paul Bernard, "L'instruction des indigènes de l'Algérie," in Arthur Rousseau, ed. *Congrès international de sociologie coloniale* (Paris, 1901), vol. 2, p. 405.

[33]See, e.g., L. P. Curtis, Jr., *Anglo-Saxons and Celts: A Study of Anti-Irish Prejudice in Victorian England* (Bridgeport, Conn., 1968), esp. pp. 43–45 and chaps. 4 and 5; Bernard Semmel, *The Governor Eyre Controversy* (London, 1962), pp. 134–36; Lynn Lees, "The Irish in London," in S. Thernstrom and R. Sennett, eds., *Nineteenth Century Cities* (New York, 1969), esp. 359–61; George Stocking, Jr., *Victorian Anthropology* (Chicago, 1987), pp. 213–14, 219, 229–30; and Maxine Berg, *The Machinery Question and the Making of Political Economy, 1815–1848* (Cambridge, Eng., 1980), pp. 142–43. As Walter Houghton has shown, this contempt for indolence was also directed on occasion against idle aristocrats; see *The Victorian Frame of Mind* (New Haven, Conn., 1957), p. 248.

was middle-class groups that dominated the commercial enterprises, missionary organizations, and bureaucratic agencies essential to the dissemination of the ideas and institutions of industrializing Europe. The sense that trading, preaching, and administering were helping to proselytize norms and to diffuse skills and technology that had come to be seen as prerequisites for civilized life provided a powerful rationale for imperialist expansion, however nasty the immediate effects of that process might be.

Perceptions of Man and Nature as Gauges of Western Uniqueness and Superiority

One of the effects of the growing dominance of the scientific world view in Western culture was the elevation of humans to a position clearly distinct from and above the rest of nature, a conception long prominent in some strains of the Judeo-Christian philosophical tradition.[34] From the Renaissance period onward, scientific discoveries and technological breakthroughs made it more and more possible for Europeans to translate anthropocentric presuppositions into the actual mastery of previously untapped sources of animate and inanimate energy and into an unparalleled capacity to reshape the natural environment to suit human needs and desires. Most Western thinkers came to see human beings, however meager their level of cultural development, as creatures sharply demarcated from all "lower" forms of animal life.[35] Descartes's view, for example, of animals as beings devoid of the capacity for thought or genuine consciousness was widely accepted by those who styled themselves scientists from the late seventeenth century onward.[36] The obsession with classification that peaked in the late eighteenth and early nineteenth centuries hardened the boundary between humans and animals and at the same time accentuated perceptions of differences within the human species. For those who embraced the Western scientific view of the world, the subjective immersion of the self in nature—as exemplified by the Indian *sadhu*, the Zen and Daoist monk, or even the medieval European mystic—was irrational and con-

---

[34]Lynn White, Jr., "The Historical Roots of Our Ecologic Crisis," *Science* 155 (March 1967), 1203–7; William Leiss, *The Domination of Nature* (New York, 1972), esp. pp. 29–35.

[35]As Keith Thomas argues in *Man and the Natural World* (Harmondsworth, Eng., 1984), pp. 93–120, at the popular level the distinction was a good deal more blurred. The boundaries were also eroded by evolutionary biology and fundamentally shaken by the appearance of Darwin's *On the Origin of Species* in 1859.

[36]Herbert Butterfield, *The Origins of Modern Science, 1300–1800* (New York, 1965), p. 136.

trary to human nature and destiny. Scientists set nature apart as an object of dispassionate inquiry. As Charles Gillispie has argued, Galileo represented the "cruel edge" of this drive for objectivity in his determination to reduce natural phenomena to primary qualities: size, figure, number, and motion. All other, more subjective attributes, such as color or taste, were relegated to secondary status.[37]

The medieval view of nature as a realm of mystery and danger, graphically depicted in written accounts and pictorial representations of the great forests that still covered much of Europe,[38] gave way in the early modern era to the view that the natural world should be studied and measured and its workings understood. The eighteenth-century travelers, artists, and naturalists who provided the main impetus for the "discovery" of the natural displayed a marked preference for tame or "gentle" nature and pastoral landscapes in which humans and domesticated animals were prominently portrayed. An appreciation for nature in its "wilder, more threatening expressions" was confined to a minority of those who savored the delights of forest and seashore.[39] This preference was shared by the Victorians, who tended to associate happiness with a "bounded human landscape."[40] Francis Bacon and his successors saw scientific inquiry and mechanical innovations as the means by which "man" would restore the mastery over nature that he "had originally been designed to possess but had lost since the fall of Rome."[41] As seventeenth-century criticisms of Bacon's assumptions illustrate,[42] before the Industrial Revolution the connections between scientific discovery and material advance were not at all obvious and eminently debatable. The increased interaction of science and technology in the era of industrialization and unprecedented material advances in such indus-

---

[37]Charles Gillispie, *The Edge of Objectivity* (Princeton, N.J., 1960), p. 44. As W. C. Dampier pointed out decades ago in *A History of Science* (Cambridge, Eng., 1929), p. 132, Galileo also moved from the position that nature had been created for man to the view that nature had its own laws—but laws that humans could discover and use to bend nature to their own ends.

[38]Kenneth Clark, *Landscape into Art* (Boston, 1961), pp. 7, 15, 37, 142; Benjamin Rowland, *Art in East and West* (Boston, 1964), pp. 75–80, 89; and for a later period, Thomas, *Man and the Natural World*, pp. 192–97.

[39]D. G. Charlton, *New Images of the Natural in France* (Cambridge, Eng., 1984), chaps. 2 and 3, esp. pp. 19, 34, 41, 46.

[40]George Levine, "High and Low: Ruskin and the Novelists," in U. C. Knoepflamacher and G. B. Tennyson, eds., *Nature and the Victorian Imagination* (Berkeley, Calif., 1977), pp. 137–52 (quoted portion p. 139).

[41]For the fullest discussions of Bacon's views on the mastery of nature, see Carolyn Merchant, *The Death of Nature: Women, Ecology, and the Scientific Revolution* (San Francisco, 1983), chap. 7; and Leiss, *Domination of Nature*, chap. 3.

[42]P. M. Rattansi, "The Social Interpretation of Science in the Seventeenth Century," in Peter Mathias, ed., *Science and Society, 1600–1900* (Cambridge, Eng., 1972), pp. 27–29.

trializing areas as England and Belgium gave much greater credibility to Bacon's predictions. By the mid-nineteenth century Robert Hamilton's contention that the resources of the earth were placed there to be used by humans to improve the quality of their lives[43] was shared by the great majority of those who lived in industrializing societies, from factory workers to politicians and intellectuals on all points of the political spectrum. In some ways this assumption corresponded to the earlier belief that nature had been created for man.[44] But in the industrial era the potential to extract resources and remake the natural world to suit human ends had grown dramatically and precariously.

Much of what Keith Thomas argues in his recent *Man and the Natural World* about the growing concern in England in the early modern period for the humane treatment of animals and a greater sensitivity to nature in general is consistent with the increasing acceptance by Western thinkers of the assumption that man was destined to be the master of nature. Because humans were superior beings, they were obliged to assume responsibility for the rational and efficient management of lower creatures and natural resources as well as protect inferior animals from wanton cruelty or excessive exploitation. As Thomas observes: "Even within a fundamentally man-centered mode of thought . . . it was possible to condemn many of the ways in which animals had been customarily treated. The beasts had been created for Man's sake, but that was no reason for ill-treating them unnecessarily."[45] With "primitives" or "savages" substituted for "animals," aspects of this argument were applied to non-Western peoples through the civilizing-mission ideology. The more extreme advocates of animal intelligence in this period, including Margaret Cavendish, Lord Monboddo, and Michel de Montaigne, were usually considered eccentric by their educated contemporaries. Their views on animal (or plant) reasoning and communication were regarded as little more than "extravagant nonsense" or "poetical fancies." Thus, they represented the more radical elements of a vocal but small minority who raised important questions about "man's place in nature" at a time when most thinkers were celebrating his mastery of the natural world through scientific discovery and technological innovation.[46]

The capacity to master the natural world through scientific investigation and the application of machine power was a vital source of the

---

[43]Robert Hamilton, *The Progress of Society* (London, 1830), p. 1.

[44]On this earlier parallel view, see Thomas, *Man and the Natural World,* esp. pp. 17–20; and White, "Historical Roots."

[45]Thomas, *Man and the Natural World,* p. 165.

[46]Ibid., pp. 128–29, 166, 172, 178–79, 242–43, 302.

optimism regarding the human condition which was shared by so many thinkers of the Enlightenment.[47] Through scientific discoveries the long-hidden secrets of the workings of the "world machine" were being revealed, while a great proliferation of technological innovations was making it possible for Europeans to tap vast reservoirs of natural energy previously not even known to humans. In the nineteenth century the benefits of understanding and mastering the forces of nature became even more apparent to European thinkers. For such influential writers as Auguste Comte and Julien Virey, breakthroughs in science and technology provided irrefutable proof of the fundamentally progressive nature of human history. For Comte, advances in scientific thought were essential to the evolution of humanity and the greater social harmony that would ensue when people concentrated on controlling nature rather than dominating each other.[48] Virey stressed the connection between the Europeans' mastery of nature, which he saw as the "true lever of power for man," and their domination of the other peoples of the globe. He proclaimed that those who explored the secrets of nature and harnessed its forces to their will would gain control over, and consequently be able to extract work and wealth from, those who remained in ignorance—and thus in poverty and misery.[49]

Similar sentiments were expressed in this same period by Thomas Carlyle, who, despite his uneasiness about the condition of English workers and the long-term effects of industrialization on the English countryside, shared the general Victorian regard for the machine as the "predominant symbol of the age's harnessing of nature."[50] In his ambivalent assessment of the "signs of the times" he boasted: "We can remove mountains, and make seas our smooth highway; nothing can resist us. We war with rude Nature; and, by our resistless engines, come off always victorious, and loaded with spoils."[51] Carlyle's contemporary, the French economist Michel Chevalier, also resorted to martial metaphors: he saw the process of development as an "honorable" and "au-

[47]Peter Gay, *The Enlightenment: An Interpretation* (New York, 1967), vol. 1, pp. 182–85; vol. 2, pp. 3–11, 25–27, 124. There were, of course, eighteenth-century thinkers who stressed decadence or the cyclic tendencies of social development; see Henry Vyverberg, *Historical Pessimism in the French Enlightenment* (Cambridge, Mass., 1958).

[48]Auguste Comte, *Early Essays on Social Philosophy*, trans. Henry Hutton (London, 1911), esp. pp. 153–56. See also John C. Greene, "Biology and Social Theory: Auguste Comte and Herbert Spencer," in Marshall Claget, ed., *Critical Problems in the History of Science* (Madison, Wis., 1959), pp. 419–46.

[49]Julien Virey, *Histoire naturelle des races humaines* (Paris, 1826), pp. vii–viii, xxiii–xxvi, xxxviii.

[50]Richard D. Altick, *Victorian People and Ideas* (New York, 1973), p. 110. See also G. M. Young, *Victorian England: Portrait of an Age* (New York, 1964), p. 7.

[51]Thomas Carlyle, *Critical and Miscellaneous Essays* (New York, 1896), vol. 2, p. 60.

dacious" struggle against "brute matter." In mastering nature, Chevalier wrote, man forced it to serve him as a "docile slave," a tool that contributed to his well-being.[52]

In 1849 an anonymous essayist writing in the Edinburgh Review declared that Europeans had "succeeded in rendering almost every quality of every various form of material substance available for the purpose of utility."[53] Some decades later Lord Lytton sought to rebut those who denigrated the material achievements of the industrial age as inimical to the "poetry of life" by pointing out that spiritual advance was impossible without material improvement. Lytton found the many ways in which once wasted or unused resources of the earth had been transformed into sources of power, beyond those any "Eastern genii" could imagine, to be the "poetry of Nature herself." He believed that an engineer's design and machines extracting the earth's resources were the highest form of human expression, the clearest demonstration of the "sublime faculties which separate man from brute creatures."[54] By century's end, popularists were celebrating the increase in man's control over nature as the greatest achievement of the century. Alfred Wallace asserted that the scientists and inventors of the "wonderful century" had done more to harness the forces of nature to man's benefit than all previous thinkers and innovators since the Stone Age. James Bryce contended that man had come to know his earthly home "thoroughly." Wallace, Bryce, and their French counterpart Edmond About saw the capacity to control and exploit the forces of nature as the key to Europe's unique progress in the modern era.[55]

For many writers, the mastery of nature as manifested in scientific discovery and technological advancement was one of the most critical gauges of human achievement. In the early 1800s Julien Virey contrasted the Europeans, who were learning to control nature, with all non-Western peoples who, though they might have developed in some areas, had in this respect not risen above the level of savages. Ignorant of the workings of nature, prescientific and preindustrial peoples were subject to its every whim. Virey regarded those who lacked the desire to investigate the "sources of life" and the "treasures of the earth" as little more than brutes, "mere cattle of the prairie."[56] For Henry Thomas Buckle,

[52]Michel Chevalier, "L'Europe et la Chine," Revue des Deux Mondes 23/4 (1840), 246.
[53]"The Progress of Mechanical Invention," Edinburgh Review 89 (1849), 57.
[54]Speeches of Edward Lord Lytton (London, 1874), vol. 1, p. 178.
[55]Alfred Russel Wallace, The Wonderful Century (London, 1899), pp. 1–3, 10–11, 151–52; James Bryce, The Relations of the Advanced and Backward Races of Mankind (Oxford, 1902), p. 5; Edmond About, Le progrès (Paris, 1864), esp. chap. 6. For a recent discussion of the place of these themes in Victorian thought, see Christine Bolt, Victorian Attitudes towards Race (London, 1971), pp. 24–25.
[56]Virey, Histoire naturelle, pp. vii–viii.

who began his much noted *History of Civilization in England* in the 1850s, the "triumph of the mind over external agents" was the most meaningful measure of civilized development. Reversing the challenge-response sequence that Arnold Toynbee would later propose, Buckle argued that because the natural environment of Europe was "tamer" and "less imposing" than that found in the tropics or the far north, Europeans were able to achieve a greater degree of control than that gained by any other culture. For Buckle, this mastery was the most important proof of the superiority of western Europeans over all other peoples.[57] In one of the most influential works in early anthropology, *Primitive Culture*, Edward Tylor attempted to identify stages in human development on the basis of the capacity displayed by various cultures in "adapting nature to man's ends." Tylor judged that this ability would be lowest among "savages," moderate among "barbarians," and highest in "modern" educated societies such as those of western Europe.[58]

In the social evolutionist thought that was so widely debated in the last decades of the century, the degree to which a culture was able to control its environment was often decisive in determining its rank on the scale of savagery and civilization. Beyond the wild appearance, unintelligible speech, and cruelty to each other of the Fuegians of Patagonia, the young Charles Darwin found the extent to which they cringed shivering and half naked before the forces of their harsh environment a major reason for questioning whether they were even fellow humans, much less the equal of Europeans.[59] The Scottish geologist David Page argued that the ability of groups to ascend the evolutionary scale was determined by the extent to which they had learned to harness the forces of nature. He also believed that there was a link between a given people's intellectual development and the extent of their control over their environment.[60] In the 1890s the prolific French writer Albert Fouillée provided numerous examples, in his less precise discussion of the gradations between civilized and savage cultures, of "primitive" peoples' inability to create the tools that would allow them to control their environment.[61]

The criteria for rating societies suggested by Benjamin Kidd in his *Social Evolution*, published in the same year as Fouillée's essay, were a good deal more muddled than Buckle's and Tylor's straightforward

[57]Henry Thomas Buckle, *History of Civilization in England* (London, 1857), vol. 1, pp. 8–9, 138–42. For analysis of the biographical and intellectual background for Buckle's views, see Stocking, *Victorian Anthropology*, pp. 112–117.

[58]Edward Tylor, *Primitive Culture* (London, 1871), p. 24.

[59]Charles Darwin, *Journal of Researches* (London, 1839), esp. pp. 234–37.

[60]David Page, *Man: Where, Whence, and Whither?* (Edinburgh, 1867), pp. 80–81, 88–89.

[61]Albert Fouillée, "Le caractère des races humaines et l'avenir de la race blanche," *Revue des Deux Mondes* 124 (1894), 81–82.

stress on environmental control. Kidd offered a pastiche of altruism, cooperation, and elite benefice which he labeled "social efficiency," but he considered the mastery of nature an end product of social efficiency and a sign of the degree to which it was present in a given society. He repeatedly lauded European scientific and technological accomplishments and concluded his study with a resounding affirmation of the "infallible" superiority of the "races" who were able to "harness and use" nature rather than allow its potential to go to waste.[62] Herbert Spencer, whose writings were even more influential than Kidd's in the last decades of the century, saw "reflective consciousness" rather than social efficiency as the key to human development. But like Kidd, Spencer believed that the degree to which a given people had been able to control the natural world through the application of this consciousness was a key measure of its advance toward civilized status.[63]

Generalized assumptions about the correspondence between the mastery of nature and overall social development frequently influenced the judgments of European missionaries, travelers, and officials in the non-Western world. Often this measure of human worth was implicit in specific observations rather than a topic of elaborate discourse. Henry Drummond, for example, sought to illustrate the lowly state of the "half animal, half children" but "wholly savage" peoples who inhabited the interior of east Africa by noting that they had done little to reshape the environment in which they lived. He maintained that it never occurred to African peoples to push aside large stones that blocked well-traveled paths or even to remove trees that fell across the way.[64] H. L. Duff also commented on the Africans' inclination to reroute a path rather than remove an obstacle. He could think of nothing that more vividly revealed the contrast between the European and African character than the image, on the one hand, of "the European engineer forcing with incredible toil his broad and certain way, stemming rivers, clearing marshes, shattering tons of earth and rock; and, on the other hand, the savage, careless of everything but the present, seeking only the readiest path and content to let a pebble baulk him rather than stoop to lift it."[65] Charles Regismanset speculated that the Africans' refusal to remove natural obstacles was rooted in their fundamentally different perception of time and space: oblivious to the loss of time, not concerned to cover the shortest distance from one point to another, the Africans built roads

[62]Benjamin Kidd, *Social Evolution* (London, 1894), pp. 324–25.
[63]See George Stocking, Jr., *Race, Culture, and Evolution* (Chicago, 1982), pp. 126–27.
[64]Drummond, *Tropical Africa*, pp. 35–36.
[65]H. L. Duff, *Nyasaland under the Foreign Office* (London, 1903), p. 292. For a similar illustration of this African "trait," see Abel Hovelacque, *Les nègres de l'Afrique sus-équatoriale* (Paris, 1889), p. 429.

that twisted and turned around trees rather than cutting a straight path through the forest growth.[66]

Goldsworthy Dickinson believed that the difference between European and Asian approaches to nature was best illustrated by Hindu sculpture, which depicted the "inexhaustible fertility, the ruthlessness, the irrationality of nature, never *her* beauty, *her* harmony, *her adaptability to human needs.*" Dickinson contrasted the European stress on the latter themes in painting and architecture with the Indian view of the individual as a mere "plaything and slave of natural forces."[67] Joseph Chailley-Bert argued that the Asian subservience to nature was reflected in such customs as cow veneration. He contended that Asian peasants would refuse to drive birds from their crops, to kill rats spreading the plague, or even to crush scorpions that had stung their own children.[68] Few nineteenth-century observers displayed the degree of cultural relativism exemplified by Michel Chevalier's remark that though the Chinese lacked the Europeans' power to dominate and remake nature, in their manners and social codes the much-maligned "Orientals" displayed a much greater capacity than the British or French to master themselves.[69]

The assumptions that it was desirable for humans to master nature and that the scientifically minded and inventive Europeans were best at doing so led many authors to the conviction that it was the destiny and duty of the Europeans to expand into and develop regions occupied by less advanced peoples. The criticisms of earlier travelers to Africa, remarking on the inefficient exploitation of forest and mineral resources, often implied that if Europeans settled or controlled these areas, their resources could rapidly be harnessed to production for the world market. Similar arguments continued to be made by European observers throughout the nineteenth century. In fact, the growing demand for raw materials in industrializing areas in Europe and North America became (despite its neglect by writers on imperialist exploitation, such as Lenin, who placed unwarranted stress on the need to export Europe's surplus capital)[70] one of the most frequently cited rationales for imperialist expansion into Africa, southeast Asia, and even heavily populated and extensively cultivated areas such as China. As Thomas Carlyle's highly

---

[66]Charles Regismanset, *Questions coloniales, 1900–1912* (Paris, 1912), p. 54. See also Cairns, *Prelude to Imperialism*, pp. 78, 111–12; and Bolt, *Victorian Attitudes*, p. 175.

[67]Goldsworthy Lowes Dickinson, *An Essay on the Civilisations of India, China, and Japan* (London, 1914), p. 12 (my italics).

[68]Joseph Chailley-Bert, *Administrative Problems of India* (London, 1910), p. 58.

[69]Chevalier, "L'Europe et la Chine," pp. 210–11.

[70]Lenin, *Imperialism: The Highest Stage of Capitalism* (New York, 1939). Critiques of the capital-outlets thesis are numerous. One of the most succinct is Mark Blaug, "Economic Imperialism Revisited," *Yale Review* 50 (March 1961), 335–49.

charged 1840 essay "Chartism" makes clear, the need to discover and exploit untapped resources was for many writers more than a mere pretext for European conquest and domination of other peoples; it was a moral obligation. Carlyle seethed with indignation over the widespread poverty and unemployment in England when overseas there existed "a world where Canadian Forests stand unfelled, boundless Plains and Prairies unbroken with the plough . . . green desert spaces never yet made white with corn; and to the overcrowded little western nook of Europe, our Terrestrial Planet, nine-tenths of it yet vacant or tenanted by nomads, is still crying, Come and till me, come and reap me!"[71] Some decades later the French writer Edmond About exhorted the youth of Europe in the name of progress and higher civilization to migrate to the interior regions of Africa, Australia, and New Caledonia and make the resources of these areas available to Europeans who had the scientific knowledge and technical skills necessary to put them to productive use.[72]

By the 1890s such writers as Arthur Girault were seeking to establish beyond all doubt Europe's need and *right* to claim and develop the resources of areas inhabited by backward peoples who were unable to make use of them. Against the arguments of those who opposed expansion because it had often led to cultural disintegration and demographic catastrophe for "primitive" peoples, Girault insisted that France and other advanced European nations had both the right and the unavoidable duty to occupy and exploit the lands of "ignorant," "powerless," "truly infantile" peoples who lacked the energy and sense of purpose to make use of the resources in their natural environment. He quoted at length the views of Rudolf von Ihering, a prominent German legal expert, who argued that it was in the interest of all humankind for Europeans to settle North America and bring its vast and fertile plains under cultivation and to "open" China and Japan to world commerce. Ihering believed that if the Europeans shrank from these great tasks, they would be defying the "workings of the natural order" and the "commandments of history."[73]

No writer surpassed Benjamin Kidd in championing European colo-

---

[71] Thomas Carlyle, *English and Other Critical Essays* (London, 1915), pp. 237–38.

[72] About, *Le progrès*, p. 131. As Numa Broc has observed, very often "scientific" expeditions overseas were little more than missions designed to scout out the resource potential of various regions; see "Les Voyageurs français et la connaissance de la Chine (1860–1914)," *Revue Historique* 276 (1986), 111–15.

[73] Arthur Girault, *Principes de colonisation et de législation coloniale* (Paris, 1895), p. 31. As Manuela Semidei has pointed out, this view was espoused in textbooks for French school children in the early twentieth century; see "De l'empire à la décolonisation: A travers les manuels scolaires français," *Revue Française de Science Politique* 14/1 (1961), 63–64.

nization and exploitation of the "great and fertile" lands of the "lush tropics." Kidd believed (incorrectly, as subsequent research has shown)[74] that tropical regions possessed fertile soil and the greater portion of the earth's resources. Despite these generous endowments, he argued, the indolence, bad government, and scientific and technological backwardness of the inhabitants meant that the great natural potential of these areas was "running largely to waste." Because only those who had achieved a high level of "social efficiency"—namely, the industrialized Europeans—were capable of fully developing the resources of tropical regions, he advocated European settlement where possible (the indigenous peoples would simply have to be pushed aside or recruited as laborers) and European conquest and tribute exaction in areas not climatically suited to "white" colonization. Echoing sentiments expressed by Carlyle more than half a century earlier, Kidd concluded that the more energetic peoples of temperate lands, who were rapidly depleting the resources and occupying all the arable land in their own zones of habitation, could no longer "tolerate" the lack of resource development in tropical areas.[75]

Kidd's obsession with the struggle over scarce and diminishing resources, which has again come to be a dominant theme in the late twentieth century, had for Europeans a more ominous side that he did not explicitly develop. Partly because of his fixation on the white/nonwhite dimensions of the global contest for resources, Kidd neglected the struggle *among* European peoples that the escalating demands of rival industrial societies helped to fuel in the late nineteenth century. Though its importance as a motive for European imperialistic expansion is often difficult to weigh against other economic considerations and strategic concerns, clearly the fear of being denied future access to key raw materials and food supplies—which the Germans graphically labeled the *Torschlußpanik*, or fear of the closing door—played a vital role in the scramble for colonial possessions during the late nineteenth century.[76] The struggle for resources was, for example, uppermost in H. F. Wyatt's mind as he attempted in the 1890s to define the "ethics of empire." For Wyatt, the seizure of colonies and the exploitation of their resources was essential to England's survival as a major power. He surmised that if England withdrew from the race for col-

[74]See, e.g., Pierre Gourou, *The Tropical World* (London, 1954).

[75]These ideas are treated at length in Benjamin Kidd, *Control of the Tropics* (London, 1898). See also *Social Evolution*, pp. 316–17, 323–24. For similar statements of this position, see Farrar, "Aptitudes of Races," p. 125; and C. H. Stigand, *Administration in Tropical Africa* (London, 1914), esp. pp. 2–5.

[76]Hans-Ulrich Wehler, *Bismarck und der Imperialismus* (Cologne, 1972), esp. pp. 437–39.

onies, it would forfeit the resources of Africa and southeast Asia to the French, Germans, or some other power with which it was engaged in "ceaseless competition for national existence."[77]

From early in the nineteenth century the notion that it was the Europeans' destiny and duty to develop the resources of the globe was included in the mixture of humanitarian sentiment, cultural arrogance, and self-serving rationalization that advocates of imperial expansion blended into the civilizing-mission ideology. Once again Julien Virey was one of the first to extol European expansion, not only because it would enrich the peoples and states of Europe itself but because it would bestow great benefits on peoples overseas. In addition to building cities and public works, Virey foretold, the Europeans would apply their "ingenious [industrial] arts" and techniques of exploiting natural resources in order to increase the wealth and well-being of the colonized peoples.[78] Even the more aggressive expansionists Benjamin Kidd and H. H. Johnston sought to depict European rule in India, Egypt, and sub-Saharan Africa as an altruistic enterprise that would result in the "steady development" of the areas colonized.[79] Edmond Ferry, an equally vociferous French advocate of imperial expansion, cited the mastery of disease—against which he believed non-Western peoples to be defenseless—to illustrate the benefits that less developed peoples could receive from European tutelage. For Ferry, disease was but one of the *forces extérieures* that only European peoples had learned to control through science and technology. His stress on the links between colonization and improved medical care and hygiene exemplified one of the most persuasive and often proclaimed arguments offered in defense of imperialism in the last decades of European global dominance.[80]

In keeping with the spirit of the civilizing mission as a whole, the uplift of colonized areas through resource development was normally seen in paternalistic, even arrogant, terms. H. H. Johnston, for example, insisted that it was intolerable for enterprising Europeans to allow "rich countries to lie idle because the natives, who cumber them often to little purpose and with little right, are too brutish and ignorant to appreciate or make use of the advantages with which their native soil has been naturally endowed."[81] Like virtually all writers who addressed these

[77]H. F. Wyatt, "Ethics of Empire," *Nineteenth Century* 242 (April 1897), 523–24.
[78]Virey, *Histoire naturelle*, pp. xxvi–xxviii.
[79]Kidd, *Social Evolution*, pp. 318–22; H. H. Johnston, "British West Africa and the Trade of the Interior," *Proceedings of the Royal Colonial Institute* 20 (1888–89), 91.
[80]Edmond Ferry, *France en Afrique* (Paris, 1905), 224–26, 231. See also Deschamps, *Les méthodes et les doctrines*, pp. 157, 166 (quoting Gallieni); and Buck, *Science and China*, pp. 21–33.
[81]Johnston, "British West Africa," p. 91.

issues, the French traveler J. B. Savigny assumed that areas such as Senegal, which he believed to be admirably suited for the cultivation of indigo and cotton, would be much more productive if Europeans supervised "native" laborers. Savigny added that the Senegalese themselves vastly preferred French supervisors to any other.[82] Nearly half a century later Edmond Ferry claimed that a Wolof leader had confided in him the hope that the Europeans, to whom God had given such great inventive capacity and technical skill, would teach his people the practical "arts" that they required to develop the resources of their lands.[83]

That Ferry's anecdote may have been more than a colonial administrator's fantasy is suggested by the fact that the European colonizers consciously sought to impress their African and Asian subjects[84] with their superior aptitude for resource exploitation and the benefits that the colonized would enjoy from the application of these skills to the hitherto untapped potential of their lands. The Kenyan nationalist leader Oginga Odinga has vividly illustrated this aspect of colonial education in his recollection of his British supervisor in a veterinary school for Africans. Odinga relates that the Englishman called him into his office and lectured him at length about the lazy Africans' inability to make use of the vast resources of their continent before the arrival of the whites. The supervisor sought to drive home his point by asking Odinga to remember that before the Europeans moved into Kenya, the Africans did not even have wheels and in fact had invented nothing whatsoever. The Englishman asserted that it would take the Africans three hundred years to reach the level of development then possessed by the Europeans.[85]

The Machine as Civilizer

More than any other technological innovation, the railway embodied the great material advances associated with the first Industrial Revolution and dramatized the gap which that process had created between the Europeans and all non-Western peoples. Powered by the steam engines that were the core invention of the industrial transformation, locomotives boldly exhibited the latest advances in metallurgy and machine-tooling. Running on tracks that reshaped the landscape across vast swaths of Europe and later the Americas, Africa, and Asia; crossing great bridges that were themselves marvels of engineering skill; serviced

[82]J. B. Savigny, *Narrative of a Voyage to Senegal in 1816* (London, 1818), p. 297.
[83]Ferry, *France en Afrique*, p. 243.
[84]As well as school children in France itself; see Semidei, "De l'empire à la décolonisation," pp. 62–63.
[85]Oginga Odinga, *Not Yet Uhuru* (New York, 1967), pp. 58–59.

in railway yards whose sheds and mounds of coal became familiar features of urban centers around the world, railways were at once "the most characteristic and most efficient form of the new technics."[86] "The railroad as a system," as Leo Marx has observed, "incorporated most of the essential features of the emerging industrial order: the substitution of metal for wood construction; mechanized motive power; vastly enlarged geographical scale; speed, rationality, impersonality, and an unprecedented emphasis on precise timing."[87]

From the time of its introduction into England in the 1820s, the railroad was regarded as the "great wonder-worker of the age."[88] In his biography of George Stephenson, whose engineering genius was responsible for the construction of the first successful commercial railway, Samuel Smiles reported that the Liverpool-Manchester line drew spectators from throughout the country. "To witness a railway train some five-and-twenty years ago," Smiles observed at midcentury, "was an event in one's life."[89] In the decades after its introduction the railway became the most pervasive and compelling symbol of the new age. The laying of thousands and then tens of thousands of miles of tracks in the middle decades of the century was paralleled by the rapid growth in the number of travelers making use of trains. In England alone their numbers increased from five or six hundred per week in the late 1820s to nearly 360,000 per annum in 1871.[90] No invention rivaled the railway in capturing the imagination of poets, novelists, and social commentators.[91] William Thackeray suggested that railway tracks provided the "great demarcation line between past and present," while Tennyson exclaimed, after viewing a passing train, "Let the great world spin forever down the ringing grooves of change."[92] But words do not convey

[86]Lewis Mumford, *Technics and Civilization* (New York, 1963), p. 109.

[87]Leo Marx, "Closely Watched Trains," *New York Review of Books*, March 15, 1984, p. 28.

[88]"Progress," *Edinburgh Review* 89 (1849), 61.

[89]Samuel Smiles, *The Life of George Stephenson: Railway Engineer* (Boston, 1858), pp. 350 (quoted portion), 186, 266–67, 274–76, 278. For other reactions from these early years, see Humphrey Jennings, *Pandaemonium, 1660–1886* (New York, 1985), pp. 172–76.

[90]Samuel Lilley, "Technological Progress and the Industrial Revolution, 1700–1914," in Carlo Cipolla, ed., *The Fontana Economic History of Europe* (Glasgow, 1974), vol. 3, p. 207; L. C. B. Seaman, *Victorian England: Aspects of English and Imperial History* (London, 1973), p. 29.

[91]Herbert L. Sussman, *Victorians and the Machine: The Literary Response to Technology* (Cambridge, Mass., 1968), pp. 1–4, 9, 76, 166; Myron F. Brightfield, "The Coming of the Railroad to Early Victorian England, as Viewed by the Novels of the Period (1840–1870)," *Technology and Culture* 3/1 (1963), 44–72.

[92]Thackeray quoted in Altick, *Victorian People*, p. 75; and Tennyson in Mumford, *Technics*, p. 184.

the Victorians' exhilaration at the power of the railway as wonderfully as J. M. W. Turner's *Rain, Steam, and Speed*, which was painted in 1844 at the height of the British railway mania. The locomotive racing unimpeded through a swirling storm proclaims that the Europeans have devised a machine that allows them to challenge the elements themselves.

When railways were introduced in North America in the 1830s and India in the 1850s, many European observers fixed upon them as the key symbol of the superiority, material as well as moral, that Western societies had attained over all others. The great engines and thousands of miles of tracks provided pervasive and dramatic evidence of European power and material mastery. Railways (and the steamboat in heavily forested, riverine areas such as that which provides the setting for Conrad's *Heart of Darkness*) made it possible for Europeans to open vast stretches of "hinterland" and "undeveloped wilderness" to colonization, settlement, and economic exploitation. Most of these areas had been little affected by European activities during the early centuries of expansion which, except in the New World, had been concentrated in coastal and island regions. Trains and steamboats led to more direct and effective European rule in such areas as India, Egypt, and Vietnam and furthered colonial expansion throughout much of Africa, southeast Asia, and Australia. Officials and their families could be stationed in interior districts and troops moved about quickly to places where the "natives" showed signs of unrest. Perhaps the most impressive use of the railway for troop movement was the construction in the late 1880s of four hundred miles of track across the Nubian desert from Wadi Halfa to the Blue Nile to transport and supply the expeditionary force led by Lord Kitchener against the Mahdist state in the Sudan.[93]

European officials and travelers delighted in the awe and, at times, outright terror that the dirty, noisy, and obviously powerful locomotives instilled in the peoples they sought to rule, convert, and recruit as laborers. At midcentury Harriet Martineau recorded an early encounter: "Under the Western Ghauts [mountains in southwest India] the villagers come out at the sound of the steam whistle, and the babies gasp and cry when the train rushes by; and nobody denies that the railway is a wonderful thing."[94] More than half a century later Elspeth Huxley observed that even the slow and antiquated Uganda mail train was an object of

[93]Winston S. Churchill, *My Early Life* (London, 1983), p. 175. For an able analysis of the complex forces that led to the construction of the railway, see David Levering Lewis, *The Race to Fashoda* (New York, 1987), chap. 6.

[94]Harriet Martineau, *British Rule in India* (London, 1857), pp. 340–41. See also *Illustrated London News*, June 8, 1854, p. 4; and Daniel R. Headrick, *The Tentacles of Progress* (Oxford, 1988), p. 66.

awe for the Kikuyu of Kenya: "In fact our train made something of a triumphal progress, with long pauses to allow people to admire at close quarters a creature so strong and inexplicable, that brought to remote places a flavour of adventure, a whiff of the mystery of unknown lands."[95]

In addition to the complexity, scale, and power that rendered locomotives such superb physical and symbolic manifestations of the European colonial presence, the railway also served as a reminder of more subtle sources of European dominance and the attributes that nineteenth-century thinkers believed set Western peoples off from all others. In the speed and regularity of their comings and goings, trains (and steamships) proclaimed the Europeans' mastery of time and space and demonstrated their capacity for precision and discipline. As the nineteenth century progressed, improvements in steam engines and railway carriages, and ultimately the shift to diesel or electric power, gave impressive evidence of the insatiable appetite for experiment, innovation, and technological change that had become defining characteristics of industrial European and North American societies.

As early as the 1830s, European colonial administrators and missionaries came to view railroads, steamships, and Western machines in general as key agents in their campaigns to revive "decadent" civilizations in Asia and uplift the "savage" peoples of Africa. Lord William Bentinck, the reform-minded Governor-General of India in this period, viewed the steamboat as a "great engine of moral improvement" in a land that at the time of his arrival he considered "cursed from one end to the other by the vice, the ignorance, the oppression, the despotism, the barbarous and cruel customs that have been the growth of ages under every description of Asiatic misrule." Bentinck judged that British plantations, factories, and mines, where Western technology was concentrated, would serve as the best "schools of instruction" through which the British might elevate the level of civilization in India. A decade later an editorial in *The Times* of London boasted that a combination of good government and the introduction of Western technology, particularly through railway construction, made it inevitable that the British would become "the greatest benefactors the Hindoo race had known." The "science and steadiness of the north," *The Times* predicted, would galvanize the "capabilities of the East" and result in the full development of India's resources and a great improvement in the condition of its peoples.[96]

[95]Elspeth Huxley, *The Flame Trees of Thika* (Harmondsworth, Eng., 1983), p. 248.
[96]Bentinck and the editorial are quoted in George Bearce, *British Attitudes towards India, 1784–1858* (Oxford, 1961), pp. 161–62, 219–20.

Though all of Bentinck's successors shared his faith in the power of technology to effect a "complete moral revolution" in India,[97] none acted on this conviction as vigorously as the Marquis of Dalhousie, who was Governor-General of India in the early 1850s. Drawing on the considerable knowledge of railways that he had gained in the previous decade while serving on the Board of Trade in England, Dalhousie oversaw the introduction of both the telegraph and the railway into India and the inauguration of the subcontinent's first unified postal system. By the end of his governor-generalship in 1856, more than two hundred miles of track had been completed in an Indian railway network that would become one of the largest in the world by the century's end. Though these improvements were undoubtedly prompted in part by military and political considerations, Dalhousie was convinced that only a large influx of Western technology could shake India from its lethargy and alleviate the poverty and backwardness of its masses.[98] Building on Bentinck's metaphor, Dalhousie declared in 1856 that railways, telegraphs, and a uniform postal system were the "three great engines of social improvement" in India. In line with views frequently expressed in the eighteenth century and dominant in the nineteenth, Dalhousie viewed commerce and the expansion of Indian production for domestic and overseas markets as the key to prosperity and beneficial social change in the subcontinent. Improved transport would make commercial growth possible because foodstuffs and industrial crops such as cotton could be profitably extracted from the Indian interior for the first time, while industrial Britain would find a market for its mass-produced consumer goods. Britain's exports, from kerosene lamps to corrugated iron roofing, would in turn stimulate peasant production for the market.[99]

Dalhousie and his contemporaries also envisioned numerous indirect improvements that would result from the introduction of Western technology into India. Perhaps the most discussed was the hope that railways would accelerate the breakdown of the caste barriers that many British observers considered responsible for India's stagnation and economic decline. If high-caste Brahmins wished to travel by rail, they would have to rub elbows with low-caste farmers and laborers, thus making a shambles of notions about pollution through touch and

[97]*Ibid.*, p. 163, quoting William Cabell, a secretary at the Board of Control of the East India Company.

[98]W. W. Hunter, *Marquess Dalhousie* (Delhi, 1961), chap. 10.

[99]S. C. Ghosh, *Dalhousie in India, 1848–56* (New Delhi, 1975), pp. 2 (quoted portion), 69–70, 137; and Hunter, *Dalhousie*, pp. 135–37. For expressions of similar views by Dalhousie's contemporaries and predecessors, see Bearce, *British Attitudes*, pp. 162, 218–20; and Headrick, *Tentacles of Progress*, pp. 59–60.

encouraging social intercourse between the members of different caste groups.[100] Dalhousie's biographer Edwin Arnold, among others, viewed Western technology as a means of subverting the religious beliefs that underlay the caste system and Hinduism itself:

> Those who have travelled on an Indian line, or loitered at a Hindoo railway station, have seen the most persuasive missionary at work that ever preached in the East. Thirty miles an hour is fatal to the slow deities of paganism, and a pilgrimage done by steam causes other thoughts to arise at the shrine of Parvati or Shiva than the Vedas and Shastras inculcate. The Hindoo sees many villages and hills now beside his own; he travels, that is, he learns, compares, considers and changes his ideas.[101]

In addition to undermining Indian adherence to what most of the British in India regarded as grotesque rituals and degrading superstitions, railways were considered a powerful way to introduce the Indian multitudes to the values and attitudes that the British believed were responsible for England's prosperity and global power. W. A. Rogers, an officer in the Indian Civil Service, noted some of the lessons in the late 1870s:

> Railways are opening the eyes of the people who are within reach of them in a variety of ways. They teach them that time is worth money, and induce them to economise that which they had been in the habit of slighting and wasting; they teach them that speed attained is time, and therefore money, saved or made. They show them that others can produce better crops or finer works of art than themselves, and set them thinking why they should not have sugarcane or brocade equal to those of their neighbours. They introduce them to men of other ideas, and prove to them that much is to be learnt beyond the narrow limits of the little town or village which has hitherto been the world to them. Above all, they induce in them habits of self-dependence, causing them to act for themselves promptly and not lean on others.[102]

Dalhousie not only equated technological advance with progress and both in turn with civilization but found technology a source of aesthetic

---

[100]E. Davidson, *The Railways of India* (London, 1868), p. 3; John Kaye, *The Indian Mutiny of 1857–58* (London, 1906), vol. 1, pp. 139–41; W. A. Rogers, "The Domestic Prospects of India," *Proceedings of the Royal Colonial Institute* 1 (1869), 111–12.

[101]Edwin Arnold, *The Marquis of Dalhousie's Administration of British India* (London, 1862), vol. 2, pp. 241–42. For a later expression of the same hope, see Charles E. Trevelyan, *Christianity and Hinduism* (London, 1882), p. 9.

[102]Rogers, "Domestic Prospects," pp. 112–13. See also Headrick, *Tentacles of Progress*, pp. 86–87.

pleasure. On one occasion he wrote of the "beautiful symmetry" of telegraph lines.[103] But despite his subsequent reputation for running roughshod over Indian traditions, his infatuation with technology did not completely blind him to the cultural accomplishments of his Indian subjects. On one occasion he intervened to save the ruins of a palace at Rajmahal that a railway engineer had scheduled for demolition to make way for his tracks; at another point he forbade the construction of a trestle that would have passed within yards of the Taj Mahal.[104] His restraint, however, was not apparent to his Indian subjects, who were well aware of the challenges that Western technology posed for their ancient customs and beliefs. Ironically, one of the most striking measures of the success of Dalhousie's policies was the great rebellion they helped to precipitate in May 1857, just months after he had departed from India. Virtually all historians of the Indian Mutiny of 1857–58 have agreed that it was in part a reactionary attempt by the princes and Hindu priests to put a stop to the bewildering changes that Dalhousie's technological innovations had set in motion.[105] Characteristically, Dalhousie had the last word: the telegraph lines and railroads he had done so much to promote played crucial roles in the suppression of the rebellion.

Decades later, when the Mutiny was a distant memory but Indian nationalists were growing more and more critical of British rule, British administrators and missionaries reaffirmed their sense of mission and legitimacy with frequent references to the "most glorious gifts"[106] that British engineers and entrepreneurs had bestowed on India in the form of the latest Western technological innovations. Lord Curzon (Governor-General, 1899–1905), who was fond of machine metaphors for all manner of human endeavors from university administrations to trade guild organizations, sought to rebut Indian accusations that the British had drained India of its wealth and resources with reminders of the "progress and prosperity" that the introduction of European science and industry had brought to the subcontinent.[107] S. J. Thomson ex-

---

[103]Quoted in Bearce, *British Attitudes*, p. 220.

[104]Ghosh, *Dalhousie in India*, p. 138.

[105]See Kaye, *Indian Mutiny*, vol. 1, pp. 139–41, for a contemporary statement of this view. For twentieth century writers who have adopted this interpretation, see Michael Adas, "Twentieth Century Approaches to the Indian Mutiny of 1857–58," *Journal of Asian History* 5/1 (1971), 1–19.

[106]The phrase is Lady Wilson's; see *Letters from India* (London, 1911), p. 253; and James Samuelson, *India Past and Present* (London, 1890), p. 185.

[107]See, for examples, *Speeches on India* (London, 1904), pp. 9, 12, 40–41; or Thomas Raleigh, *Lord Curzon in India* (London, 1906), p. 312. Similar themes were pushed by British and French educators to convince the increasingly restive "natives" of the benefits of European rule. Lord Lugard, e.g., thought that African students should be taught about the ways in which British invention and industry had improved the lot of all

pressed similar views; years of service in India had convinced him that engineers were the "true missionaries of progress and enlightenment." He declared that railways and electricity had done more than all the Western schools put together to convince the Indians of the advantages of European ideas and knowledge.[108]

The most resounding late Victorian affirmation of the role of technology in Britain's civilizing mission in India was Benjamin Kidd's in his discussion of the importance of altruism among the qualities that he believed were conducive to "social evolution." Dismissing the suggestion that Great Britain had developed India solely for its own advantage, Kidd listed the many benefits that the Indians had enjoyed as a result of British efforts to "share" their superior technology:

> There has been for long in progress in India a steady development of the resources of the country which cannot be paralleled in any other tropical region of the world. Public works on the most extensive scale and of the most permanent character have been undertaken and completed; roads and bridges have been built; mining and agriculture have been developed; irrigation works, which have added considerably to the fertility and resources of large tracts of the country, have been constructed; even sanitary reform is beginning to make considerable progress. European enterprise too, attracted by security and integrity in the government, has been active. Railways have been gradually extended over the Peninsula. . . . The cotton industry of India has already entered on friendly rivalry with that of Lancashire. Other industries, suited to the conditions of the country, are in like manner rising into prominence, without any kind of artificial protection or encouragement.[109]

G. O. Trevelyan, a civil servant who was famed for his satirical sketches of British life in India, summed up a good deal more laconically the widespread sentiment that the railway was a key agent of Europe's civilizing mission: he simply observed that in India all signs of civilization disappeared beyond one hundred yards on either side of the railway track.[110]

---

humankind; see *Dual Mandate in British Tropical Africa* (London, 1922), p. 452. See also Gail P. Kelly, "The Presentation of Indigenous Society in the Schools of French West Africa and Indochina, 1918–1938," *Comparative Studies in Society and History* 26/4 (1984), 537–38.

[108]S. J. Thomson, *The Real Indian People* (Edinburgh, 1914), pp. 315–16.

[109]Kidd, *Social Evolution*, pp. 318–20.

[110]G. O. Trevelyan, *The Competition Wallah* (London, 1864), p. 26. Some decades later Winston Churchill made a similar comment, substituting telegraph lines in upper Egypt for Indian railroads; see Frederick Woods, ed., *Young Winston's Wars* (London, 1972), p. 78.

In Africa, the spread of commerce had been widely associated with the advance of civilization since the days of the campaign for the abolition of the slave trade.[111] In the late nineteenth century the explorers, traders, and missionaries who led the way into the interior of Africa and whose activities were often the prelude to colonial annexations[112] continued to see commercial expansion as a vital ally in the struggle to civilize the "Dark Continent." They also linked the potential for commercial development to the introduction of Western communications technology.[113] "There is no civilizer like the railway," proclaimed H. H. Johnston, colonial official and one of the most influential spokesmen for the British colonial party in Africa. Johnston viewed railways and steamships, which could open the ports of "barbarous countries," as the chief means of greatly expanding trade and the extraction of natural resources as well as a way to "sap race prejudice and dissolve fanaticism."[114] Similar sentiments were expressed by officials in all parts of the continent. Steamboats were considered the key to civilizing the interior of West Africa; telegraphs and roads were deemed essential to the moral and material uplift of the peoples of southern Africa; and the road from Dar-es-Salaam to Lake Nyasa was greeted by Frederic Elton, the British consul at Mozambique, as "the most promising sign that the march of civilization has commenced" in East Africa.[115] Fred Shelford, a consulting engineer for the Colonial Office, provided a detailed list of the various ways in which railways, by improving the technical skills of the Africans, contributed to their uplift:

> The education of the native in the practical arts and crafts is a most important effect of the introduction of the iron horse, although

[111]For examples, see Jean-Baptiste Durand, *Voyage au Sénégal* (Paris, 1802), pp. 369–70; S. M. Golberry, *Travels in Africa* (London, 1803), vol. 1, pp. 244ff.; and J. Corry, *Observations upon the Windward Coast of Africa* (London, 1827), pp. 81–82, 117. As Philip Curtin points out in *The Image of Africa* (Madison, Wis., 1964), pp. 428–31, however, by the 1840s English humanitarian reformers had begun to question the conviction that trade was necessarily an agent of civilization.

[112]For examples of various aspects of these linkages, see the many essays of C. W. Newbury; Anthony Hopkins, *An Economic History of West Africa* (New York, 1973), chap. 4; and R. O. Oliver, *The Missionary Factor in East Africa* (London, 1952).

[113]One of the most prominent advocates of commerce as a civilizer was David Livingstone: see *Missionary Travels and Researches in South Africa* (New York, 1858), p. 34; and *Expedition to the Zambesi*, pp. 6–9. For discussion of the links between improved transport and the spread of commerce, see Viscount Mountmorres, "The Commercial Possibilities of West Africa," *Proceedings of the Royal Colonial Institute* 38 (1906–7), 236–37; and Ajayi, *Christian Missions*, pp. 16–19.

[114]Johnston, "British West Africa," p. 91.

[115]Respectively, M. E. Mage, *Voyage dans le Soudan Occidental* (Paris, 1968), p. 663; Donald Currie, "Thoughts on the Present and Future of South Africa, and Central and East Africa," *Proceedings of the Royal Colonial Institute* 7 (1876–77), 388; and J. Frederic Elton, *Travels and Researches among the Lakes and Mountains of East and Central Africa* (London, 1829), p. 21. See also Ajayi, *Christian Missions*, pp. 16–17.

railways share this influence with mining operations and other industries. The native if left to himself will learn nothing and will aspire to nothing but the simple husbandry of his forefathers, which supplies him with his food and a small surplus of products for exchange for clothing, gun and gunpowder. But when railway construction is begun he is called upon to assist in surveying, clearing of forest upon a large scale, excavation of cuttings in earth, and blasting of cuttings in rock, building of embankments, excavation of bridge foundations, construction of masonry or concrete bridges, and erection of steelwork, erection of station buildings, workshops, quarters and telegraph, laying of permanent way and ballast, each of which, together with many other branches of the work, must educate him and advance him in the scale of civilisation.[116]

This enthusiasm for technology as an agent of social change was elevated to a general principle by H. H. Kol in an address to an international colonial congress in Paris in 1900. Nothing, Kol declared, would put the peoples of backward lands on the path to civilization more rapidly than the introduction of machines to stimulate the "forces of production." He saw the new technology, and the division of labor along Western lines which would result from its diffusion in the colonies, as forces capable of eradicating all manner of abuses found in "primitive" societies, from cannibalism and human sacrifice to polygamy.[117] Georges Hardy, whose writings and administrative decisions had far-reaching effects on French educational policy in West Africa in the first decades of the twentieth century, viewed machines and Western scientific knowledge as both the basis for European dominance and the keys to transforming not just the customs but the very psychology of African peoples. In his major treatise on colonial education, *Une conquête morale*, he outlined some of the foreseeable results of the incontestable superiority of the Europeans: "Thus, the colonized peoples are enlightened and and their horizons expanded; the prohibition which seems to weigh down the blacks is lifted; ambition becomes acceptable and finds regular and reliable outlets; fantasies of dark forces no longer entangle the skein of human endeavors at every instant; nature is clearly set against man, no longer as a legendary dragon, but as a storehouse that should be discovered and can be exploited; the peaceful conquest of the riches of nature is bound to begin."[118]

---

[116]Fred Shelford, "The Development of West Africa by Railways," *Proceedings of the Royal Colonial Society* 35 (1903–4), 251.

[117]In Rousseau, *Congrès international*, vol. 1, pp. 187, 361–62.

[118]Georges Hardy, *Une conquête morale: L'enseignement en A.O.F.* (Paris, 1917), p. 359.

Because most of China was not ruled directly by European colonizers, and its preindustrial technological endowment was so impressive, there were significant differences between European approaches to the introduction of Western technology into the Qing empire and those adopted in India and Africa. In the minds of most European observers, the roles that Western machines and techniques could play in improving the condition of the Chinese people were also somewhat different from those envisioned for other areas. Through most of the early centuries of contact and roughly the first half of the nineteenth century, Europeans involved in China sought to induce the Chinese to adopt Western technology by demonstrating its superiority over the tools and machines the Chinese themselves possessed. For the most part this approach was frustrated by the stubborn refusal of the Chinese to realize, or at least to admit, that their technology was in fact inferior and that the adoption of Western innovations would be advantageous.

The Reverend Jacob Abbott vented some of these frustrations in his 1835 account of English relations with the Qing empire. Drawing on John Barrow's recollections, Abbott described in some detail the clocks, intricate firearms, and celestial machines that the Macartney mission had presented to the emperor and Qing officials in the hope of "forcing upon them a comparison [with] the highest efforts of the arts and sciences in their own country." Abbott surmised that this attempt at stimulating technological change through diplomatic channels was too subtle for the Chinese, who in their "self-conceit" saw not weapons and machines that were clearly superior to their own but tribute proffered by barbarians who had come from remote lands to honor their emperor.[119] Within five years of Abbott's observations, the British, having lost patience with diplomacy and demonstration, resorted to warships and an expeditionary force to impress upon the Chinese the obvious fact of their technological superiority. The precipitant, of course, of the Opium War that resulted (as its name so graphically illustrates) was not the reluctance of the Chinese to import Western technology but their refusal to engage in commerce on terms that the British found consistent with their highly vaunted principles of free trade. In this instance, free trade meant that the Chinese must allow the importation of unlimited quantities of the Assamese opium that had disrupted the Qing economy and was sapping the strength of the military and the bureaucracy.[120] Whatever the causes of the war, J. Elliot Bingham and many of his contempo-

---

[119]Jacob Abbott, *China and the English* (New York, 1835), pp. 163–70.

[120]The best discussion on the causes of the conflict remains Hsin-pao Chang, *Commissioner Lin and the Opium War* (Cambridge, Mass., 1964).

raries stressed the critical role of military technology in "opening" China to the "great work" of restoring the decaying empire through commerce and missionary proselytization.[121]

Further clashes between the European powers and the beleaguered Chinese in the last half of the nineteenth century only served to underscore Europe's technological superiority and expand the opportunities for missionaries and traders to undermine the crumbling edifice of Confucian civilization. In the face of the rapidly deteriorating condition of the Chinese masses, European diplomats and missionaries sought to justify their activities with the argument that only Western civilization could provide an alternative system for the moribund Confucian order. Though they stressed the importance of European political institutions and modes of social and economic organization, the "China helpers," as Jonathan Spence has so fittingly labeled them,[122] frequently singled out the introduction of Western technology and scientific learning as essential ingredients in their formulas for the revitalization of Chinese society. As early as the 1850s, British observers were recommending a "taste of Western machinery" as a prescription for the floundering empire's ills. An anonymous essayist in the *Edinburgh Review* suggested that only Western inventions and science could lift China from the barbarism to which it had sunk; the Europeans themselves, he said, could never have risen to their current state of civilization without them.[123] A decade later the Marquis de Courcy marveled at the "inconceivable greatness" Chinese civilization might have attained if European industrial technology had been combined with the Chinese aptitude for practical invention and imitation, as well as Chinese patience and eagerness for profit.[124]

In the 1870s the Protestant missionary Arthur Smith condemned Chinese resistance to the construction of railways near or through ancient cemeteries as the futile struggle of the forces of superstition and backwardness against the harbingers of the modern age.[125] By this time, locomotives and Western machines were so regularly touted as instru-

---

[121]J. Elliot Bingham, *Narrative of the Expedition to China* (London, 1842), vol. 2, pp. 393–94; see also "The Chinese Empire," *Edinburgh Review* 15 (1831), 311.

[122]Jonathan Spence, *The China Helpers: Western Advisors in China, 1620–1960* (London, 1969).

[123]Captain Fishbourne, *Impressions of China* (London, 1855), p. 430 (quoted phrase); "Relations of England with China," *Edinburgh Review* 98 (1853), 112–13.

[124]Marquis de Courcy, *L'empire du milieu* (Paris, 1867), pp. 444–45.

[125]Arthur Smith, *Chinese Characteristics* (Shanghai, 1890), p. 160. Smith was apparently unaware of the prolonged and strident resistance that had greeted the introduction of railways into England itself. See Smiles, *George Stephenson*, pp. 168–73, 201–3, 291–97, 305–17.

ments of China's salvation, that English art critic John Ruskin, who had little fondness for the industrial order, was moved to ridicule the faith that machines could civilize China. To do so, he related an incident that had occurred during his travels in the area around Shanghai:

> We were sailing on the river in a steam launch, which was making the air impure with its smoke, and whistling as steam launches are wont to do. The scene was appropriate to the conversation, for we were among a great forest of great junks—most quaint and pictur-esque they looked—so old-fashioned they seemed that Noah's Ark, had it been there, would have had a much more modern look about it. My friend to whom the launch belonged, and who is in the machinery line himself, gave his opinion. He began by giving a significant movement of his head in the direction of the uncouth-looking junks, and then pointing to his own craft with its engine, said he did not believe much in war [as a means of civilizing China]; and the missionaries were not of much account. "This is the thing to do it," he added, pointing to the launch; "let us get at them, with this sort of article, and steam at sixty pounds on the square inch; that would do it; that's the thing to civilize them— sixty pounds on the square inch."[126]

Probably unaware of and certainly undaunted by Ruskin's barbs, the China helpers, who found growing numbers of Americans and Japanese in their ranks in the last decades of the century, laid more and more stress on the role of technology in China's regeneration. In a manner worthy of the modernization theorists a century later, James Wilson, a general in the United States Army, set forth the measures that he felt the Chinese had to take if they were to rescue their country from chaos and foreign conquest. They must, he argued, replace their "dry husks of worn-out philosophy" with Western scientific thinking and adopt West-ern approaches to practical problems. In addition, they must learn from Westerners how to build railroads, dig mines, produce iron and steel, and extract the untapped resources of their vast land. Wilson recom-mended the introduction of railways and steamships, above all other measures, to incease "human comfort" in China. These great inven-tions, he predicted, would teach the Chinese the means by which West-erners had learned to "annihilate" the obstacles of time and space and harness the forces of nature to human needs.[127]

---

[126]John Ruskin, *Works* (London, 1903–8), vol. 28, p. 105.

[127]James H. Wilson, *China: Travels and Investigations in the Middle Kingdom* (New York, 1887), pp. 311–12. See also J. H. Gray, *China* (London, 1878), vol. 1, pp. 16–17.

Though linked to aspirations to improve the condition of the peoples of the colonized world, the late Victorian advocacy of the diffusion of Western technology in Africa and Asia was couched in terms that accentuated the Europeans' sense of dominance. The spokesmen for Europe's civilizing mission stressed the dependence of Africans and Asians on Europeans for the machines, the capital, and the entrepreneurial and engineering skills that were essential to the technological transformation of colonized societies. Again, the nature and extent of the dependence varied according to the society in question. Because they were short on finance capital and lacked the technical skills needed to manufacture engines or rails, Lord Curzon deemed the Chinese "incapable of constructing a great [railway] line except by foreign assistance." He mused that the Chinese determination to construct the railway on their own was bound to "postpone such an enterprise until the 'Greek Kalends'"[128]—hence forever, since the Romans had calends but the Greeks did not. The Indians were considered wanting not only in technology and engineering skills, but also in initiative. H. H. Johnston, though speaking on West Africa, could not resist an aside on the Indian situation and a swipe at those who dared to question the propriety of the civilizing mission:

> Left to themselves, is it credible that the native inhabitants of India—that is to say the various Negritos, Dravidian, Mongol and Aryan tribes who arrived there before we did—would in the year 1889 have carried out an almost perfect topographical survey, have criss-crossed India with railways, have constructed canals and dams and reservoirs, have built hospitals and orphanages and universities, have established steam navigation between every port, and generally brought up the mean of the country to the level of a civilized European state? I think not, and although I fancy there are still a few ignorant, narrow-skulled fanatics existing in England, whose intelligence conceives little beyond the limits of their local vestry, who still cherish the notion that it is kindest and best to leave the uncivilized and the savage to wallow in their half-animal existence, I scarcely think there can be anyone in my audience tonight who doubts that India has been vastly benefited by our hundred years of rule.[129]

As for the Africans, the capacity even to have conceived of a railway was deemed impossible.[130]

---

[128]G. N. Curzon, *Problems of the Far East* (London, 1894), p. 344.

[129]Johnston, "British West Africa," pp. 91–92.

[130]T. J. Aldridge, "Sierre Leone up to Date," *Proceedings of the Royal Colonial Institute* 40 (1908–9), 48.

Rudyard Kipling's short story "The Bridge-Builders" superbly illustrates the many ways in which the Europeans' mission to diffuse industrial technology enhanced their global dominance. The railway bridge itself, spanning the mighty Ganges, which for millennia had been one of the main sources of Indian civilization, symbolizes the British capacity to reshape the Indian environment in order to increase its productivity. The strength of the bridge is tested by "Mother Gunga," the goddess of the river, with a great flood intended to sweep away the nearly completed structure. But the goddess finds that she is able to rip away only a "handful of planks" and rages because the British have "chained my flood, and the river is not free any more."[131] The bridge's great towers, "loopholed for musketry and pierced for big guns," stand as a reminder to the subject peoples of the military force that provides the ultimate sanction for British imperial authority. The entire structure—"one mile and three-quarters in length; a lattice-girder bridge, trussed with the Findlayson truss, standing on seven-and-twenty brick piers"—will facilitate the movement of travelers and products for market which, as we have seen, the British believed necessary to their efforts to regenerate Indian civilization.

Kipling's tribute to the engineering genius of the Victorian Englishman also suggests the hierarchy, based on degrees of technical expertise, that the Europeans believed their mission of technological diffusion necessarily imposed on colonized societies. The bridge is designed and its construction supervised by two Europeans, Findlayson and Hitchcock, who are "wholly absorbed in a purposeful refashioning of the material environment and whose ethic of work, responsibility and personal honour through achievement, matches this concern."[132] Europeans "borrowed from the railway workshops" are responsible for such skilled jobs as riveting and fitting. A mix of Europeans and Eurasians serve as foremen of the labor gangs. The laborers are, of course, all Indians. An exceptional Indian, a former seaman named Peroo, has risen to the position of chief assistant to the British engineers because of his "knowledge of tackle and the handling of heavy weights" and his ability to supervise the Indian laborers. In his loyalty to Findlayson and his ingenuity at carrying out the construction tasks that are required ("No

---

[131]Unless otherwise indicated all quotations in the following discussion are taken from the 1914 edition of Kipling's "The Bridge Builders" (1894) in *The Day's Work* (New York, 1898).

[132]Benita Parry, *Delusions and Discoveries: Studies on India in the British Imagination, 1880–1930* (Berkeley, Calif., 1972), p. 237. Parry's stimulating discussion of this story (pp. 228–38) focuses on Kipling's use of the bridge and the flood as metaphors for British and Indian world views respectively. As Martine Loutfi has pointed out (*Littérature et colonialisme*, pp. 115–16), the theme of the European engineer as civilizer was also popular in French literature on the colonies in this period.

piece of iron was so big or badly placed that Peroo could not devise a tackle to lift it"), Peroo exemplifies the qualities that Kipling thought fit for assistants from among the colonized peoples employed in technical fields. They could be skilled but had to be clearly subordinate. They must be willing to acknowledge their dependence on Europeans for the mastery of the underlying principles and scientific knowledge that have made possible the wonders of technological innovation which have multiplied in their lands as a result of European rule.[133]

Kipling cannot resist contrasting the hardworking, amiable Peroo with the aristocratic Indian playboy, the Rao of Baraon. The Rao Sahib has acquired an English education, a taste for such Western pastimes as billiards, and contempt for his own people's traditions. Though he treats the English engineers more or less as his equals, the Rao is busily squandering the revenues extracted from his hardworking subjects and is completely ignorant of Western technology. He has a "new toy," a steam-driven pleasure boat with "silver-plated rails, striped silk awning, and mahogany decks," but is unable to pilot the craft. The boat rescues Finlayson, but much of the time it is just "horribly in the way as when the Rao [comes] to look at bridge works." At the end of the story the faithful Peroo, who is "well known to the crew," pilots the Rao's pleasure boat "craftily up-stream."

Displacement and Revolution: Marx on the Impact of Machines in Asia

Karl Marx and Frederick Engels would have dismissed Dalhousie's and Arthur Smith's science- and machine-based designs for the improvement of colonial peoples as mere expressions of the "profound hypocrisy" they believed inherent in "bourgeois civilization."[134] As their fragmentary writings on India and China make clear, Marx and Engels believed that Europe's colonial enterprises were motivated only by "the vilest interests" and that the techniques of the colonial overlords were "stupid" and brutal.[135] But Marx's views (at least those contained

[133]At one point in the story Kipling appears to admit that it is possible for the colonized, or at least clever Indians like Peroo, to acquire this knowledge: when Finlayson suggests that someday Peroo "will be able to build a bridge in [his] own fashion," Peroo (rather impertinently) replies that he will build a "sus-sus-pen-sheen" bridge—not one with many trusses like Finlayson's.

[134]Karl Marx, The First Indian War of Independence, 1857–1859 (Moscow, n.d.), p. 38. Although Marx's writings in this volume and the companion collection On Colonialism (Moscow, n.d.) have been combined and edited by Shlomo Avineri in Karl Marx on Colonialism and Modernization (Garden City, N.Y., 1968), the page numbers I cite refer to the separate editions.

[135]Marx, Indian War of Independence, pp. 21 (quoted portions) and esp. 75–80.

in his early writings) on the long-term effects of European imperialism were a good deal more ambivalent than his charges of colonial misrule would suggest. His depiction of the woeful effects of mercantile and industrial capitalism on Asian textile workers and peasants was set within a larger argument that stressed the progressive and constructive roles these forces would ultimately play in the Asian context. In making that larger argument, Marx espoused views on non-Western societies, the power of technology to effect social change, and the proper relationship of humans to nature almost identical to those held by improvers like Dalhousie and Smith or Bentinck and Edmond Ferry. Thus, though Marx scornfully dismissed the suggestion—implicit in the civilizing-mission ideology—that the colonizers acted out of a concern for the well-being of their colonial subjects, he concluded that the unintended effects of their domination over non-Western peoples was revolutionary and would ultimately be beneficial.

The research Marx undertook, after his arrival in London in 1849, on the nature and impact of British colonialism in India and China led him to conclude that the stages of development followed by Asian societies were quite distinct from those he had delineated for Europe. For the familiar slave-feudal sequence which, he argued, had followed the communal stage in the West, Marx substituted a single, prolonged phase of development; he termed it the "Asiatic mode" and distinguished a number of variations.[136] This conception of an alternate path of Asian historical development is very much in evidence in the articles that he and Engels wrote for the *New-York Daily Tribune* in the mid-1850s, generally regarded as their authoritative pronouncements on colonialism and Asian affairs. Though Marx was more concerned with the chronological and dialectical relationships between stages of development and modes of production than in ordering them hierarchically, he clearly saw the Asiatic as backward and oppressive. He rejected the notion (which by that time few of his contemporaries entertained) that India had enjoyed a golden age before the coming of the British. Therefore, though he and Engels were extremely critical of the British handling of the great Muti-

---

[136]Marx's most detailed exploration of the "Asiatic mode" is contained in the manuscript of the *Grundrisse der Kritik der Politischen Ökonomie*, which he completed in 1857–58 but which was not published until the late 1930s. The relevant portions first appeared in English translation (by Jack Cohen) under the title *Precapitalistic Economic Formations* (London, 1964) with an excellent introduction by Eric Hobsbawm. Numerous authors, of whom perhaps the best known is Karl Wittfogel, have elaborated on Marx's original formulations. For a fine summary of their views and an attempt to distinguish Marx's own approach, see Umberto Melotti, *Marx and the Third World* (Atlantic Highlands, N.J., 1972). For a consideration of Marx's contribution to a larger literature on Asiatic despotism, see Perry Anderson, *Lineages of the Absolutist State* (London, 1974), pp. 474–76.

ny of 1857–58 and traced its origins to colonial misrule, they had little sympathy for its Indian leaders, whom they regarded as reactionaries seeking to turn back the historical clock.[137]

Marx concurred with the view that had been widely held since the mid-eighteenth century that Asiatic societies, burdened by bloated and despotic but highly centralized governments, had stagnated for centuries and fallen far behind the West technologically, economically, and politically. Marx identified village India, where the overwhelming majority of the subcontinent's population resided, as the essential base for "Oriental despotism" and the source of the "undignified, stagnatory, and vegetative" existence that he believed the Indian masses had endured for millennia. He condemned the village for its "barbarian egotism," for the "passive sort of existence" it sustained, and for being an "unresisting tool of superstition." Like so many of his contemporaries, Marx deplored the ways in which these forces had "subjugated man to external circumstances instead of elevating [him] to be the sovereign of circumstances." He was repulsed by the Indians' "brutalizing worship of nature, exhibiting its degradation in the fact that man, the sovereign of nature, fell down on his knees in adoration of Hanuman, the monkey, and Sabbala, the cow."[138]

Though Marx did not attempt an assessment of the techniques and tools of Indian handicraft manufacture, he viewed its destruction as part of the general demise of artisan production that he, and Engels before him, had seen as one of the central effects of the introduction of steam power and the factory system in Britain beginning in the late eighteenth century.[139] Marx argued that British machine-made textiles had first driven Indian handicraft-manufactured cloth from the markets of Europe, where it had been a valued import in the early centuries of European expansion. Then, as the subcontinent came under colonial control, British exports had captured much of the Indian market as well. Massive unemployment among textile workers and the rapid decline of India's

[137]See Marx, *Indian War of Independence*, esp. pp. 41–45, 66–69, 75–80, 89, 103, 116, 154–60. Many historians of India who claim to be writing in the Marxist tradition have in fact viewed the rebellion in very different ways than did Marx and Engels; see, e.g., Adas, "Indian Mutiny," pp. 8–13.

[138]Marx, *Indian War of Independence,* pp. 18–21.

[139]Engels developed these views at length in his *Condition of the Working-Class in England,* first published in German in 1845. Marx's most extensive discussion of industrial technology and its effects is contained in vol. 1, pt. 4, chap. 15 (esp. secs. 3–8) of *Das Kapital,* originally published in German in 1867; I have used the S. Moore and E. Aveling translation, published as *Capital: A Critique of Political Economy* (New York, 1906). For insightful discussions of Engels's views and their impact on Marx, see John M. Sherwood, "Marx, Malthus, and the Machine," *American Historical Review* 90/4 (1985), 837–65; and Berg, *Machinery Question,* esp. pp. 315–26.

major textile-producing centers followed. Perhaps more critically in Marx's view, factory imports "broke up the hand-loom and destroyed the spinning-wheel" which, with hand-tilling, had served as the economic pillars of India's village communities.[140] Marx believed that the "opening" of China, as well as Persia and Asia Minor, to Western trade had caused similar dislocations and massive suffering on the part of the textile workers and their dependents.[141]

Even though Marx's account of the destruction of Asian handicraft production amounted to little more than a few pages, his themes were taken up in the late nineteenth and twentieth centuries by numerous disciples, as well as non-Marxist nationalists and European critics of imperialism. Until recent decades Marx's vision was widely accepted by those concerned to draw up balance sheets on the impact of colonial rule.[142] Because this part of his argument has dominated discussions of his views on colonialism, it would appear that his stance on the introduction of Western technology into Asian societies directly contradicted the belief of mid-nineteenth-century missionary and administrative improvers that machines were a key instrument in the campaign to civilize backward peoples. But though Marx was more sensitive than many of his contemporaries to the potential disruption and adversity that technological change could unleash in a colonized society, he shared their conviction that machines could break down the barriers that he (and they) believed had held Asiatic peoples in superstition and poverty for millennia. Marx's thinking on the long-term effects of the introduction of machines and machine-made goods in Asia took into account a multiplicity of factors that belie his critics' charge of narrow technological determinism.[143] He saw technological innovation as one facet of a broader process of market expansion and transformation, of the "opening up" of India and China to new forms of entrepreneurial investment and management, and of the introduction of new modes of political and social hegemony. Within this larger context, Marx, like so many of his

[140]Marx, *Indian War of Independence*, pp. 17–20; *On Colonialism*, p. 280; and *Capital*, p. 471.

[141]Marx, *On Colonialism*, pp. 16–19.

[142]Perhaps the best introduction to the debate touched off by Morris D. Morris's revisionist challenges in the mid-1960s can be found in the exchanges in *Indian Social and Economic History Review* 5 (1968). For a critical look at these patterns in the Chinese context, see Albert Feuerwerker, "Handicraft and Manufactured Cotton Textiles in China, 1871–1910," *Journal of Economic History* 30 (1970), 338–78.

[143]On the centrality of the charge of technological determinism in writings critical of Marx, see Donald Mackenzie, "Marx and the Machine," *Technology and Culture* 25 (1984), 473–78; and Nathan Roseberg, "Marx as a Student of Technology," *Monthly Review* 23/8 (1976), 58–62.

contemporaries, stressed the roles of the railway, the telegraph, and the steamship in undermining the foundations of Asiatic despotism and the Asiatic mode of production, while building the basis for the Asians to advance to higher stages of development.[144]

Marx saw new modes of communication as the means of unifying India politically for the first time. He argued that in combination with the expansion of India's internal and the international market, they were breaking down the stultifying isolation of India's villages. Railways and newspapers bolstered the position of new Indian elites who held private property in land, which Marx termed "the great desideratum of Asiatic society," and who flocked to English-language schools to be "endowed with the requirements for government and imbued with European science." He was struck by the consensus among British colonizers regarding the Indians' "particular aptitude for accommodating themselves to entirely new labor, and acquiring the requisite knowledge of machinery." Marx viewed railways as the most essential technology for India's regeneration because

> when you have once introduced machinery into the locomotion of a country, which possesses iron and coal, you are unable to withhold it from its fabrication. You cannot maintain a net of railways over an immense country without introducing all those industrial processes necessary to meet the immediate and current wants of railway locomotion and out of which there must grow the application of machinery to those branches of industry not immediately connected with railways. The railway system will therefore become, in India, truly the forerunner of modern industry.[145]

Marx also predicted that modern industry would "dissolve" the Indian caste divisions that had been such "decisive impediments to Indian progress and Indian power." He was confident that industrial technology would allow the Indians to break out of the pattern of cyclic stagnation which had dominated their long history and permit them to exercise "a scientific domination of natural agencies." In so doing, they would advance to the bourgeois stage of development, thereby establishing the conditions for the "great social revolution."

Marx had far less to say about the effects of industrial technology on China, but he found it "gratifying" that "the bales of calico of the English bourgeoisie" had within decades smashed China's "barbarous

[144]The following discussion is based on Marx's 1853 essay "The Future Results of British Rule in India," in *Indian War of Independence*, pp. 33–40.

[145]Ibid., p. 37.

and hermetic isolation" and brought the Manchu empire to the brink of revolution.[146] He also intimated that the upheavals in China and India might fuel revolutionary movements in Europe itself. But he premised his analysis of the impact of colonialism in Asia on the assumption that Europe had advanced to a higher stage of development than either India or China and that the same technological advances that had been central to Europe's transformation would be vital for the regeneration of Asian societies. As was the case in his far more extensive writings on European development, Marx blamed the capitalist context in which they were operated for the exploitation and degradation of the working classes caused by the introduction of machines in Asia. At the same time he acknowledged the potential of machines for human liberation and improvement.[147]

As numerous commentators have observed, Marx's views were undoubtedly Eurocentric.[148] He shared the convictions of numerous European intellectuals during the first wave of industrialization that social progress could best be measured by inventions and that by this standard Asians had made little headway. Peter Hudis and others have suggested that in later writings and personal correspondence Marx tempered his Eurocentric critiques of Asian institutions and his sanguine view of the potential of technology to regenerate Asian societies.[149] But his views from the 1850s have long been considered his definitive pronouncements. They have buttressed radical nationalist critiques of "traditional" Asian cultures and contributed to the neglect of non-Western alternatives to industrialization along European lines. Marx's views from the 1850s also strongly influenced the "social engineering" approach to development that dominated the thinking of the Bolshevik leaders in Russia, who were the first revolutionaries to seize power in his name and seek to spread rebellion in the colonized world.

Time, Work, and Discipline

The Europeans' consistently low regard for non-Western timekeeping devices and perceptions of time became more pronounced in the era

[146]Marx, *On Colonialism*, pp. 14, 16.

[147]As P. D. Anthony and others have suggested, Marx attacked the effects of industrialization in its capitalist guise, not industry itself; see Anthony, *The Ideology of Work* (Bristol, 1977), p. 85; and Albert J. La Valley, *Carlyle and the Idea of the Modern* (New Haven, Conn., 1968), p. 225.

[148]See, e.g., Melotti, *Marx*, p. 115; and H. C. d'Encausse and S. R. Schram, *Marxism in Asia* (London, 1965), pp. 7–10.

[149]Peter Hudis, "The Third World Road to Socialism: New Perspectives on Marx's Writings from the Last Decade," *South Asia Bulletin* 3/1 (1983), 38–52. See also Melotti, *Marx*, p. 118.

of industrialization, when what were regarded as African and Asian deficiences in dealing with time were cited as evidence of the fundamental differences that set Europeans off from the rest of humankind. As manifestations of these deficiences, essayists and colonial policymakers pointed to the supposedly inherent lack of punctuality exhibited by non-Western peoples, their improvidence and lethargic work habits, and their apparent indifference to time "lost" or "wasted" in gossip, meditation, or simply daydreaming.

These views were nurtured by the rise and spread of the factory system in the late eighteenth and early nineteenth centuries. Ever larger numbers of Europeans, from managers and business executives to laborers and office clerks, became involved in work routines that were regulated by clocks and whistles. The "fit and start character" of preindustrial labor, in which work and play were rather indiscriminately mixed,[150] was ill suited to the new industrial order. Punctuality was at a premium; tardiness was "paid for" by fines or loss of employment; and machines set the pace at which men, women, and children labored. Time became a commodity that could be "saved," "spent," or "wasted." Laborers sold it; entrepreneurs bought it. Work time and leisure time were clearly demarcated, and there emerged an industrial work ethic that stressed time thrift, human subordination to machine rather than natural or personal rhythms, and productivity rather than individual skills or expression.[151]

In the nineteenth century the impact of the factory system on temporal perception was extended by the mass production of clocks and watches, which made possible their widespread purchase by the lower middle classes and the better-paid elements of the working classes. In addition, the rapid growth of the middle, skilled working, and urban service classes plus the great expansion of national and international market and transportation networks meant that the new secular and mechanized time sense became a dominant feature of day-to-day life throughout much of western Europe and parts of North America.[152]

[150]Robert Darnton, The Literary Underground of the Old Regime (Cambridge, Mass., 1982), pp. 162–66. The extent to which this pattern disappeared with the coming of industrialization can be overstated. My own limited stint on a Detroit assembly line in the 1960s revealed that there was a good deal more "goofing off" and idle talk than the Taylorist vision of automated production would allow.

[151]E. P. Thompson, "Time, Work-Discipline, and Industrial Capitalism," Past and Present 38 (1967), 56–97; Mumford, Technics, pp. 16–18, 196–98, 269–73; and Sebastian De Grazia, "Time and Work," in Henri Yakes, ed., The Future of Time (Garden City, N.Y., 1971). Again, not all Europeans were caught up in these patterns. For a discussion of the varieties of European time by class and locale, see Georges Gurevitch, The Spectrum of Social Time (Dordrecht, 1966).

[152]David Landes, Revolution in Time (Cambridge, Mass., 1983), pp. 227–36; Altick, Victorian Ideas, pp. 77, 96–97; Sussman, Victorians and the Machine, p. 93; and Thompson, "Time," pp. 66–69.

New modes of transport, especially the railway, also accelerated—at a steadily increasing rate—the very pace at which Europeans lived their daily lives. At midcentury the obscure Victorian novelist George Whyte-Melville observed:

> Ours are the screw-propellor and the flying-express—ours the thrilling wire that rings a bell in Paris even while we touch the handle in London—ours the greatest possible hurry on the least possible provocation—we ride at full speed, we drive at full speed—eat, drink, sleep, smoke, talk, and deliberate, still at full speed—make fortunes and spend them—fall in love and out of it— are married, divorced, robbed, ruined, and enriched all *ventre à terre!* nay, time seems to be grudged even for the last journey to our long home.[153]

Work schedules, business appointments, and railway timetables regulated the rhythm of life in industrializing societies. Inventors, investors, business managers, and eventually efficiency experts such as Frederick Taylor strove for technological and organizational advances that would render production more time-efficient and transportation and communications more rapid. Educational systems, which affected ever larger portions of the populations of northwestern Europe, inculcated the virtues of time thrift, saving and investing, and punctuality that became vital to success in the capitalist environment. New religious sects such as the Methodists reinforced the earlier admonitions of the Calvinists and other time-conscious religious movements against wasting or losing time, and exhorted the faithful to "walk circumspectly . . . redeeming the time; saving all the time you can for the best purposes; buying up every fleeting moment out of the hands of sin and Satan."[154] Thus, by the early nineteenth century, the Europeans who explored, colonized, and sought to Christianize Africa and Asia were setting out from societies dominated by clocks, railway schedules, and mechanical rhythms. They "went out" to cultures still closely attuned to the cycles of nature, to societies in which leisure was savored, patience was highly regarded, and everyday life moved at a pace that most Western intruders found enervating if not downright exasperating.

Some societies were more enervating and exasperating than others. Like the Polynesians, Amerindians, and other "primitives," African peoples had not only failed to devise astronomical instruments and time-

---

[153]Quoted in Brightfield, "Coming of the Railroad," p. 72.
[154]Quoted in Thompson, "Time," p. 88. See also De Grazia, "Time and Work," p. 470.

keepers of any sort but, nineteenth-century observers insisted, they lacked any sense of time apart from that dictated by nature's rhythms. Africans marked the passage of time only by the cycle of the seasons (wet and dry), the transitions from day to night, birth to death, and, at best, the waxing to the waning of the moon. Because they had not developed writing and little was known of their oral traditions, most Europeans wrote of the Africans as if they had no history and assumed that they were not concerned to record their past. To drive home this point to his readers, Edmond Ferry claimed that the peoples of the Sudan had no verb forms to express the past tense and in fact made no distinctions between past, present, and future.[155] Lionel Decle asserted that not only were Africans unable to record their past; they were incapable of remembering even quite recent events.[156] H. L. Duff compared Africans to "intelligent animals" because of their presentist orientation. He contended that the development of their societies had been stunted by an inability to see beyond their most immediate and basic needs or to plan for the future. Duff and numerous other observers linked this lack of time sense to the Africans' notorious improvidence—their tendency to squander all of their earnings or to consume whatever food was available with little concern for future needs.[157] By the late nineteenth century this lack of "prevoyance" or "ability to anticipate" was considered a clear sign of the primitive state of African societies.[158]

European observers pointed out that the Africans' poorly developed sense of time was reflected in the vague and naturebound manner in which they measured its passage. African modes of time-reckoning were considered crude even by travelers such as Mungo Park and Frederick Forbes, whose impressions of African peoples were generally favorable. Forbes, for example, believed that the Africans had little more than the rumblings of their stomachs to tell them the time of day.[159] Later explorers noted that the Africans measured time only by

[155]Ferry, *France en Afrique*, p. 228. See also Farrar, "Aptitudes of Races," p. 125; Samuel Baker, *Ismailia: A Narrative of an Expedition to Central Africa* . . . (London, 1874), vol. 2, p. 315; and John Barrow, *Travels into the Interior of South Africa* (London, 1806), vol. 1, p. 110.

[156]Lionel Decle, *Twelve Years in Savage Africa* (London, 1898), p. 510.

[157]H. L. Duff, *British Administration in Central Africa* (London, 1903), pp. 227, 230–31, 276, 292. See also Ferry, *France en Afrique*, p. 223; Hovelacque, *Les nègres*, pp. 424, 431–32; A. Cureau, *Les sociétés primitives de l'Afrique équatoriale* (Paris, 1912), pp. 63–65; Decle, *Savage Africa*, p. 510; J. G. Wood, *The Natural History of Man* (London, 1868), vol. 1, p. 23; Léon Fanoudh-Siefer, *Le mythe du nègre et de l'Afrique noire dans la littérature française* (Paris, 1968), pp. 163–64; and William B. Cohen, *The French Encounter with Africans* (Bloomington, Ind., 1980), p. 24.

[158]Quoted phrases taken respectively from Ferry, *France en Afrique*, p. 237; and W. Bagehot, *Physics and Politics* (London, 1872), p. 19.

[159]Frederick Forbes, *Dahomey and the Dahomeans* (London, 1851), vol. 1, p. 38; and Mungo Park, *Travels in the Interior Districts of Africa* (London, 1799), vol. 1, p. 271.

such imprecise gauges as changes in the seasons and the moon—though Richard Burton believed that they possessed no names for the moon's different phases.[160] David Livingstone summed up the late Victorian view in describing an itinerant Portuguese pedlar named Sequasha, who carried among his wares a variety of cheap American clocks. Livingstone judged these a "useless investment" because the merchant had no hope of finding buyers in a part of Africa where "no one cares for the artificial measure of time." To support his impromptu market analysis, Livingstone reported that Sequasha had complained to him of a heavy fine he had been forced to pay to a local "chief" who had been frightened by his clocks and had angrily denounced them as evil products of witchcraft.[161]

Travelers to Africa in this period almost invariably registered their frustration at having to deal with African guides, bearers, and provisioners who were apparently utterly unconcerned about the loss or waste of time. E. D. Young, who led an expedition that set out to find Livingstone in 1867, lamented that time was nothing to the African; his bearers could not be prodded to move faster than a slow walk, much less hurry, despite the obvious urgency of his mission.[162] John Buchanan echoed Young's sentiments and registered his "typical traveller's complaint" that the Africans cared nothing about delays of days or even weeks as long as they had food and drink.[163] Though clocks and watches were often described as objects of wonder and amusement for the Africans, European writers considered timekeeping devices as such completely meaningless to their guides and bearers. They were contrivances of a world of "railways, steamers, and telegraphs" which was impossible for the primitive African even to imagine. For explorers like Young, the Africans' inability to comprehend the workings and uses of clocks was yet another manifestation of the chasm in social development that separated Europeans from the peoples of the "dark contintent." Young averred that the Europeans had risen far above the "natives" by centuries of cumulative change, during which the Africans had "altered very little" from the time when the two "races" were "afloat in the Ark together."[164] For European travelers, settlers, and missionaries, timekeeping devices were also tangible links to the more "advanced" societies they had left behind in order to brave the dangers and discomforts

[160]Richard Burton, *A Mission to Galele, King of Dahomey* (London, 1864), vol. 2, pp. 20–22. See also Decle, *Savage Africa*, p. 85.

[161]Livingstone and Livingstone, *Expedition to the Zambesi*, pp. 345–46.

[162]E. D. Young, *The Search for Livingstone* (London, 1868), p. 248.

[163]John Buchanan, *The Shire Highlands* (London, 1885), p. 171. See also Livingstone and Livingstone, *Expedition to the Zambesi*, pp. 104, 281.

[164]Young, *Search for Livingstone*, p. 248. See also Drummond, *Tropical Africa*, pp. 56–57.

of "savage Africa." Robert Moffat, David Livingstone's co-worker, captured this association vividly when he wrote in his journal soon after arriving in Matabeleland, "Today we have unpacked our clock and we seemed a little more civilized."[165]

Most European observers found the Indians as frustrating to deal with as the Africans, given the equally pronounced Hindu indifference to delay and punctuality and the languid pace at which they went about their daily affairs. Nevertheless, most European writers readily conceded a significant difference: the ancient Hindus had devised numerous instruments of celestial observation, invented intricate systems of measurement and enumeration, and worked out remarkably elaborate methods of calculating the flow of history, which they traced eons into the past. In the late eighteenth century, the Orientalists William Jones and H. T. Colebrooke had described Indian modes of reckoning time in considerable detail. Though some writers refrained from sweeping judgments about the accuracy of Indian techniques or subordinated these to favorable expositions of Indian philosophy as a whole, other writers were less neutral. William Robertson, who generally admired Indian scientific achievement, dismissed the "wild computations" that the Hindus called history as extravagant and repugnant. John Gilchrist thought Hindu timekeeping imperfect, inaccurate, and full of absurdities, adding that the "better sorts [of Asiatic] peoples" found their own systems so inadequate that they were eager to adopt European methods.[166] In this period European assessments of Indian astronomical calculations were equally negative. Writing in the 1790s, Q. Craufurd termed Hindu speculations about the age of the earth "extravagant." A half-century later, mild reproach had turned to ridicule. In his infamous "Minute on Indian Education," Macaulay ridiculed Indian astronomy as little more than fantasy that "would move laughter in girls at an English boarding school" and mocked Hindu history which "abounded in kings thirty feet high and reigns thirty thousand years long."[167]

[165]Robert Moffat, *Missionary Labours and Scenes in Southern Africa* (London, 1842), p. 36.

[166]William Robertson, *An Historical Disquisition concerning the Knowledge Which the Ancients Had of India* (London, 1791), pp. 36, 329, 360; John Gilchrist, "Account of the Hindustanee Horometry," in W. Jones, ed., *Supplemental Volumes to the Works of William Jones* (London, 1801), vol. 2, pp. 863, 868. See also G. de la Galaisière le Gentil, *Voyage dans les mers de l'Inde* (Paris, 1779), p. 236; and *Lettres édifiantes et curieuses,* ed. Pierre Du Halde (Paris, 1707–73), pp. 33–36.

[167]Q. Craufurd, *Sketches Chiefly Relating to the History, Religion, Learning, and Manners of the Hindoos* (London, 1790), p. 222; Thomas B. Macaulay, *Four Articles on Education* (London, 1871), p. 93. See also C. Trevelyan, *Christianity and Hinduism*, p. 9; and Robert Knox, *The Races of Man* (London, 1862), pp. 123–24.

By the last decades of the nineteenth century, European officials and travelers tended to focus on the more mundane problems that arose from the Hindu approach to time. John Crawfurd, for example, decried the Indians' inability to manufacture the simplest of timepieces and linked this to their general lack of aptitude for mathematics and the sciences.[168] In the 1890s Charles Elliot remarked that the introduction of modern communications and transport had made the "comparative stagnation" of Indian society all too obvious to his "progressive" countrymen and a good deal more intolerable than it would have been for their counterparts in the preindustrial era. George Trevelyan complained that planning and carefully made schedules were futile in the Indian milieu. He told of working out a most detailed timetable, which hinged on his making an early morning train; the next day he awoke to find all his "coolies" still asleep, the train long departed, and his carefully laid plans undone.[169] Another traveler, Gustave Le Bon, also cited a railway scene to illustrate both the Hindus' disregard for punctuality and their methods of coping with the routinized, machine-patterned Western sense of time. When railways were first introduced into India, he reported, prospective passengers—having learned that the trains would not wait for them to drift in—had adjusted not by consulting schedules and showing up on time but by arriving at stations two or three hours before their trains were scheduled to depart. Thus, Le Bon observed, the Indians, including those who had received a Western education, did nothing to correct for their lack of punctuality but "simply in the language of the algebrist changed the signs."[170]

For a handful of European observers, the pace of Asian life stirred nostalgic longings for a time when Westerners themselves lived a less frenetic existence. In his memoirs Leonard Woolf intertwined reminiscences of his youth with the familiar theme of traveling back in time while living among non-Western peoples, in this case the Sinhalese:

> In describing my childhood I said that in those days of the eighties and nineties of the 19th century the rhythm of London traffic which one listened to as one fell asleep in one's nursery was the rhythm of horses' hooves clopclopping down London streets in broughams, hansom cabs, and four-wheelers, and the rhythm, the

---

[168]John Crawfurd, "On the Physical and Mental Characteristics of the European and Asiatic Races of Man," *T.E.S.L.* 5 (1867), 68.

[169]Charles Elliot, *Laborious Days* (Calcutta, 1892), p. 2; G. Trevelyan, *Competition Wallah*, pp. 258–59.

[170]Gustave Le Bon, *Les civilisations de l'Inde* (Paris, 1887), pp. 186–88. Arthur Koestler describes a similar pattern followed by Russian peasants in the prerevolutionary era; see *Darkness at Noon* (New York, 1965), pp. 224–25.

tempo got into one's blood and one's brain, so that in a sense I have never become entirely reconciled in London to the rhythm and tempo of the whizzing and rushing cars. . . . But in Ceylon, in the jungle road between Anuradhapura and Elephant Pass, and again in my last three years in the south of the island, I had gone straight back to the life and transport of the most ancient pastoral civilizations, in which the rhythm of life [had] hardly altered or quickened. . . . I am glad that I had for some years, in what is called the prime of life, experience of the slow-pulsing life of this most ancient type of civilization.[171]

Though British poets and novelists might pine for a tranquil England where time stood still,[172] few Europeans overseas found this pace anything but exasperating. Most travelers and administrators viewed the Indians' poorly developed sense of time only as a source of everyday inconvenience and annoyance. Most colonizers would have heartily agreed with Lord Curzon's conviction that Indians operated according to a time sense which was not only different from but doggedly contrary to that which the British sought to establish in the subcontinent.[173] Similar sentiments are standard fare among present-day Western visitors to India—from tourists growing indignant over the hours spent in queues for railway tickets to Peace Corps volunteers struggling to adjust project deadlines at variance with the rhythms of village life.

In some ways nineteenth-century comments on conceptions of time in Chinese culture were merely extensions of earlier critiques of Chinese horological and astronomical instruments. John Barrow's discussion of the inability of the Chinese to predict the timing of eclipses and the inaccuracy of their timekeepers added nothing to what the Jesuits had reported in greater detail and with more insight centuries earlier.[174] But as the nineteenth century proceeded, the catalogue of stock complaints that developed concerning the incompatibility of Western and Chinese attitudes toward time bore a strong resemblance to the one compiled for the Indians.

The 1890s accounts of China by Emile Bard, a French merchant, and

[171]Leonard Woolf, *Growing* (New York, 1961), p. 31.

[172]Martin Weiner, *English Culture and the Decline of the Industrial Spirit* (Cambridge, Eng., 1981), pp. 51–64.

[173]Curzon, *Problems of the Far East,* p. 4.

[174]John Barrow, *Travels in China* (London, 1804), p. 288. For examples of later criticisms, see D. J. MacGowan, "Modes of Keeping Time among the Chinese," *Chinese Repository* 20/7 (1851), 426–33; William H. Milne, *Life in China* (London, 1857), p. 209; and Léopold Saussure's introduction to Alfred Chapuis, *La montre chinoise* (Neuchâtal, 1919), p. 15.

Arthur Smith, a missionary who resided in China for decades, covered between them the main points of the European critique. Bard charged that time—their own or that of others—had absolutely no value for the Chinese, noting that in four years of residence in China he had never known a Chinese to keep an appointment on time or apologize for being late. He complained of the lengthy visits, apparently for the most trivial of reasons, that Chinese notables inflicted on busy European officials and travelers. He found evidence of the Chinese *méprise du temps* in their irregular and lackadaisical work habits, their indifference to the inefficiency of their tools, and their bemused incomprehension of European impatience in situations where time was being "wasted."[175] Smith opined that the Chinese sense of time was similar to that of other non-Western peoples, which a fellow missionary had labeled "antediluvian." This meant that the Chinese had no sense of punctuality, could not be made to understand that time wasted equaled money lost, had only the vaguest notions of the time of day or even their own ages, and "spent" far too many profitless hours at feasts and theatrical performances, which, Smith claimed, could go on for days.[176]

By the end of the century, most European observers would have concurred with J. Dyer Ball's estimate that the Chinese not only were unable to develop their own instruments to measure the passage of time accurately but appeared unwilling or unable to make use of European clocks and watches. Ball claimed that Western-made timepieces abounded in the treaty ports but that there was no time standard established to synchronize them and, in any case, the Chinese ignored even the watches they wore and the clocks they had in their shops.[177] Alfred Fouillée captured the sense of frustration and contempt that the apparent Chinese indifference to time aroused in most Europeans when he remarked that the chief contribution of the Orientals to the *économie du temps* was the prayer wheel. He mused that if the Chinese were to dominate the globe, prayer wheels would be substituted everywhere for scientific investigation and mechanical invention.[178]

In the last half-century, textual study and field research have undermined the nineteenth-century conviction that African and Asian peoples had very poorly developed conceptions of time.[179] What nineteenth-

[175]Emile Bard, *Les Chinois chez eux* (Paris, 1899), pp. 29–34.
[176]A. Smith, *Chinese Characteristics*, pp. 74–77, 82.
[177]J. Dyer Ball, *Things Chinese* (London, 1892), pp. 387–89.
[178]Fouillée, "Le caractère des races," p. 91.
[179]For some of the pioneering works, see Heinrich Zimmer, *Myths and Symbols in Indian Art and Civilization* (New York, 1966); Joseph Campbell, ed., *Man and Time* (New York, 1957); Rodney Needham, "Time and Eastern Man," *Occasional Papers of the Royal Anthropological Institute*, no. 2 (London, 1965); and E. E. Evans-Pritchard, *The Nuer* (Oxford, 1940).

century observers did get right, however, was that no matter how complex and sophisticated non-Western modes of time perception might be, they were very different from the secular, clock-oriented time sense that had become dominant in industrializing Europe and North America. On this as on other issues, for almost all nineteenth-century Europeans, different meant inferior. Many non-Western peoples had no devices for measuring time or charting the movements of the planets, and even those who did, most notably the Indians and Chinese, relied on contrivances that were unreliable and inaccurate by the standards then current in Europe. From the European perspective, African and Asian cultures promoted values that were antithetical to time thrift, punctuality, routinization, and other attitudes and patterns of behavior that were believed essential to the successful functioning of "advanced" capitalist societies. Above all, African and Asian peoples—often even those who had had a Western education,[180] continued to adhere to patterns of life regulated by nature's clock, by biological and natural rhythms, as well as religious-ritual cycles. Thus, the ways in which non-Western peoples perceived and marked time provided European writers with further examples of the failure of Asians and Africans to rise above and reshape nature to human purposes.

As time came to be oriented to the regular beat of machines and viewed as a commodity to be economized or squandered, European attitudes toward work altered in ways that further emphasized the contrasts between the industrializing West and the rest of the world. The introduction of bells and clocks had already begun to change European work patterns, particularly in urban centers specializing in textile manufacturing, centuries before the industrial era. Long before the factory system became a dominant feature of western European economies, merchants and members of some religious groups (Calvinists, for example) had also firmly committed themselves to the virtues of time thrift and hard work, and the harassment of idlers. The philosopher David Hume praised the incessant activity and hard work of the commercial classes, while Voltaire complained of the excessive number of religious holidays that kept laborers from their productive tasks.[181] New attitudes

---

[180]See, e.g.,the incident related by Paul Mus (in John T. McAlister, Jr., ed., *The Vietnamese and Their Revolution* [New York, 1970], pp. 104–5) in which a French-educated Vietnamese schoolteacher reveals his doubts about the heliocentric nature of the solar system; or Georges Hardy's account of a similar response among African students (*Une conquête morale*, p. 6).

[181]Jacques Le Goff, *Time, Work, and Culture in the Middle Ages* (Chicago, 1980), pp. 35ff., 44–47; Landes, *Revolution in Time*, pp. 70–78; Thompson, "Time," p. 87; Anthony, *Ideology of Work*, pp. 43–44; Gay, *The Enlightenment*, vol. 2, pp. 45–51.

toward time and work were central elements in the new sensibility that emerged among the European and North American middle classes as a consequence of the rise of capitalist economies. The growing importance of contractual agreements and savings and investment rendered self-discipline and foresight essential attributes for economic success and social approbation. A premium was placed on manipulative and problem-solving skills, on forecasting future trends, and on long-term calculations and planning.[182] Time became a factor to be controlled by the ascendant middle classes, whom the nineteenth-century essayist William Greg characterized as the "energetic, reliable, improving element of the population."[183]

The demands of industrial production intensified and spread the new work patterns spawned by clock time. Great expenditures of capital on plant and equipment, fuel, and large numbers of laborers could be lost if factories stood idle or were underused or if workers arrived late, left early, or were distracted by chatter and pranks. The substitution of wage for piecework payment of labor made it essential to compartmentalize work and leisure and closely supervise and regulate workers on the job.

As Michel Foucault has shown, conscious and cumulative efforts to control, serialize, and coordinate bodily operations were in evidence in military, educational, and penal endeavors as early as the seventeenth century.[184] Techniques for increasing efficiency and imposing discipline in one sphere spread to and shaped methods in the others. The early influence of these patterns on reactions to non-Western societies can be seen in the accounts of numerous eighteenth-century writers who drew attention to the contrast between the small, highly trained professional European armies of the day and the ill-disciplined, cumbersome, and poorly organized forces of Asian rulers.[185] By the early decades of the nineteenth century, those techniques of control, discipline, and organization which applied to labor had given rise to a "gospel" of work that was espoused by thinkers as disparate as the French utopian Henri Saint-Simon, the Scottish essayist Thomas Carlyle, and the English art critic

---

[182]For a stimulating exploration of these connections, see Thomas Haskell, "Capitalism and the Origins of the Humanitarian Sensibility," *American Historical Review* 90/3 (1985), esp. 550–53, 557–60.

[183]Quoted in John C. Greene, *Science, Ideology, and World View* (Berkeley, Calif., 1981), p. 108.

[184]Michel Foucault, *Discipline and Punish* (Harmondsworth, Eng., 1982), pp. 135–41.

[185]For examples, see Abbé Groslier, *A General Description of China* (London, 1788), vol. 1, pp. 14–15; Daniel Defoe, *The Farther Adventures of Robinson Crusoe* (London, 1925), p. 267; François Bernier, *Travels in the Mogol Empire* (Westminster, Eng., 1891); and J. C. Noble, *A Voyage to the East Indies in 1747–1748* (London, 1762), p. 143.

John Ruskin.[186] Although it was advocated most vociferously by industrial entrepreneurs and other parvenus of the ascendant middle classes, the new work ethic was disseminated at all levels of European industrial societies by best-selling tracts, epitomized by Smiles's *Self-Help*;[187] by religious teachings, particularly those of the Methodists in England;[188] and by educational institutions that affected ever larger numbers of the laboring classes as the century passed.

In the early centuries of overseas expansion, numerous European visitors commented on the irregular work patterns and lackadaisical attitudes of African and Asian peoples. In the nineteenth century, passing remarks gave way to an ubiquitous and rather predictable set of responses relating to hard work and its rewards or, conversely, sloth and its ill consequences. Important distinctions were normally made between the Chinese, who were depicted as hardworking in most accounts,[189] and the Africans and Indians. Exceptions were also made, mainly by retired district officers who were more likely to be aware of peasant work routines, for specific peoples such as the Baganda in East Africa and the Sikhs of northwest India.[190] But in most accounts Africans and Indians were pictured as incurably lazy and negligent.[191] Some authors and colonial officials, such as Arthur Phayre and Lord Elgin, explicitly ranked different peoples in terms of their potential for hard work. Typically the Chinese were preferred as laborers, with the Indi-

[186] Frank E. Manuel, *The New World of Henri Saint-Simon* (Cambridge, Mass., 1956), pp. 237, 239, 241; Ghita Ionescu, *The Political Thought of Saint-Simon* (Oxford, 1976), chap. 4; Saint-Simon, "Lettres d'un habitant de Genève à ses contemporains (1803)," in *Oeuvres* (Paris, 1868), vol. 1, pp. 55–57; William F. Roe, *The Social Philosophy of Carlyle and Ruskin* (New York, 1921), pp. 54, 93, 147; and La Valley, *Carlyle*, pp. 198, 203–5, 207.

[187] Anthony, *Ideology of Work*, pp. 77–78; Asa Briggs, *Victorian People* (Harmondsworth, Eng., 1954), chap. 5; Samuel Smiles, *Self-Help* (London, 1859), and his biographies of George Stephenson and other Victorian engineers.

[188] E. P. Thompson, *The Making of the English Working Class* (New York, 1966), pp. 355–74.

[189] M. G. Mason, *Western Concepts of China and the Chinese, 1840–1876* (New York, 1939), p. 140; Charles Gutzlaff, *China Opened* (London, 1838), vol. 2, p. 3; "De Guignes' Voyage à Pékin et Chine," *Edinburgh Review* 14 (1809), 161, 166. Some travelers complained of Chinese indolence; see, e.g., John L. Nevius, *China and the Chinese* (London, 1904), p. 278.

[190] For examples, see Cairns, *Prelude to Imperialism*, p. 80; Bolt, *Victorian Attitudes*, p. 199; Joseph Thompson, *To the Central African Lakes and Back* (London, 1881), vol. 1, p. 106; Livingstone and Livingstone, *Expedition to the Zambesi*, pp. 40, 139; A. St. H. Gibbons, "Marotseland and the Tribes of the Upper Zambezi," *Proceedings of the Royal African Institute* 29 (1897–98), 264–65; and T. F. Victor-Buxton, "Missions and Industries in East Africa, *Journal of the Africa Society* 8 (1908–9), 283.

[191] Roger Mercier, *L'Afrique noire dans la littérature française* (Dakar, 1962), pp. 41, 52, 108, 123, 134, 207–9; Fanoudh-Siefer, *Le mythe du nègre*, pp. 44–45, 161–65; Curtin, *Image of Africa*, p. 224.

ans a poor second and Africans least favored of the three. Africans were in turn preferred to the Polynesians, Amerindians, and "aborigines" of Australia, who were presumed by many authors late in the century to be on their way to extinction.[192]

For many European authors, the generally accepted capacity of the Chinese for hard work was offset by their refusal to adopt new, more efficient work patterns and more advanced foreign machines and tools. The French merchant Bard sought to sum up the deficiencies that he believed made the Chinese laborer far less productive than his European counterpart, despite the former's well-known capacity for sustained exertion. Bard stressed the "miserable" quality of Chinese tools and machines. He contended that it would never occur to a Chinese worker to use a wheelbarrow (which the Chinese themselves had invented) to transport bricks from one point on a construction site to another; instead, the laborer would make numerous time-consuming trips to transfer the bricks by hand. According to Bard, a Chinese worksite as a whole resembled an anthill: lots of activity; little being accomplished. He also commented on the irregular working hours of the Chinese, their abrupt comings and goings, and their lack of concern about the time it took to perform a particular task. He complained that his Chinese neighbors worked far into the night and slept through much of the day.

Bard's greatest annoyance, however, arose from the stubborn refusal of the Chinese to adopt foreign ways and tools that were so obviously superior to those they had used since ancient times. He recounted his efforts to persuade his Chinese gardener to use a long-handled hoe rather than a small knife for weeding the extensive flowerbeds about the merchant's residence. Having himself purchased and demonstrated the effectiveness of the Western-style hoe, Bard delighted in the gardener's obvious admiration for the new tool. But the Frenchman was startled and chagrined when, during his morning stroll on the following day, he came upon the gardener contentedly weeding with his little knife.[193] For Bard and numerous European writers, the lesson was clear: strong attachment to customary tools and time-honored work patterns had blocked and would continue to block Chinese acceptance of newer ones

[192]For Elgin, see George W. Cooke, *China: Being "The Times" Special Correspondence from China in the Years 1857–58* (London, 1858), p. xxii; for Phayre, see Government of India, *Secret and Political Correpondence*, Range 201, vol. 15 (20 February 1857), no. 842. See also Bolt, *Victorian Attitudes*, pp. 179, 197; Francis G. Hutchins, *The Illusion of Permanence* (Princeton, N.J., 1967), p. 65; George Campbell, *Modern India* (London, 1852), p. 59; Henry Guppy, "Notes on the Capabilities of the Negro for Civilization," *Journal of the Anthropological Society* 2 (1864), ccx–ccxi; and Hovelacque, *Les nègres*, pp. 424, 428–30.

[193]Bard, *Les Chinois*, pp. 29–32. See also Gutzlaff, *China Opened*, p. 213.

from the West, even though the latter were demonstrably more efficient.

European observers found African and Indian tools even more primitive and inefficient than those of the Chinese, who were acknowledged to have excelled in invention and toolmaking in earlier times. Indians and Africans, in contrast to the industrious but inefficient Chinese, were generally depicted as indolent and highly resistant to all forms of sustained bodily exertion. Nineteenth-century travel literature and official and missionary documents abound in descriptions of African males lounging about, ceaselessly chatting, drinking, and carousing, or fast asleep during the daylight hours that the Europeans slated for peak activity. The fact that the heat of the tropics, especially at midday, dictated a different work routine from that in northern Europe (which experienced colonial officials and missionaries readily acknowledged and themselves adjusted to) was rarely discussed in popular accounts of the "dark continent," which did much to shape European attitudes at all social levels. Explorers and other commentators concentrated instead on what were intended to be amusing stories to document the Africans' addiction to sloth and their elaborate devices to avoid hard work. Some missionaries told how "their boys" believed that God had made just two good things, sleep and Sunday, when there was no work. Richard Burton reported that the Africans regarded only death as a greater evil than hard work. Numerous mid-nineteenth-century writers lamented the ruin that the sloth of the freed "Negroes" had brought to the once prosperous plantation colonies of Jamaica and Haiti. Henry Guppy even charged that slave rebellions had no objects beyond "lust and ease." Some Europeans were particularly angered by their perception that African women labored long and hard while their fathers and husbands exchanged gossip and sampled the local brew.[194]

Even those rare observers who challenged the stereotype of the lazy African conceded that Africans were incapable of the punctuality, regularity, and work discipline expected of European laborers.[195] Ignorant of the seasonal and daily work rhythms of African peoples, the patterns of labor division by sex, and the ravages of parasitic disease, as well as oblivious to the negative impact of colonialism on the traditional male activities of hunting and war, nineteenth-century European writers

[194]Cairns, *Prelude to Imperialism*, p. 79; Fouillée, "Le caractère des races," pp. 80–81; Guppy, "Capabilities of the Negro," pp. ccx–ccxi; and James Hunt, "On the Negro's Place in Nature," *Memoirs of the Anthropological Society of London* (1865), 40–47. The classic statement of the slothful ex-slave of the West Indies is, of course, Thomas Carlyle's, in his "Occasional Discourse on the Nigger Question" (1849).

[195]E.g., Viscount Mountmorres, "Commercial Possibilities," p. 226.

rarely discarded the ethnocentric blinders that made the myth of the lazy African one of the most pervasive and harmful fabrications of the era of Western dominance.

In India, although some District Officers were well aware of the long and arduous hours worked by peasants and laborers, the impression conveyed by most nineteenth-century works on the subcontinent was perhaps best captured by G. O. Trevelyan, who wrote in the 1860s of the lazy, languid Bengalis "contented to glisten and bask in the sun for days and weeks together."[196] As early as 1801, a majority of the English magistrates in the province of Bengal characterized the Indians as indolent in response to a survey ordered by Governor-General Lord Wellesley. James Mill concurred with this judgment (though he had never seen an Indian peasant or laborer) and thereby established the stereotype of the phlegmatic Hindu which was mindlessly repeated by most European writers until well into the twentieth century. Like Africans, Indians were considered incapable of sustained effort, improvident, and resistant to any work beyond that required for their minimum subsistence. For British officials, who believed that they went out to India to labor in the "heat and dust" so that the subcontinent might be developed,[197] the Indians' inability to appreciate their rulers' devotion to the "sacredness of toil"[198] was a constant source of frustration and complaint.

Explanations for Indian and African indolence varied somewhat, but by the last decades of the nineteenth century a related myth—that of the lush tropics, which had been favored by a number of eighteenth-century writers—was widely accepted. Earlier authors such as James Mill blamed centuries of despotic rule—under which hard work gained the individual little advantage and was likely to increase his chances of being exploited by the state—for the languid dispositions and aversion to "honest" toil of African and Asian peoples.[199] Later observers, from travelers and missionaries to colonial officials and merchants, singled out environmental factors (which some argued had become racially in-

[196]G. Trevelyan, *Competition Wallah*, pp. 255–56. The following summary of dominant European attitudes toward Indian work patterns is based on a variety of sources, including Ainslee Embree, *Charles Grant and British Rule in India* (London, 1962), p. 150; James Mill, *History of British India* (London, 1826), vol. 1, pp. 412–13; Le Bon, *Les civilisations de l'Inde*, pp. 181–85; and Louis Figuier, *The Human Race* (London, 1872), p. 342.

[197]A. J. Greenberger, *The British Image of India* (London, 1969), pp. 31–32; and Lewis D. Wurgaft, *The Imperial Imagination: Magic and Myth in Kipling's India* (Middletown, Conn., 1983), pp. 76–77, 101–2, 110, 160–61. This view is epitomized by E. M. Forster's character Ronny in *A Passage to India*. For French variations on this theme, see Loutfi, *Littérature et colonialisme*, pp. 25–29, 33–34.

[198]G. Trevelyan, *Competition Wallah*, p. 257.

[199]James Mill, *History*, vol. 1, pp. 412–13. For Africa, see Frederic Hornemann, *Travels in the Interior of Africa* (London, 1802), p. 70.

grained) rather than bad government as the chief cause of laziness. The widespread view that the tropics were a zone of abundant rainfall, fertile soil, and luxuriant plant growth led many to conclude that ready food and shelter and easy living had much to do with "native" sloth.[200]

Some authors, following the example of Chateaubriand, conjured up visions of tropical paradise,[201] but most pictured the consequences of lush environments and easy living in much grimmer terms. They saw environmental differences as the key to the contrast between the energetic and active Europeans and the phlegmatic and passive tropical peoples.[202] According to this view, in their struggle to survive in the harsher climate of the temperate zones, European peoples were forced to work longer and harder to alter the natural world about them. As a consequence, their bodies grew stronger and more vigorous, their endurance greater, and their material expectations higher. More work and inventiveness resulted in ever higher levels of social development and economic productivity, culminating in the scientifically oriented, industrialized nation states that had come to rule the globe. Work meant progress; industry produced prosperity; felt needs drove the Europeans to dominate the earth.

The expansionist implications of the work ethic were stressed by Samuel Smiles, who celebrated the energy and perseverance of such conquerors as Robert Clive and Warren Hastings, and by Gustave Le Bon, who viewed the "absence of energy" among the Hindus as the chief reason why sixty thousand Europeans could rule over 250 million Indians.[203] The bodily discipline that the Europeans displayed in their work, military training, and athletic competitions was seen as a key factor in their dominance over less controlled, less well-organized, less goal-oriented peoples.[204]

[200]Curtin, *Image of Africa*, p. 224; John Howiston, *European Colonies in Various Parts of the World* (London, 1834), vol. 1, p. 18; Drummond, *Tropical Africa*, p. 56; Cohen, *French Encounter*, p. 211; and Le Bon, *Les civilisations de l'Inde*, p. 181. Occasionally, authors in this era challenged the lush tropics myth and the lazy "native" stereotype that went with it; see Emile Baillauds, "The Problem of Agricultural Development in West Africa," *Journal of the Africa Society* 18 (1906), 119.

[201]Mercier, *L'Afrique noire*, p. 196.

[202]This notion has persisted well into the twentieth century. See, e.g., Ellsworth Huntington, *The Mainsprings of Civilization* (New York, 1935); or Kelly, "Presentation of Indigenous Society," p. 537, for examples from twentieth-century French textbooks.

[203]Smiles, *Self-Help*, pp. 44, 50, 58–59, 235–36, 244, 360; Le Bon, *Les civilisations de l'Inde*, pp. 184–85. For discussions of this theme in various strands of Victorian thought, see Altick, *Victorian People and Ideas*, p. 171; Briggs, *Victorian People*, pp. 126, 133, 136; Cairns, *Prelude to Imperialism*, p. 76; Michael Ruse, *The Darwinian Revolution* (Chicago, 1979), p. 155; and H. F. Wyatt, "The Ethics of Empire," *Nineteenth Century* 242 (1897), 523.

[204]Foucault, *Discipline and Punish*, pp. 135–38, 142, 150–53, 160; Smiles, *Self-Help*, chap. 9; Briggs, *Victorian People*, pp. 134, 141; Ferry, *France en Afrique*, p. 232; J. Berncas-

Conversely, peoples who lived in lush, tropical environments, where demands for food and shelter were minimal and easily met, had become soft, lazy, uninnovative, passive, and destined to be subjugated—if not exterminated.[205] Some authors blamed indolence and lack of vigor for many of the flaws associated with various African and Asian peoples, ranging from underdeveloped languages and thick skulls to technological backwardness and (despite the obvious contradiction of the lush tropics argument) famines and poverty.[206] Other writers saw a clear correspondence between capacity for hard work and level of civilization.[207] On the eve of the great Indian rebellion of 1857–58, which was in part a reaction against the changes wrought by the aggressive and energetic British colonizers in India, Harriet Martineau summarized the Victorians' sense of the challenges that the combination of European technology and dynamism posed for the ancient cultures of Asia: "The great fundamental condition of goodness of every sort—patient slowness—seems to the Hindoo to be overthrown by our [British] inventions. Immutability, patience, indolence, stagnation have been the venerable things which the Hindoos hated the Mussulmans for invading with their superior energy; and now what is Mussulman energy in comparison with ours, judged by our methods of steaming by sea and land, and flashing our thoughts over 1,000 miles a second."[208]

The implications of the widely held belief in the lazy "natives" were direct and vital, particularly for the colonized peoples of Africa and southern Asia. European officials had to find ways to entice or compel indolent peoples to provide cheap and docile labor. In the best of circumstances, consumer incentives could be employed. Education was also seen as a major means "to patiently instruct the natives first of all

---

tle, *A Voyage to China* (London, 1850), vol. 2, p. 189; and Q. Garfield Jones, "Athletics Helping the Filipino," *Outing* 64/5 (1914), 585–92.

[205]Charles Pearson, *National Life and Character* (London, 1893), pp. 32–33; Alfred Russel Wallace, "The Origins of Human Races and the Antiquity of Man Deduced from the Theory of 'Natural Selection,'" *Journal of the Anthropological Society* 2 (1864), clxiv–clxv.

[206]Ferry, *France en Afrique*, pp. 228, 243; Loutfi, *Littérature et colonialisme*, pp. 64–65; Harry H. Johnston, "England's Work in Central Africa," *Proceedings of the Royal Colonial Institute* 28 (1896–97), 42; Hornemann, *Travels*, p. 70; and Edward Bowdich, *Mission from Cape Castle in Ashanti* (London, 1819), p. 12.

[207]Ferry, *France en Afrique*, pp. 237, 244–47; Elton, *Travels and Researches*, pp. 19–20; J. K. Tuckey, *Narrative of an Expedition to the River Zaire* (London, 1818), p. 369; Theodore Waitz, *Introduction to Anthropology* (London, 1863), p. 337; Houghton, *Victorian Frame of Mind*, p. 250. Carlyle and Ruskin went so far as to assert that work was an attribute that separated man from beast; see the introduction to Knoepflamacher and Tennyson, *Nature and the Victorian Imagination*, p. xx.

[208]Martineau, *British Rule*, pp. 340–41.

that idleness is the eighth deadly sin."[209] But the widespread resort to coercive measures, ranging from hut and head taxes to corporal punishment and hostage taking, was often thought necessary. These measures did much to stimulate African and Asian resistance to European demands, however, and resistance—whether of the "footdragging" variety or flight—in turn reinforced European assumptions about lazy "natives." European forced-labor techniques also made a mockery of the civilizing mission, at least for those who, like Joseph Conrad, paused to test ideology against practice. In *Heart of Darkness*, Conrad describes perhaps the most horrific effects of the European insistence on putting the "natives" to work. It was the Belgians' drive to "civilize" the "savage" peoples of the Congo region by forcing them to deliver quotas of ivory and palm oil which led, Conrad believed, to the line of poles before the hut of Mistah Kurtz, each topped by a human head "black, dried, sunken with closed eyelids—a head that seemed to sleep at the top of that pole, and, with the shrunken dry lips showing a narrow white line of teeth, was smiling, too, smiling continuously at some endless and jocose dream of that eternal slumber."[210]

The need to make efficient laborers out of what were perceived to be indolent, pleasure-addicted, sensuous peoples fueled arguments for the reimposition of slavery in Africa and inspired innumerable discourses on the techniques of supervision and control. Textbooks used in colonial schools urged "native" school children to reject the slothful ways of their ancestors and embrace the work ethic of their conquerors.[211] Though indolence, drunkenness, and improvidence were recognized as vices that were far from being eradicated among the working classes of Europe itself, great advances were believed to have been made in regulating their work patterns. Insofar as possible (and European observers differed considerably on the extent to which the imposition of labor discipline overseas was feasible), Africans and Asians had to be compelled to be punctual, to work according to time clocks, and to learn that amusement was not to be mixed with work, that labor and leisure were distinct activities.[212]

[209]Johnston, "British West Africa," p. 98; Jean and John Comaroff, "Christianity and Colonialism in South Africa," *American Ethnologist* 13/1 (1986), 3, 12–14.

[210]Joseph Conrad, *Heart of Darkness* (New York, 1950), p. 133.

[211]Kelly, "Presentation of Indigenous Society," pp. 536–39; and the fourth section of Chapter 5, below.

[212]Foucault, *Discipline and Punish*, pp. 149–51, 154; Cairns, *Prelude to Imperialism*, p. 205; Wilhelm Junker, *Travels in Africa* (London, 1890–92), vol. 1, p. 97; Ferry, *France en Afrique*, pp. 237, 244; Burton, *Mission to Galele*, vol. 1, pp. 18, 204; Guppy, "Capabilities of the Negro," pp. ccx–ccxi; Mercier, *L'Afrique noire*, pp. 23, 53; and Hardy, *Une conquête morale*, pp. 206–11.

Space, Accuracy, and Uniformity

The widely held belief that the Europeans had learned to measure, regulate, and make productive use of time to an extent inconceivable to any non-Western people was complemented by the equally firm conviction that Westerners had also mastered space to an unparalleled degree. Traveling by rail and steamship or using the telegraph and later the telephone, Europeans could cover distances in hours or days that would have taken months or even years to traverse in the preindustrial era. Technological innovations that allowed Western businessmen and colonizers to overcome geographic barriers and in effect shrink space became the most prominent manifestation of yet another aspect of perception in which Europeans believed themselves to differ from and excel all other peoples. Signs of the Africans' and Asians' inferior capacity for space perception, noted by European observers in the early centuries of expansion, continued to be described by nineteenth-century writers. Numerous travelers and diplomats commented, for example, on the lack of perspective and proportion in Asian painting.[213] But most nineteenth-century observers concentrated less on Asian artistic preferences than on what they viewed as stronger evidence of non-Western deficiencies in space perception and measurement. As representatives of a civilization that had been exploring and mapping the earth for centuries, and had in the process developed increasingly sophisticated ways to chart locations, record distances, and survey sites, European visitors were quick to note the poor quality or complete lack of African and Asian maps and the "ridiculous" notions or "feeble" understandings of non-Western peoples about geography. John Barrow's dismissal of Chinese mapmaking and navigational treatises as "nonsense" is especially significant in this regard because he contrasts his own judgments with those of early Jesuit missionaries who, he asserts, lavished praise on the Chinese for their skills in these endeavors.[214]

From the late eighteenth century on, European visitors to African and Asian societies also became a good deal more sensitive than earlier travelers had been to the irregularities of non-Western urban configurations and the human landscape in these areas more generally. Nineteenth-

---

[213]See, e.g., Barrow, *Travels in China*, pp. 323–24; Julia Corner, *China* (London, 1853), p. 221; Farrar, "Aptitudes of Races," p. 123; J. D. Ball, *Things Chinese* (London, 1892), pp. 23, 29; and esp. F. S. Feuillet de Conches, "Les peintres européens en Chine et les peintres chinois," *Revue contemporaine* 25 (1856), 235–36.

[214]Barrow, *Travels in China*, p. 39. On these aspects of spatial perception, see Samuel W. Williams, *The Middle Kingdom* (New York, 1879), vol. 2, p. 153; James Mill, *History*, vol. 1, p. 65; Francis Galton, *Narrative of an Explorer in Tropical South Africa* (London, 1853), p. 88; Cameron, *Across Africa*, p. 284; and Georges Pouchet, *De la pluralité des races humaines* (Paris, 1858), pp. 109–10, 295.

century observers, whether missionaries, merchants, explorers, or colonial officials, went out to Africa and Asia from urban centers in which planning and linear orientation had been growing in importance since the Renaissance.[215] The elaborate geometrical designs for cities and fortifications pioneered by Italian planners in the late fifteenth and sixteenth centuries were extended in the Baroque era under the aegis of absolutist rulers seeking to proclaim their power through dramatic vistas and architectural grandeur. Broad, straight avenues and orderly block layouts were complemented by formal, geometrically ordered gardens. Symmetry reigned, and new urban space was regimented. The growing traffic of carts and wagons in urban areas, reflecting both the mercantile expansion and conspicuous consumption of the aristocracy which were hallmarks of the era, gave added impetus to the drive for regularity and linearity. In the eighteenth and nineteenth centuries, canals, paved roads, and railway lines carried these trends to the countryside and imposed a sense of pattern on human movement in both town and country. The ascendancy of the commercial and later the industrial bourgeoisie was reflected in the widespread adoption of the gridiron approach to urban planning, which favored rectangular layouts over natural contours. The drive to regularize city design and accommodate the flow of vehicular traffic culminated in the middle decades of the nineteenth century in Joseph Paxton's schemes for a Great Victorian Way in London, which was never built, and Baron Haussmann's plans for the renovation of Paris, which to the chagrin of many of his contemporaries were largely carried out. Haussmann, dubbed the "Attila of the straight line" by one of his most vehement critics,[216] oversaw the demolition of much of what remained of medieval Paris to make way for broad avenues, orderly block grids, and thoroughfares for modern transport.

European travelers could not help being disturbed by differences in the ways non-Western peoples arranged their living space. Visitors to Africa might admire the orderly layout of cities such as Benin, but they considered them exceptional. Explorers who had little more than contempt for most of the peoples they encountered lavished praise on the few who were found to have "clean and orderly" streets and well-

---

[215]This discussion of shifts in urban design is based on the relevant portions of Robert Dickinson, *The West European City* (London, 1951); Lewis Mumford, *The City in History* (New York, 1961), esp. pp. 418–60 ; Pierre Lavedan, *Histoire de l'urbanisme: Renaissance et temps modernes* (Paris, 1941), esp. pp. 9–34, 70–118, 277–323; and Leonardo Benevolo, *The Origins of Modern Town Planning* (London, 1967).

[216]Victor Fournel in his anti-Haussmann polemic, *Paris nouveau et Paris futur* (1865), cited in Françoise Choay, *The Modern City: Planning in the Nineteenth Century* (New York, 1969), p. 14.

built houses.[217] Europeans regarded most African towns, particularly those south of the savanna zone, as jumbles of ramshackle buildings thrown together without apparent design, and they dismissed African roads as meandering footpaths. Robert Knox contrasted the controlled and symmetrical landscapes of "civilized" European-settled areas in colonies such as South Africa with the "chaos" of African-occupied areas, which showed little sign of having been transformed by humans.[218]

Comments on the disturbing lack of regularity and the crowded disorder of Indian and Chinese cities were standard fare in accounts of Asia written in this period.[219] But many authors praised the construction of ancient roadways, were comforted by the semblance of order provided by the irrigation networks in rural areas, and (more rarely) noted the intricately planned layout of the palace centers of Asian potentates.[220] Europeans found the urban quarters of Asian commoners, on the other hand, randomly arranged and claustrophobically crammed with buildings, people, and domestic animals. Even Chinese gardens that had delighted early travelers to Asia and served as focal points for the eighteenth-century rage for *chinoiseries* were now considered poorly planned. One traveler went so far as to conclude that "nothing shows more completely the [Chinese] deficiency in science than their wretched attempts at landscape in which every object is most absurdly arranged and miserably out of proportion."[221]

Many writers in this period viewed seemingly chaotic human landscapes and ethnocentric maps—featuring Asian kingdoms drawn greatly out of proportion to their actual size—as symptoms of much deeper flaws in the space perception of non-Western peoples. Africans, for example, were thought hopeless when it came to measuring a distance

[217]See, e.g., Cameron, *Across Africa*, p. 133; and Paul Du Chaillu, *A Journey to Ashango-Land* (New York, 1967), pp. 258–59.

[218]Knox, *Races of Man*, pp. 542–43; J. Thompson, *Central African Lakes*, vol. 1, pp. 43, 73; Elton, *Travels and Researches*, p. 30; Cairns, *Prelude to Imperialism*, p. 78; Richard Freeman, *Travels and Life in Ashanti and Jaman* (London, 1898), pp. 214–15; and Duff, *Nyasaland*, pp. 282–83.

[219]*Chinese Repository* 1 (1833), 158–59, and 2 (1834), pp. 436–37; Ball, *Things Chinese*, p. 8; and Melchior Yvan, *Six Months among the Malays and a Year in China* (London, 1855), pp. 305, 307, 328, 358. For similar reactions in the eighteenth century, see, e.g., Pierre Sonnerat, *A Voyage to the East Indies and China between 1774 and 1781* (Calcutta, 1788), p. 190; John Grose, *A Voyage to the East Indies* (London, 1766), vol. 1, p. 111; and Kelly, "Presentation of Indigenous Society," p. 538.

[220]E.g., as late as 1799, George Staunton (*An Authentic Account of an Embassy . . . to the Emperor of China* [London, 1799], pp. 19–20), a prominent member of the McCartney mission, contrasted the broad avenues of Beijing with the cramped and narrow streets of London and other European cities.

[221]Berncastle, *Voyage*, vol. 2, p. 162. See also P. J. Marshall and Glyndwr Williams, *The Great Map of Mankind* (London, 1982), p. 81.

or drawing a straight line. Some travelers claimed that Africans could not distinguish a rectangle from a triangle; others averred that they paid no attention to direction when traveling. H. L. Duff, ever willing to pronounce judgment on African shortcomings, reported that they could not tell a crooked from a straight line; that an African servant found it impossible to place a centerpiece in the middle of a dining table; and that an African gardener asked to make a straight border would invariably end up with a zigzag edge.[222]

European assessments of Chinese spatial perception differed somewhat from their views on Africans in this regard. There was something of a consensus that the Chinese had attained a level of "practical mediocrity"[223] in measuring distances and determining the dimensions of buildings and public works but that their sense of space was a good deal less refined than that of the Europeans. The missionary Arthur Smith, for example, pointed out that Chinese measures of distance were variable. He noted that the Chinese *li*, which he equated with the English mile, was considerably shorter on an imperial highway than on a country road, and much longer on a mountain pathway. Thus, according to Smith, Chinese calculations of distance were vitally affected by the character of the space in question: by whether it was flat or hilly, by the time of day or night, and by the weather. This approach, of course, made a shambles of attempts to establish the uniform standards of measurement so dear to the Europeans.[224]

The French merchant Emile Bard concurred with Smith's assertions about Chinese measurements, adding that standards for length varied according to the item in question or the locality in which the measurement was being made. Bard also reported that the Chinese did not hold to the principle, self-evident to any sensible European, that the whole was equal to the sum of its parts. He related being "struck dumb" when he learned that a journey said to be forty *li* long proved in reality to be made up of two eighteen-*li* segments. When he questioned his guide about the discrepancy, the latter sought to alleviate his puzzlement by pointing out that "four nines are forty, are they not?" This exchange reminded Bard of the tale of an imperial courier who was threatened with punishment for arriving late with an important message for a local magistrate. The courier pleaded that the *li* along the route he had trav-

[222]Duff, *British Administration*, pp. 282–83. See also Decle, *Savage Africa*, p. 85; Baker, *Ismailia*, vol. 1, pp. 298–9; Wood, *Natural History of Man*, p. 344; Mary Kingsley, *Travels in West Africa* (London, 1897), p. 17; Livingstone, *Missionary Travels*, p. 56; and Fanoudh-Siefer, *Le mythe du nègre*, p. 179.

[223]Williams, *Middle Kingdom*, p. 154.

[224]A. Smith, *Chinese Characteristics*, p. 81–83.

eled were very large. After having them measured, the magistrate not only agreed with the courier, but added nearly half again as many *li* to the distance of the route recorded in the imperial records.[225]

European writers frequently linked weaknesses in temporal and spatial perception to a general disregard on the part of Africans and Asians for the accuracy and precision that had come to be valued so highly in Western culture. The emphasis on experiment and empirical validation, central to the scientific approach to the material world, had resulted in the imposition of ever stricter standards of observation, measurement, testing, and recording from the seventeenth century onward. The demand for quantification and precision led in turn to great advances in scientific instrumentation, which were often associated with innovations in watch and glass manufacture, machine-tooling, and metalworking.[226] The introduction within a century of such key instruments as the thermometer, barometer, telescope, and microscope made it possible for Europeans to explore and measure the natural world to a degree unimaginable to non-Western peoples. The steadily expanding market for clocks and watches, navigational and medical instruments, and military equipment gave further impetus to refinements in machine-tooling, which culminated in the manifold innovations in metalworking essential to the process of industrialization. Advances in mathematics and a growing emphasis on the collection and use of statistics by businessmen, bureaucrats, and scientists concurrently heightened the value placed on accurate enumerations and estimates. By the early nineteenth century the drive for quantification and precision had so pervaded the industrializing societies of western Europe that European observers overseas inevitably became increasingly sensitive to the crudeness and inaccuracy of African and Asian statistics and instruments.

Although all non-Western peoples failed to live up to European standards of accuracy, none—predictably—were considered as careless and imprecise as the Africans. Some travelers, including the influential David Livingstone, remarked on the care Africans took in counting their livestock and the dazzling precision they displayed in coordinating masses of drummers.[227] But the predominant view was that Africans,

---

[225]Bard, *Les Chinois*, pp. 21–23. See also Paul Giran, *De l'éducation des races* (Paris, 1913), pp. 20–24.

[226]Landes, *Revolution in Time*, esp. pp. 98, 103–13, 128. See also Silvio Bedini and D. J. de Solla Price, "Instrumentation" (pp. 168–87), and Eugene Ferguson, "Metallurgical and Machine Tools" (pp. 264–84), all in Melvin Kranzberg and Carroll J. Pursell, Jr., *Technology in Western Civilization* (London, 1967), vol. 1.

[227]Livingstone, *Missionary Travels*, pp. 21–22, 42; Freeman, *Travels in Ashanti*, p. 99; and Duff, *British Administration*, p. 277.

like black slaves in the Americas,[228] were sloppy, prone to exaggeration, inattentive to details, devoid of uniform standards, and incapable of quantification beyond (and sometimes including) elementary counting. For nineteenth-century Europeans, the fact that Africans had not developed numbers and their alleged inability to count higher than the number of fingers on both hands strikingly confirmed the conviction that they were incapable of accuracy or enumeration. Francis Galton provided one of the fullest discussions of these shortcomings in his 1850s account of his travels in South Africa. He concluded by comparing an African herdsman, struggling to relate sticks of tobacco offered in barter for some of his cattle, with his own bitch spaniel, checking and rechecking her litter of puppies to make sure they were all there. Galton surmised that the spaniel "evidently had a vague notion of counting, but the figure was too large for her brain. Taking the two as they stood, dog and Damara, the comparison reflected no great honour on the man."[229] Nearly two decades later John Wood, a successful writer of adventure stories for boys, did much to popularize Galton's equation of African and canine mathematical skills by retelling this story in his *Natural History of Man* and repeating Galton's conclusions.[230]

It was widely recognized that the Indians and Chinese had developed numbers; as we have seen, the Indians were notorious for the complexity and staggering magnitude of their calculations. But both peoples, as Arthur Smith observed, appeared to be "free from the quality of accuracy" and the "mania which seems to possess the Occidental, to ascertain everything with unerring exactness."[231] These deficiencies were widely thought to be related to the chronic dishonesty that Europeans believed afflicted colonized peoples.[232] As early as the first decades of the nineteenth century, such prominent Indianists as Charles Grant and James Mill had complained of the absence in Indian culture of an inquisitive spirit and the Indians' lack of concern for careful observation and

[228]For examples, see Gerald W. Mullin, *Flight and Rebellion* (London, 1972), esp. chap. 2; and Eugene Genovese, *Roll, Jordan, Roll* (New York, 1974), pp. 295–309.

[229]Galton, *Narrative*, pp. 133–34. See also on these themes Harmand, *Domination et colonisation*, pp. 267–68; William J. Burchell, *Travels in the Interior of Southern Africa* (London, 1822), vol. 2, p. 349; Cairns, *Prelude to Imperialism*, p. 78; Duff, *British Administration*, pp. 277–80; Fouillée, "Le caractère des races," pp. 82, 84; and Farrar, "Aptitudes of Races," p. 122. One traveler, Dr. A. Cureau, conceded that the Africans could count into the thousands but believed they had little capacity for more complex sorts of computations (*Les sociétés primitives*, pp. 86–88).

[230]Wood, *Natural History of Man*, vol. 1, pp. 344–45.

[231]A. Smith, *Chinese Characteristics*, pp. 183–84.

[232]This allegation, one of the most ubiquitous in the colonial literature, is found in many forms, from the mutterings of explorers about their lying bearers to the infamous speech of Lord Curzon in which he informed a large audience of Bengali intellectuals that truth was a peculiarly Western characteristic.

scientific investigation. They attributed the backward state of the sciences in India to the weakness of these drives in Indian culture and the paucity of accurate instruments for empirical inquiry.[233] By the last decades of the century Gustave Le Bon could confidently declare that the Indians were unequaled in their disregard for precision and that this flaw was responsible for the murky and vacillating quality of Hindu thought in all fields from religion to history. Le Bon felt that no Indian endeavor had suffered more on this account than science. The Hindus' lack of precision, he argued, in conjunction with their want of a "critical spirit," their excessive imagination, and their inability to "see things as they are," had severely hampered their scientific investigations. Le Bon judged that the Indians were capable of borrowing scientific learning from other peoples (he assumed that the ancient Indians had in fact done so from the Greeks and Arabs) but that they were unable to add to or improve upon this knowledge, which they found difficult to reconcile with their own "childish" investigations.[234]

As we have seen, many nineteenth-century writers attributed the failure of the Chinese to develop their early discoveries in science and technology to a neglect of general theory and systematic accumulation of knowledge. In the view of John Barrow and numerous later observers, these costly deficiencies were closely related to a disregard on the part of the Chinese for accuracy in any form.[235] Though they had numbers, the Chinese were thought to be hopelessly imprecise in their use of them. Thus, as Smith and Bard cautioned, Chinese statements about numbers or quantities should be regarded as little more than "informed guesses." Like their gauges of distance, their standards of weights and measures and the value of their currency were reputed to be highly variable, depending on regional differences, social situations, and calculations of personal advantage. Smith found the Chinese indifference to accuracy especially troubling because he felt that it might greatly impede Chinese efforts to master the sciences of the West. He mused that such a failure was likely to have a devastating impact on the first generation of Chinese chemists, which would "probably lose many of its number, as a result of the process of mixing a 'few tens of grains' of something, with 'several tens of grains' of something else, the consequences being an unexpected earthquake."[236]

[233]Charles Grant, "Observations on the State of Society among the Asiatic Subjects of Great Britain . . ." *Parliamentary Papers,* 1832, East India Company (8) "Report from Committee," vol. 1, app. 1, p. 26; James Mill, *History,* vol. 2, pp. 22, 41.

[234]Gustave Le Bon, *The Psychology of Peoples* (London, 1899), p. 68, and *Les civilisations de l'Inde,* pp. 192, 547–48.

[235]Barrow, *Travels in China,* p. 36. See also Yvan, *A Year in China,* p. 340.

[236]A. Smith, *Chinese Characteristics,* p. 84 (quoted portion), 78–85; Bard, *Les Chinois,* pp. 22–23, 33.

Worlds Apart: The Case of Ye Ming-chen

No other writer captured the Europeans' sense of the great differences that set them off from non-Western peoples in the perception of time and space, attitudes toward work, and the capacity for discipline, precision and organization as well as George Cooke, correspondent for *The Times* of London during the second round of wars between the European powers and China in the late 1850s. Cooke used a rather trivial encounter between the crew of a British man-of-war, aptly named the *Inflexible*, and the ex-governor of Canton, Ye Ming-chen, to illustrate the many ways in which Europeans had become superior in thought and action to the once highly esteemed Chinese. As Cooke relates the incident, Ye, who has been captured by the British during the war and is being taken into exile, has come up on deck after dinner and comfortably ensconced himself in a large bamboo chair. All about him the crew of the *Inflexible* is in motion as they work their way through a series of drills: putting out and taking in sails, manning battle stations, and firing the long rows of cannon. Delighting (as any patriotic Englishman would) in this impressive display of imperial power and martial discipline, Cooke watches Ye to see how the "Chinaman" is responding to the spectacle and whether or not the prisoner is suitably impressed with the skills of the navy that has contributed so much to the humiliating defeats suffered by the Chinese people: "Is he marking this orderly energy, this discipline, this zeal of art, this heartiness of work, this scene of a multitude in motion with one object, and is he pondering the lesson?"[237]

To Cooke's chagrin, he is not. In fact, Ye appears to be paying no attention whatsoever to the complex and energetic drill, much less contemplating these exercises as a demonstration of European superiority. Instead, the old mandarin is delighting in a wrestling match that is taking place—out of the officers' view—between a tall, brawny sailor and a smaller shipmate. As Ye rises slowly to go below for the night, Cooke exclaims with exasperation that the "Chinaman" has learned nothing, that he will remember only a smallish English sailor being dunked by a larger one. The lesson for Cooke and his readers was all too clear. Asians not only lacked the capacity for discipline and precision and the views of time and space essential to these; they were perversely indifferent or openly hostile to adopting the attitudes and patterns of behavior that had contributed so much to European dominance of the Chinese and all humankind.

In Cooke's rendering, Ye Ming-chen is more than a defeated adver-

---

[237]Cooke, *China*, pp. 424–25.

sary who foolishly refuses to learn from his conquerors. The old scholar-official embodies the values and perspectives of the ancient Chinese "Great Tradition," which had for decades been an object of derision among Europeans and was becoming a major target of their reforming zeal. Ye and the culture he sought to defend are in Cooke's view the antithesis of the ascendant Western middle-class and industrial civilization that are overwhelming them. Ye is passive, flaccid, and effete; the British are active, muscular, and energetic. He is indolent, little concerned to make the best use of his time by gathering information about his powerful enemies; the British (or most of them) are hard at work, taking advantage of the remaining daylight to improve their drill and gunnery skills. Ye appears resigned to his fate and to China's continued defeat and decay; the British strive to ensure victory, whatever the odds, and exude confidence in their ability to expand Britain's power and increase its wealth. For Ye, time is static or degenerative; for the British, it is purposeful and progressive. In his sloth and self-indulgence Ye cannot begin to appreciate the discipline and precision of the British officers and crew. He is utterly oblivious to the many ways in which the powerful battleship reflects the unprecedented control over nature that the Europeans have won through their superior science and technology.

Though Cooke's exposition is more extensive and explicit, many of his contemporaries saw similar contrasts between themselves and the peoples they encountered overseas. Unlike the effort in Europe, where the aristocratic and middle-class elites were deemed to be safely in the civilized fold, civilizing offensives in Africa and Asia had to be mounted at the level of both elite and popular culture. Besides complaining of peasant improvidence and the indolence of their servants, many writers emphasized the resistance of the elite guardians of the Great Traditions of Asia and Africa: Ye Ming-chen and the mandarins of China, the Brahmins and princes of India, the "chiefs" and *ulama* of Africa. The conviction that they possessed not only knowledge and skills that Africans and Asians could never acquire on their own but a vastly superior way of approaching both the natural and supernatural worlds was central to the European colonizers' sense of themselves and their mission in overseas societies. It was used to justify their conquest and domination; it bolstered their determination to remake other peoples and cultures in their own image.

The European colonizers were convinced that they had the right to rule and the duty to suppress and transform because, in contrast to the corrupt and unpredictable African and Asian elites who contested their claims to dominance, they were disciplined, foresightful, and reliable.

They claimed to hold power not for its own sake or their own self-aggrandizement but for the improvement of their colonial subjects as well as their countrymen in Europe. The scientific and technological advances of the civilization they championed made them confident that they could overcome natural obstacles and solve social problems that their Asian or African rivals had fatalistically accepted or allowed to fester for centuries. The antithesis between the European colonizers and the leaders of the cultures they had come to dominate was at the heart of the civilizing-mission ideology. That ideology gave the Europeans the sense of righteousness, self-assurance, and higher purpose which as much as the firepower of their Maxim guns made it possible for them to dominate most of humankind through much of the nineteenth century.

However central it was to their own sense of purpose and virtue, the European colonizers made little attempt to propagate the civilizing-mission ideology—as distinct from the work habits or the penchant for punctuality that it advocated—beyond the constricted circles of their African and Asian allies, who played vital roles in all colonial administrations. Because it was intended primarily to promote the esprit and justify the policies of the colonial overlords rather than win the consent of the masses of colonized societies, the civilizing mission was an ideology of dominance with a more limited scope than those aimed at achieving cultural hegemony in Gramsci's sense of the term.[238] Most of the colonized knew little or nothing of the concepts of work, time, and contract that shaped European thinking, even though all felt the effects of the policies that resulted. Few of the colonized, and virtually none beyond the Western-educated elite, had any appreciation for the theoretical breakthroughs and technical accomplishments that were responsible for the European dominance. But nearly all were only too aware of the Europeans' unprecedented capacity for the application of scientific knowledge to production, communications, and, above all, military technology. The Europeans were able to maintain their rule over Africans and Asians mainly because the colonizers possessed an overwhelming advantage in the instruments of coercion at their disposal—though the actual application of force in most colonial settings was far less frequent than much of the literature on imperialism would lead one to assume.[239] For the great majority of the colonized, the legitimacy of

[238]See Antonio Gramsci, *Selections from Prison Notebooks*, ed. and trans. Q. Hoare and G. Smith (New York, 1971), pp. 12–14, 55–63, 275–76.

[239]The generally infrequent resort to violence resulted chiefly from the preference of the colonized for indirect methods of resistance and protest rather than confrontations involving force. See the articles by James C. Scott, Michael Adas, and Ann Stoler in Scott and B. Kerkvliet, eds., *Everyday Forms of Peasant Resistance in South-east Asia* (London, 1986); and Allen Isaacman, Stephen Michael, et al., "'Cotton is the Mother of Poverty': Peasant Resistance to Cotton Production in Mozambique, 1938–1961," *International Journal of African Studies* 13/4 (1980), 581–615.

European imperial regimes—insofar as it existed at all—depended on the preservation (and at times the creation) of "traditional" elites and indigenous rituals and symbols of authority.[240]

Only among the Western-educated elites were serious efforts made to inculcate an understanding of and adherence to the principles of the civilizing mission. This remarkable attempt by an alien elite to establish cultural hegemony over their Western-educated subordinates on the basis of alien norms and values met with mixed responses. Very often European officials and missionaries were frustrated by the refusal of the educated and professional classes among the Africans and Asians to accept fully the attitudes and patterns of behavior that so many Victorian observers had come to regard as quintessential virtues of the civilized. As we shall see in the next chapter, European explanations for this failure varied from allegations of racial incapacity to suspicions of political machinations. But some colonizers noted the extent to which Western-educated Africans and Asians, especially those who worked with Europeans in government and business, had imbibed Western preferences and standards of judgment. In George Orwell's novel of colonial life, *Burmese Days*, for example, one of the main characters is an Indian physician named Veraswami, who, at the outset at least, fervently embraces the tenets of the civilizing mission. Appalled by the cynical critique of British imperialism delivered by the English protagonist Flory, Veraswami launches into a defense of the colonizers that strikingly demonstrates his willingness to concede superiority to those who have excelled technologically and gained such an impressive mastery over nature:

My friend, it iss pathetic to me to hear you talk so. It iss truly pathetic. You say you are here to trade? Of course you are. Could the Burmese trade for themselves? Can they make machinery, ships, railways, roads? They are helpless without you. What would happen to the Burmese forests if the English were not here? They would be sold immediately to the Japanese, who would gut and ruin them. Instead of which, in your hands, actually they are improved. And while your business men develop the resources of our country, your officials are civilizing us, elevating us to their level, from pure public spirit. It is a magnificent record of self-sacrifice.[241]

---

[240]On the preservation and creation of symbols of legitimacy in the colonial setting, see the essays by Bernard Cohn and Terence Ranger in E. Hobsbawm and Ranger, eds., *The Invention of Tradition* (Cambridge, Eng., 1983); and Richard Fox's study of the Sikhs' "martial tradition" in *Lions of the Punjab: Culture in the Making* (Berkeley, Calif., 1985).

[241]George Orwell, *Burmese Days* (New York, 1963), pp. 36–37.

That Orwell's ficticious rendering of an Indian's internalization of the civilizing-mission ideology had some basis in fact is suggested by the results of a survey of Indian migrants living in central Africa in the 1960s. Floyd and Lilian Dotson found that the Indians considered the Africans "illiterate and incomprehensible savages" who were lacking in culture, lazy and without foresight, childlike in their thinking and incapable of logical deductions, and self-indulgent and morally reprobate.[242] The success of the civilizing mission as an ideology of cultural hegemony at the elite level is equally strikingly illustrated by numerous reminiscences in which African and Asian leaders admit to a youthful emulation of the ideas and behavior of the colonizers.[243] But as these same memoirs testify, African and Asian intellectuals and emerging political leaders came increasingly to reject the attitudes and assumptions underlying the civilizing-mission, dismissing them as justifications for the exploitation and oppression of colonized peoples. It is significant that counterhegemonic responses on the part of African and Asian nationalists (ironically fashioned in part on the basis of alternative European ideas) gained widespread adherence in the very decades when the civilizing-mission ideology and the assumptions about European technological and scientific superiority in which it was rooted were increasingly challenged by European intellectuals themselves. This convergence gave added impetus to the Afro-Asian assault on the ideas and assumptions that were central to European justifications for their dominance and to their efforts to persuade colonized peoples that their subordination was both necessary and beneficial.[244]

[242]Floyd and Lilian Dotson, *The Indian Minority of Zambia, Rhodesia, and Malawi* (New Haven, Conn., 1968), pp. 262–68, 320.

[243]See, e.g., Ndabaningi Sithole, *African Nationalism* (London, 1969), pp. 95–101, 157–66; Odinga, *Not Yet Uhuru*, pp. 30–37; Nirad C. Chaudhuri, *The Autobiography of an Unknown Indian* (Bombay, 1951), pp. 118–30, 187–202; and Jawaharlal Nehru, *Toward Freedom* (New York, 1942), pp. 27–28, 30–39.

[244]For European critiques, see Chap. 6, below. I plan to explore Asian and African intellectual responses to European technological supremacy in a companion study to this volume.

CHAPTER 5

The Limits of Diffusion: Science and Technology in the Debate over the African and Asian Capacity for Acculturation

B Y THE last decades of the nineteenth century there was a fairly strong consensus among those who wrote on colonial issues and those who actually implemented colonial policies on the way tasks were apportioned and goals set as defined by the doctrines of the civilizing mission. Europeans would invent, finance, and command; Africans and Asians would acculturate, labor, and obey. Industrial Europe would serve as the international source of money and machine capital and cheap consumer goods; Africa and Asia would provide market outlets for Europe's manufactures and raw materials for its factories.

This vision of the most efficient and mutually beneficial ordering of the global political economy was challenged by critics of imperialist expansion and indeed often bore little relation to the disappointing realities of colonial markets and resources.[1] But it was seldom the source of the heated debates and bitter polemic that were aroused by the more fundamental and long-term implications of the civilizing-mission ideology. These centered on the belief that the industrially and scientifically advanced Europeans were obliged to serve as teachers of the backward and superstition-bound peoples of the non-Western world. This respon-

---

[1]For a survey of these controversies in the British empire, see A. P. Thornton, *The Imperial Idea and Its Enemies* (New York, 1968); for the French, Raoul Girardet, *L'idée coloniale en France de 1871 à 1962* (Paris, 1972). Beginning with J. A. Hobson's *Imperialism*, first published in London in 1902, there have been numerous critiques of the thesis that European nations, as distinct from individual enterprises, gained great economic benefits from their overseas empires. But as Eric Wolf's *Europe and the People without History* (Berkeley, Calif., 1982) demonstrates in considerable detail, the global economic order that the imperialists advocated was closely approximated by exchanges between Europe and its dependencies in the half-century before World War I.

sibility was accepted by virtually all major thinkers on colonial policy issues. It was enshrined in Kipling's poem "Kitchener's School," which is perhaps the supreme expression of the conqueror-as-tutor ideal:

Knowing that ye are forfeit by battle and have no right to live,
He begs for money to bring you learning—all the English give. . . .
They terribly carpet the earth with dead, and before their cannon
    cool,
They walk unarmed by twos and threes to call the living to school.[2]

But serious divisions emerged in the ranks of European social thinkers and colonial policymakers over the extent to which African and Asian peoples were capable of acquiring the technological skills and mastering the scientific ideas that were the ultimate source of Western global dominance.[3] The positions adopted by those engaged in the resulting controversy were invariably shaped by the degree to which advocates of various policies adhered to the proposition that an individual's potential for acculturation was determined by his or her race.

Contrary to the impression given in much of the recent literature on nineteenth-century European colonization, in which there is a tendency to reduce European interaction with Africans and Asians to stereotypes of racist exclusivism and condescension, European responses to racial thinking varied widely in this era. To begin with, the meaning of the term "race" itself was often vague; its application changed significantly from one writer to another and even within the same work of a single author. The term "race" was used rather routinely to indicate divisions within humanity as a whole, but there was little agreement on which divisions were appropriate or even how many different races could be distinguished.[4] In addition, as contemporary critics of racist thinking such as R. P. Lesson and Theodor Waitz pointed out, opinions varied

[2]*Rudyard Kipling's Verse* (New York, 1938), pp. 231–33. For additional examples of these sentiments, see Girardet, *L'idée coloniale,* pp. 78–79; and esp. contemporary works such as Georges Hardy's appropriately titled study of colonial education, *Une conquête morale: L'enseignement en A.O.F.* (Paris, 1917).

[3]There was also debate over the degree to which these skills and secrets *ought* to be diffused, since their acquisition by non-Western peoples might well undermine Western dominance. With the emergence of Japan as an industrial power and the rise of Asian and African nationalist movements in the late nineteenth century, these doubts grew in importance, even as the controversies over the abilities of colonized peoples peaked in intensity.

[4]Perhaps the most extensive nineteenth-century critique of this aspect of racist thinking can be found in the important but neglected work of the Haitian diplomat Antenor Firmin, *De l'égalité des races humaines* (Paris, 1855). See also Theodor Waitz, *Introduction to Anthropology* (London, 1863); A. de Quatrefages, *Histoire générale des races humaines* (Paris, 1871); and Jean Finot, *Race Prejudice* (New York, 1970).

widely among nineteenth-century writers as to which characteristics should be decisive in deciding racial boundaries.[5] Not surprisingly, there was little consistency in the Victorians' usage when it came to the term "race." References can be found in nineteenth-century works to the human race; the Hindu or Islamic race; primitives as a race; the British and French or Anglo-Saxon and Celtic races; the African or European race; the Bantu or Tamil race; the Aryan and Semitic races; and of course the Negro and Caucasian, or Negroid and Caucausoid, races. As these examples suggest, the problems of agreeing upon the number of races and how to distinguish them were compounded by the fact that the same group could be classified in several different ways.

Different classifications could result in conflicting, even blatantly contradictory assessments of the racial attributes and potential of a given group. For example, linguistic and archeological evidence uncovered over the course of the century convinced many thinkers that Hindu civilization in northern India had been developed by a branch of the Aryan "race" (not language family, as present-day scholars would argue).[6] This view clashed with the tendency of some writers to classify all Indians as part of the "brown" or "black" race, and especially with the decidedly racist attitudes that dominated British social interaction with the Indians from the last decades of the nineteenth century. The notion that the northern Indians or Hindus represented a branch of the Aryan race that had been degraded by centuries of life in an enervating climate and constant intermarriage with lesser racial types provided a way to reconcile the two views. But it did little to resolve questions regarding the Indians' historic achievements and future potential.[7]

A survey of nineteenth-century works dealing with racial categories suggests that only a minority of writers used the term "race" to differentiate between and rank human groups on the basis of hereditary biological differences. Though most Europeans clearly considered themselves superior to African or Asian peoples, until the last decades of the

[5]R. P. Lesson, *Voyage médical autour du monde* (Paris, 1829), pp. 154–56; and Waitz, *Anthropology*, pp. 213–29.

[6]See Leon Poliakov, *The Aryan Myth: A History of Racist and Nationalist Ideas in Europe* (New York, 1977), esp. chaps. 9 and 10, which deal in some depth with the problems posed by the northern Indians. For sample nineteenth-century affirmations of the Aryan-Indian racial link, see Georges Cuvier, *The Animal Kingdom* (London, 1827), pp. 98–99; Arthur de Gobineau, *Essai sur l'inégalité des races humaines* (Paris, 1853), vol. 1, pp. 362–63; and William Lawrence, *Lectures on Physiology, Zoology, and the History of Man* (London, 1819), p. 484.

[7]On the varying approaches to the Aryan "problem," see Joan Leopold, "British Applications of the Aryan Theory of Race to India, 1850–1870," *English Historical Review* 89/4 (1974), 578–603; and Christine Bolt, *Victorian Attitudes towards Race* (London, 1971), pp. 190, 198.

century their conviction of superiority at the level of ideas, as distinct from that of social interaction, was based primarily on cultural attainments rather than physical differences. There is, of course, in European writings of this period a narcissistic preference for white skin and straight hair, but the notion that biological factors were responsible for Europe's achievements and global dominance was not widely argued or accepted until the latter part of the century. Because Mendel's investigations were known to only a handful of botanists, and thinking on genetics was muddled in the extreme,[8] the connections between innate physical characteristics and moral or intellectual capabilities remained ill defined even in the most systematic of the racist tracts. Though the numbers of those who sought to prove scientifically that human abilities varied according to race increased steadily from the first years of the century, their findings had only a marginal impact on policies toward colonized peoples until the 1860s and 1870s. From the works of J. C. Prichard in the early 1800s to those of Jacques Novicov at century's end, outspoken criticisms were directed against the techniques and conclusions of the scientific racists throughout this period.

The fact that virtually all nineteenth-century European thinkers accepted without question the assumption that Europeans were technologically and scientifically superior to all other peoples but did not necessarily draw from this the conclusion that Europeans were racially superior suggests that we may need to reevaluate the place of racism in its more restricted sense in the intellectual discourse of this era. Though many authors who made racist arguments paid little attention to or altogether ignored European scientific and technological advances,[9] racism was more often ideologically subordinate to a more fundamental set of European convictions that arose out of their material accomplishments. In addition, at the level of intellectual exchange as opposed to popular sentiment, scientific and technological gauges antedated racial distinctions as important proofs of European superiority. They have also (despite setbacks) shown more staying power in the twentieth century, when national liberation and civil rights struggles and Nazi atrocities have done much to discredit racist arguments.

Questions relating to the capacity of Africans and Asians to adopt Western science and technology tended to be debated at two levels: among the physicians, anthropologists, and social theorists who developed, or disputed, the racist theories that peaked in sophistication in this era; and among administrators and missionary spokesmen who shaped

[8]Vitezslav Orel, *Mendel* (Oxford, 1984), pp. 94–100; Robert Olby, *Origins of Mendelism* (Chicago, 1985), chap. 6.
[9]See, e.g., Georges Pouchet, *De la pluralité des races humaines* (Paris, 1858).

educational policies in the colonies. For most of the first half of the century, there was little interaction between the two groups. The impact of racist theory on colonial decision-making was negligble, and colonial officials and missionaries were only marginally involved in developing instruments to measure racial differences with scientific precision and supplying information about overseas societies to confirm theories based on the assumption of innate racial differences. In the last decades of the century, when both theories of racial supremacy and scientific and technological gauges of human worth were widely accepted by European politicians and intellectuals, the interplay of ideas became more intense. In lectures, memoirs, and newspaper articles, explorers and former colonial administrators recounted personal experiences and missionaries and anthropologists provided field evidence that supported racist theories. These racist ideas in turn played a major role in late nineteenth-century debates over colonial policy. The following discussion gives special attention to the impact, or lack of it, of racist thinking on British and French educational policies in their colonial possessions. As we shall see, a tautological relationship developed: scientific and technological achievements were frequently cited as gauges of racial capacity, and estimates of racial capacity determined the degree of technical and scientific education made available to different non-Western peoples.

The First Generations of Improvers

Race was not an issue in the controversy over higher education in India that pitted Orientalists against Evangelicals and Utilitarians in the 1820s and 1830s. The victory of the pro-Anglicizing Evangelicals and Utilitarians was sealed by the 1835 resolution of Governor-General Lord William Bentinck, committing the British to "the promotion of European literature and science among the natives of India." Bentinck's decision, one of the most momentous in the history of European colonization, was based on the assumption (which none of the British officials engaged in the debate had challenged) that Indians were intellectually able to master the English language and Western learning, including advanced mathematics and science. In fact, the British desire to supplant what they viewed as hopelessly antiquated and superstition-ridden Indian views of the natural world had been one of the main objectives of the Evangelical and Utilitarian factions from the outset of the campaign to win increased government support for English education for Indians, particularly at the university level. Their assaults on Sanskrit and Persian learning were culturally arrogant, but they were free of the racial prejudice that is often assumed to have been synonymous with British rule in nineteenth-century India.

The original impetus for the drive to win government support for the use of English in higher education can be traced to Charles Grant's 1792 critique of Indian civilization and his official efforts to enact a far-reaching program of reform in the British-controlled portions of the subcontintent. As President of the Board of Trade in Calcutta and later Chairman of the Court of Directors of the East India Company, Grant was in a superb position to push his proposals for administrative and legal reform. He was also one of the more prominent members of the Evangelical clique that was active in the late eighteenth and early nineteenth centuries in a wide range of reformist causes, including the campaign to abolish the slave trade. He viewed moral uplift as the sine qua non of his program to rescue the Indians from the degraded state to which they had fallen. Therefore, Christian education was central to his proposals for reform in India, though he also envisaged a major role for science and technology. Grant was confident that instruction in the revealed truths of Christianity and in the advanced sciences and technology that the British had developed through their "great use of reason in all subjects" would be sufficient to weaken the Indians' attachment to what he viewed as repulsive customs and "superstitious chimeras," and to arouse them from centuries of lethargy and passivity.[10] The mere sight of the wonderful machines devised by the British, Grant asserted, would convince the Hindus of the superiority of European "natural philosophy" and attract them to English-language schools. Through English education and Christian instruction, Indian youths could be awakened to a spirit of invention, a desire for improvement, and a taste for British manufactured goods. Having mastered the "principles of mechanics" devised by European thinkers and engineers over the centuries, educated Indians would be able to work for improvements in agricultural production and the "useful arts" which would benefit the whole population of the subcontinent.[11]

Though Grant did refer to the racial origins of British moral superiority over the Indians, his estimate of the potential for Indian improvement was decidedly nonracist. Reflecting the views of most eighteenth-century thinkers, he held that an enervating climate and centuries of

[10]Charles Grant, "Observations on the State of Society among the Asiatic Subjects of Great Britain . . ." *Parliamentary Papers*, 1832, East India Company (8) "Report from Committee," vol. 1, app. 1, p. 61. For the importance of this theme in Grant's writings as a whole, see Ainslie Embree, *Charles Grant and British Rule in India* (London, 1962), pp. 118, 151.

[11]Grant, "Observations," pp. 62, 66, 68.

despotic rule were the main causes of Indian character flaws and India's backwardness. He shared the confidence of most Enlightenment thinkers that "all branches of the human race" could be "improved through reason and science." Hindu inferiority, he argued, arose from "moral causes" not "physical origin[s]." The Bengalis were as capable as the ancient Britons of raising themselves up from the lowly state to which they had fallen, and Grant had no doubt about their desire to do so. Presumably on the basis of personal conversations, he insisted that Indians who had been in close contact with Westerners conceded the "immense superiority" of European thought and technology. He inferred that they would welcome the widespread introduction of both European ideas and machines into the subcontinent. He was convinced that these would serve as key agents of the process of social and intellectual regeneration which he had come to see as a central purpose of British rule.[12]

Interestingly, Grant was anxious to assuage the concern not of those who doubted the Indian capacity to master Western learning but of those who feared that the Indian acquisition of it would result in a rebellion against British authority—similar to that which had recently deprived the British of their valuable colonies in North America. Grant reasoned that the lethargic and servile Indians were unlikely to respond in the same way as the energetic and aggressive Americans, who had been nurtured by a temperate climate and English freedoms.[13]

In the early 1800s, Grant found numerous allies in his campaign for government-supported education in India. After 1813, when Christian proselytization was allowed for the first time in areas controlled by the East India Company, missionaries such as William Carey and John Samuel founded private colleges in which instruction in English and Western sciences was promoted.[14] Carey envisioned a somewhat elementary course of instruction in the sciences for Indian youths. Samuel's approach stressed the practical and applied over the theoretical. The college that he proposed in 1813 resembled a technical school in its approach to the sciences, and he made much of the need for rigorous physical education to inculcate the virtues of industry, activity, and

---

[12]Ibid., pp. 20–22, 31–35, 62–63.

[13]Ibid., pp. 77–79.

[14]William Carey, Joshua Marshman, and William Ward, *College for the Instruction of Asiatic Christian and Other Youth in Eastern Literature and European Science* (London, 1819), esp. pp. 4–8; John Samuel, *On Indian Civilization: Proposals for Establishing a Native School Society* (London, 1813), esp. pp. 38–44; and George Smith, *The Life of William Carey* (London, 1885), pp. 295ff., 381–82.

bodily fitness among the Bengalis, whom the British regarded as languid, soft, and passive.[15]

The privately financed efforts of Western missionaries and prominent Englishmen to establish English-language colleges in the early 1800s were bolstered by a growing demand for English education on the part of Indian leaders themselves. Hindu notables and British improvers[16] joined forces in 1817 to found the Calcutta School Book Society and the Hindu College, which were aimed in part at the dissemination of Western scientific thought in English as well as in Indian languages. The special interest in English eduation evinced by the Kayasthas and similar caste groups whose members had traditionally served as scribes and government officials, and the rapidly growing Indian commercial classes in port cities such as Calcutta arose from more than an opportunistic desire to secure positions in the colonial bureaucracy. As the 1823 petition by Raja Rammohan Roy to Governor-General Lord Amherst made clear, English education was viewed as a means of unlocking the secrets of the Western sciences that had contributed so much to the advancement of European societies. Roy, who was among the boldest of the early Indian leaders favoring sweeping reforms of Indian society and extensive Anglicization, charged that the perpetuation of the government's policy of confining its support to Sanskrit (and Arabic-Persian) education would serve only to "keep [India] in darkness." If, however, the object of govern ment was the "improvement of the native population," Roy declared, it should "promote a more liberal and enlightened system of instruction, embracing Mathematics, Natural Philosophy, Chemistry, Anatomy and other useful sciences . . . which the nations of Europe have carried to a degree of perfection that has raised them above the inhabitants of other parts of the world."[17]

Of the forces joined in the early campaign for English education in India, the Utilitarians ultimately played the most critical role in both England and India. Through his relentless critiques of virtually all aspects of Indian society, James Mill helped to prepare the way for English-language instruction, even though he opposed the ultra-Anglicizing policies advocated by other Utilitarians.[18] His lengthy com-

[15]A view later enshrined in Thomas B. Macaulay's famous essay "Lord Clive"; see *Miscellaneous Works* (New York, n.d.), vol. 5, pp. 37–38.

[16]Though the Victorians rarely applied the label "improvers" to themselves, I find it most apt in view of their frequent references to the importance of improving societies—both their own and those colonized overseas—and of improvement in general.

[17]Quoted in Charles O. Trevelyan, *On the Education of the Indian People* (London, 1838), p. 66. See also R. C. Majumdar and V. N. Datta, "English Education," in Majumdar, ed., *British Paramountcy and Indian Renaissance*, vol. 10, pt. 2, of *History and Culture of the Indian People* (Bombay, 1965), pp. 34–42.

[18]See John Clive, *Macaulay: The Shaping of a Historian* (New York, 1973), p. 384.

mentary on the dismal condition of the "arts" and sciences in the subcontinent made it clear that the introduction of Western learning and
technology were essential features of his wide-ranging plans for reform
in India. His official dispatches too left no doubt that he thought the
Indians intellectually able to acquire Western scientific learning and
adopt Western technology.[19] The hopes of Mill and other Utilitarians
that their programs to "improve" India's laws, political economy, customs, and learning would be translated into government legislation
were finally realized when Lord William Bentinck was appointed
Governor-General in 1829. For the next half-decade Bentinck strove to
make good on a promise he had made just before leaving for India:
during his tenure, Bentinck had assured Mill, Jeremy Bentham would
be the real Governor-General of India.[20]

The reforms initiated or enacted under Bentinck ranged from the
abolition of such social abuses as *sati* (the ritual immolation of the widows of high-caste Hindus) to extensive changes in the land tenure systems in some areas.[21] The decision to shift the emphasis of government
support for Indian higher education from indigenous learning to
English-language instruction and European writings was undoubtedly
one of the most significant improvements adopted. The issue had been
raised in government circles much earlier. Though the East India Company had provided small sums in support of English instruction in
private schools long before Bentinck's arrival, the first explicit official
acknowledgment of the need to promote Western learning in India came
in the bill by which Parliament renewed the Company's charter in 1813.
Under the new charter a small sum of money was set aside for "the
introduction and improvement of [the] knowledge of the sciences" in
India.[22] But the money was never spent. In 1821 the Court of Directors
of the East India Company in London, in reply to the Committee on
Education's report from India, stated bluntly that it was "worse than a
waste of time" to continue to teach the sciences as they were "found in
Oriental books." The Directors urged that the education department
work for the introduction of "useful" learning, which they believed

[19]Majumdar and Datta, "English Education," p. 44; and George D. Bearce, *British
Attitudes towards India, 1784–1858* (London, 1961), pp. 65–69, 73–74, 161.

[20]Majumdar and Datta, "English Education," p. 45.

[21]The best account of Bentinck's reforms and his governor-generalship as a whole can
be found in John Rosselli, *Lord William Bentinck: The Making of a Liberal Imperialist, 1774–
1839* (London, 1974). For broader reformist initiatives, see Eric Stokes, *The English
Utilitarians and India* (Oxford, 1959).

[22]Majumdar and Datta, "English Education," p. 10; and Percival Spear, "Bentinck and
Education," *Cambridge Historical Journal* 6/1 (1938), 79.

could be found only in Western and not in Hindu or Muslim works. But again, little was done to follow up on their suggestions.[23]

The question of government promotion of English education at the expense of instruction in the indigenous languages became a major issue only in the late 1920s. At that time reform-minded members of the Committee for Public Instruction (established in 1823) launched an open assault on the long-standing policy of government financing for instruction in Sanskrit and Arabic and for the publication of works in Indian languages. The improvers stressed the dated and deficient state of Indian scientific knowledge in their efforts to rebut the defense of Indian learning by the Orientalist members of the committee.[24] They also pointed out that the Europeans' superior scientific and technical knowledge could be properly transmitted to Indian students only through classes taught in English. It is important to note that though the Orientalists found much more of value in Indian scientific treatises than the Utilitarians or Evangelicals did, even such staunch defenders of Arabic and Sanskrit education as Henry Prinsep freely admitted that in this area European learning was vastly superior to Indian. As John Clive has persuasively argued in a recent work, the long-accepted view that the Orientalists were opposed to English-language instruction in Western subjects considerably distorts their actual position.[25] The debate between the Orientalists and the improvers over educational policy centered on questions of preference and pace. The Orientalists wanted to introduce Western learning more slowly and preserve a higher degree of government support for instruction in the Indian classics. They favored grafting Western learning, which they conceded was undoubtly superior in many areas, onto the trunk of Indian knowledge. The improvers wanted to plant a new tree imported from the West and let the banyan of ancient Indian wisdom wither as the new growth flourished.

By 1835 the quarrel between the two factions was stalemated. Bentinck called upon Thomas Babington Macaulay, the newly appointed President of the Committee for Public Instruction, to break the deadlock. Macaulay responded in February 1835 with his infamous "Minute on Education," which arrogantly dismissed the totality of Asian learning as equivalent only to the works on a "single shelf of a good European library."[26] He strongly recommended that the East India

[23]C. Trevelyan, *Education of the Indian People,* pp. 75–76.

[24]Ibid., pp. 37, 55, 72; Alexander Duff, *New Era of the English Language and English Literature in India* (Edinburgh, 1837), pp. 35–36; Spear, "Bentinck and Education," pp. 81–82; and H. Woodrow, *Macaulay's Minutes on Education in India* (Calcutta, 1862).

[25]Clive, *Macaulay,* esp. pp. 355–59, 367, 371, 380.

[26]In W. N. Lees, ed., *Indian Musalmans and Four Articles on Education* (London, 1871), pp. 91–93.

Company transfer its support from Indian-language education and publications, which contained "neither literary or scientific information," to English-language instruction. In doing so, he stressed the "immeasurable" superiority of English works in fields in which "facts are recorded and general principles investigated." Macaulay noted that the English language "abounded" in works containing "full and correct information respecting every experimental science which tends to preserve the health, to increase the comfort or expand the intellect of man." With Rammohan Roy's petition in mind, he bolstered his arguments by observing that the Indians themselves had clearly indicated a preference for English education over Sanskrit or Arabic. They clamored for English instruction, he contended, despite the fact that they had to pay for it in private schools, whereas they could be subsidized to master their traditional learning in government-supported institutions.[27]

Like Grant's "Observations" nearly half a century earlier, Macaulay's advocacy of a predominance of English-language instruction and Western learning in government-supported higher education was premised on the assumption that Indian students were intellectually able to appreciate English literature and to comprehend the findings of European scientists and mathematicians. Macaulay saw India as comparable to Russia, where the introduction of Western writings had been instrumental in transforming a backward and "barbaric" land into a more forward-looking and civilized one. His famous prediction that government support for English education would result in the creation of "a class of persons, Indian in blood and color, but English in taste, in opinion, in morals and in intellect" provides one of the most straightforward expressions of assimilationist sentiment available in official English documents. His expectation that English-educated Indians would go on to enrich vernacular learning with "terms of science borrowed from the Western nomenclature" reflected both his confidence that Indians would be able to master Western knowledge and his conviction that the Europeans' superior understanding of natural phenomena ought to filter down and enrich the lives of those at all levels of Indian society.[28]

Although it is now generally agreed that the "Minute" was not as decisive as was once thought in prompting Bentinck's decision to shift government support from Indian- to English-language education,[29]

---

[27]Ibid., pp. 91–92, 95–97.

[28]Ibid., pp. 94–95, 101–2; and Stokes, *Utilitarians and India*, p. xiii.

[29]See, for examples, Majumdar and Datta, "English Education," pp. 45–48; Kenneth Ballhatchet, "The Home Government and Bentinck's Education Policy," *Cambridge Historical Journal* 10/2 (1951), 228–29; Spear, "Bentinck and Education," pp. 83–84; and esp. Clive, *Macaulay*, pp. 360–70.

Macaulay had cogently, if abrasively, summarized the views of the improvers. In the previous months he had also worked behind the scenes for the advancement of those views. Charles Trevelyan, Macaulay's brother-in-law and disciple, hailed Bentinck's Resolution as a decisive triumph for the improvers. Citing ancient India's achievements in mathematics and the sciences, Trevelyan argued that the decision for English education would provide the British with opportunities to repay the "ancient debt of civilization" which Europe owed to Asia. He was thrilled by the prospect of sciences "cradled in the East and brought to maturity in the West" being disseminated among all peoples of the globe. He viewed the scientific and literary revelations that English education would make available to the Indians as agents of regeneration that would arouse in the colonized peoples a "desire for improvement" and lead them to abandon the "tradition of servility" and the "prostration of the mind" which he believed had been characteristic of Asiatic peoples.[30]

In the decades after the enactment of Bentinck's Resolution, the furtherance of English education came to be regarded as one of the key objectives of Britain's imperial mission in India. When the Home Government wavered following Bentinck's departure from India, fearing that his decision might drive the holy men and pundits of India to foment rebellion, his successor, Lord Auckland, vigorously defended Bentinck's policies. Auckland reminded the Directors in London that only government-supported English education could satisfy the growing Indian demand for instruction in European literature and Western sciences.[31] In the 1840s the dissemination of Western scientific knowledge remained one of the chief rationales for the extension of English education in India.[32] Missionaries as outspoken as Alexander Duff predicted that English education would lead to the "demolition" of Indian "idolatry and superstition." It is significant that Duff regarded the "truths of Western science and history, not Christian revelation," as the main weapons with which Indian errors and falsehoods would be combatted.[33]

Even Orientalists such as H. H. Wilson admitted to the benefits of English education, stressing the increasing facility it offered in teaching Western sciences to Indian students. Wilson felt that Indian youths were

[30]C. Trevelyan, Education of the Indian People, pp. 168, 194.

[31]Ballhatchet, "Bentinck's Education Policy," pp. 224–27.

[32]F. Boutras, "Enquiry into the System of Education Most Likely to Be Generally Popular in Bihar and the United Provinces," in Government of India, Selections from Educational Records, pt. 2, 1840–59 (Calcutta, 1922), extracts 6 and 10.

[33]Duff, New Era, pp. 38, 40.

able to gain a command of Western literature and science that was "rarely equalled by [the students of] any schools in Europe itself."[34] More than a decade later George Campbell, who subsequently rose in the Indian Civil Service to the rank of Lieutenant-Governor of Bengal, commented on the fine intellect of Indian students. He was particularly impressed by their skills in mathematics and their aptitude for "figures and the exact sciences," which he confessed were not typical of Englishmen.[35]

The champions of English education in India had proceeded on the assumption that there were no innate intellectual obstacles to the dissemination of Western scientific learning among the peoples of the subcontinent. They also shared the conviction, so rudely enunciated by Macaulay, that there was little worthwhile in the Indians' own scientific and mathematical treatises. A number of officials and writers, especially those belonging to the shrinking Orientalist faction, disagreed with this latter view. None challenged it as effectively as Lancelot Wilkinson, a Sanskrit scholar and the Assistant Resident in the princely state of Bhopal. After an intense study of Indian scientific texts, Wilkinson concluded that there was a great deal of value in some of them, particularly the *Siddhantas*, and much that was in accord with Western scientific thought. He proposed that these texts, printed in the Indian vernaculars, be used to introduce Indian students at an early age to the "first principles" of scientific learning in a variety of fields from astronomy to trigonometry.

Wilkinson believed that Indian learning could carry instruction in science to the point where Copernicus and Galileo revolutionized it and thus lead Indian students on to the study of the "pure and unadulterated truths" of Western science. He also saw the use of the *Siddhantas* and other Indian texts as a way of interesting Indian priests and scholars in Western scientific learning and, through them, teachers reaching large numbers of Indian students. Wilkinson's students at the Sehore School, where he had put his approach to the test, convinced him that young Indians were "eager" and "prepared" for the serious study of Western sciences and mathematics.[36] One student in particular had a "wonderful talent" for mathematics; in a remarkable display of pedagogic modesty, Wilkinson confessed that the boy was able to work algebraic problems

[34]Quoted in J. R. Covin, "Note on Education," in Government of India, *Selections from Educational Records*, pt. 1, *1781–1839* (Calcutta, 1920), p. 171.

[35]George Campbell, *Modern India* (London, 1852), pp. 59–60.

[36]Wilkinson noted that an approach similar to his had met with equal success in Ceylon; see "On the Use of the Siddhantas in the Work of Native Education," *Journal of the Asiatic Society of Bengal* 34/3 (1834), 511.

that he himself could not solve. He also noted the youth's interest in chemistry and averred that with proper instruction he could become another Hippocrates.[37]

Wilkinson's approach to scientific education was commended by a number of Company officials,[38] and it helped to sustain the Orientalists' appreciation for Indian scientific learning in an age in which the British increasingly denigrated Indian thought and customs. The synthesis of Indian and Western knowledge that Wilkinson envisioned served as a model, in the years after the great Indian rebellion of 1857–58, for those who advocated a retreat from the policies of sweeping reform and the wholesale substitution of European for Indian learning which had been pursued since the 1830s.[39] Though Wilkinson, like the Orientalists more generally, clearly considered Western scientific attainments superior to Indian, his willingness to acknowledge and make use of Indian discoveries exemplified the spirit that made possible major British contributions to the late nineteenth-century rediscovery of ancient Indian civilizations.

Beginning in the 1840s, growing emphasis was given to English-language instruction in the applied sciences and technical training. A chair in engineering and the natural sciences was established at the medical school at Calcutta in 1844. By 1856 a separate college of engineering had been founded. Open to Indians, Eurasians, and Europeans, it was designed to produce civil engineers proficient in Western surveying techniques and architectual design. The "Education Dispatch of 1854" explicitly linked the "advance of European knowledge" in India to the economic development of the subcontinent and its growing potential as an outlet for British manufactured goods and source of raw materials for British industry. The authors of the Dispatch predicted that English education would "teach the natives of India the marvelous results of the employment of labor and capital" and "rouse them to emulate us in the development of the vast resources of their country."[40]

The growing awareness of the scientific and technological backwardness of China in the era of industrialization did little to diminish the

---

[37]Ibid., pp. 504–19, passim., and *A Brief Notice of the Opinions of the Late Mr. Lancelot Wilkinson* (London, 1853), esp. pp. 7–10.

[38]Covin, "Note on Education," pp. 174–75.

[39]Arthur Mayhew, *The Education of India, 1835–1920* (London, 1926), p. 26. The principal of Benares College, Dr. Ballantyne, established a course of scientific instruction similar to Wilkinson's in the middle of the century. See Donald McLeod, "Speech to the Native Nobility of Lahore and Umritsur," in *Short Essays and Reviews on the Educational Policy of the Government of India* (Calcutta, 1866), pp. iii–iv.

[40]Government of India, *Selections*, pt. 2, p. 365 (quoted portions), 338–40.

confidence of most European observers that the Chinese were capable of adopting, if they chose to do so, the technology and scientific ideas of the West. The majority view that China had once been a great civilization persisted throughout the nineteenth century. Evidence of the technological accomplishments of that civilization—its bridges and canals, sailing vessels and water mills—served as reminders of the great feats of invention and engineering that the Chinese had once performed and could again achieve. As we have seen, most Western observers linked China's stagnation to the policies of a succession of despotic regimes and the Chinese people's excessive reverence for past precedents and established custom.

Though a number of writers in the mid-nineteenth century suggested that the servility and passivity of the Chinese might be racial traits and thus difficult to overcome, most missionaries and diplomats in this period emphasized Chinese xenophobia, not innate intellectual deficiencies, as the major barrier to Chinese borrowing. The three missionaries who produced the brief 1838 essay "Intellectual Character of the Chinese," for example, declared that without exception the "natural [mental] endowment" of the Chinese was equal to that of "any other people on earth."[41] An anonymous essay published a half-decade earlier had deplored the cultural arrogance of the Chinese because it had blocked the introduction of Western learning. But the author insisted that the potential for spreading the science and technology of the West was greater in China than in any other Asiatic society. He noted that the Chinese already had a substantial scientific tradition to build upon, a large number of literate individuals, and a highly adaptable language. He proposed exposing the Chinese to European inventions and establishing a society to introduce them to the superior scientific knowledge of the West as the most effective way of rousing the Chinese from their lethargy and making them aware of their backwardness.[42]

In the reports of the Medical Missionary Society in the late 1830s and the 1840s, there were repeated references to the need to instruct young Chinese in Western medicine, chemistry, and inductive reasoning and to bring to their attention the practical applications of Western mathematics and mechanics. Members of the society, who earnestly professed their faith in the unity of mankind, foresaw little difficulty in disseminating European learning and technology once Chinese political restrictions had been removed. As early as the 1830s the missionary Peter Parker had worked to turn his brightest students into competent

[41]*Chinese Repository* 7 (1838), 1–2.
[42]Ibid., 3 (1834), 508–9.

medical doctors, and his efforts were periodically though haphazardly taken up by other missionaries throughout the middle decades of the century.[43] Though opportunities for educating the Chinese were greatly restricted until after the British and French military incursions of the late 1850s, missionary opinion in general was reflected in the views of Evariste Huc, one of the most respected and widely quoted visitors to China in the mid-nineteenth century. Noting the abundance of China's resources and the intelligence and industriousness of its people, Huc concluded that the Chinese "nation" would industrialize rapidly once its leaders committed it to the acquisition and use of Western ideas and machines.[44]

Except for enclaves such as Sierre Leone, where freed slaves were settled under European sponsorship, British or French control over the education of Africans in the first decades of the nineteenth century was almost as limited as their access to the Chinese. The available evidence, however, suggests that both missionary and official opinion regarding the intellectual capacity of the Africans was only marginally, if at all, affected by the thinking of those who sought to demonstrate racial or biological reasons for the perceived inferiority of the Africans. The predominant view among administrators and missionaries at work in the coastal enclaves where Europeans exercised some control was rooted in the Enlightenment assumption that all branches of the human family possessed equivalent mental and physical faculties and were thus inherently improvable. Of course, the "imperial humanitarians" were certain that the Africans were decidedly inferior to the Europeans, but they stressed cultural rather than biological deficiencies in attempting to explain African backwardness.[45] Therefore, though missionaries such as John Philip believed that "the speculations of science" and the "pursuits of literature" were "above the comprehension" of the "untutored savage" and accordingly stressed religious education, they generally concurred with Philip's view that it was not the quality of the African's mind that created the obstructions but rather the "objects it is used to contemplating."[46]

[43]Ibid., 7 (1838), 37–44; 11 (1842), 338–39; 14 (1845), 476–77; and Peter Buck, *American Science and Modern China* (Cambridge, Eng., 1980), p. 33.

[44]E. R. Huc, *The Chinese Empire* (London, 1855), vol. 1, pp. 410–11.

[45]See, for examples, T. J. Barron, "James Stephen, the 'Black Race' and British Colonial Administration, 1813–47," *Journal of Imperial and Commonwealth Studies* 5/2 (1977), 139; Philip Curtin, *The Image of Africa* (Madison, Wis., 1964), pp. 138–39; S. M. Golberry, *Travels in Africa* (London, 1803), vol. 1, pp. 244–46, and vol. 2, pp. 371–72; and Edward Bowdich, *Mission from Cape Castle to Ashanti* (London, 1819), pp. 7–9.

[46]John Philip, *Researches in South Africa* (London, 1828), vol. 2, pp. 356–57. For similar conclusions among French officials, see Georges Hardy, *L'enseignement au Sénégal de 1817 à 1854* (Paris, 1920), p. 57.

Perhaps the fullest insights in this period into European estimates of the African potential for adopting Western technology and for education in European sciences can be found in the views expressed by French colonial officials and educators with regard to the proper instruction for Africans living in the French commercial enclaves of Gorée and St. Louis in Senegal. In contrast to the deliberations over educational policy in India, the role of the colonial administration and the language of instruction were not central to the policy disputes in Senegal. In the first half of the nineteenth century, virtually all administrators and missionaries in areas colonized by France adhered to the policy of assimilation. This approach rested on two convictions that were somewhat contradictory: the belief that all human groups have similar intellectual capacities, which we have seen was dominant in the Enlightenment and Revolutionary eras; and a strong sense of French cultural superiority over Africans or Asians. In the towns of St. Louis and Gorée, where the doctrines of assimilation in their various guises were most fully applied, it was assumed that young Africans who attended French schools could be transformed into full citizens of France through their mastery of the French language, French literature, and other aspects of French culture.[47]

Disagreement over educational policy in this period focused on developing a curriculum appropriate for African students. The place of science and technical education in that curriculum was a major issue through much of the early nineteenth century. In fact, the struggle between those who promoted scientific learning for Africans and those who opposed it—the latter insisting upon an emphasis on religious and moral instruction—reveals a great deal about French attitudes toward African peoples they had long considered among the most advanced on the continent. Like the efforts to spread Western science and technology in India and China, the deliberations and decisions of the French with regard to Senegal also illustrate the major ways in which perceptions formed in earlier centuries continued to shape varying European views and policies toward different colonized peoples.

From the beginning of the debate over educational policy in Senegal, the question of scientific instruction was inextricably bound up with the broader issue of the level at which education in French schools for Africans ought to be pitched.[48] The initial view, espoused by Jean Dard,

[47]For a general history of assimilation in Senegal, see Michael Crowder, *Senegal: A Study of French Assimilation Policy* (London, 1967); and for approaches to assimilation in the empire as a whole, see the citations in n. 149 below.

[48]This account is based primarily upon Hardy, *L'enseignement au Sénégal,* and Joseph Gaucher, *Les débuts de l'enseignement en Afrique francophone: Jean Dard et l'École Mutuelle de Saint-Louis du Sénégal* (Paris, 1968).

headmaster of the École Mutuelle de Saint-Louis from 1816 to 1819, was highly favorable to the Africans. Dard, who strove to learn enough of one of the local African languages (Wolof) to speak it fluently in the classroom, was extremely optimistic about the ability of Senegalese students to respond to his self-appointed mission to sow the first seeds of the exact sciences—"les premiers germes des sciences exactes"—in Africa. He commented on the excellent memory of the Senegalese and their capacity for the study of nature, once they had been instructed in its general laws. He reported that his students were particularly fond of mathematics, botany, agriculture, and mineralogy. He was persuaded by his successes that Africans would prove able students in all areas of Western learning.[49]

In the years after his departure from Senegal, Dard's effusive estimate of his students' achievements was challenged by both French officials and his pedagogical rivals. Most of his critics charged that Dard's students had not even mastered French, much less the many sciences he claimed to have taught them. As a result, most of the French were much less sanguine about the Africans' capacity for scientific instruction than Dard had been. Though his techniques were strongly defended by the Baron Jacques Roger, Governor of Senegal in the mid-1820s, both of Roger's predecessors (Fleuriau and Lecoupé) sided with Dard's critics, who claimed that he had greatly inflated both his own and his students' achievements. Roger's successor as governor, H. B. Gerbidon, concurred with the majority view. Proceeding on the assumption that the Africans' intelligence was "very limited," he recommended that French educational efforts be confined to elementary instruction. But none of Dard's critics stressed intrinsic racial differences as the cause of the Africans' poor showing. Rather they saw the enervating climate and centuries of barbarism as the sources of the great obstacles to the African acquisition of Western learning—obstacles that they felt could only gradually be overcome. Their views were more or less ratified by an 1829 educational commission that advised French teachers to concentrate on the French language and to attempt only limited instruction in geography and arithmetic for even the best students.[50] Despite the intermittent efforts of some officials to introduce classes in applied mathematics for the sons of the African elite in schools in both St. Louis and Gorée,[51] instruction in mathematics and the sciences remained rudimentary for all but a handful of especially gifted youths who were sent to Europe for their higher education.[52]

[49]Gaucher, *Les débuts*, pp. 54 (quoted portion), 58–59, 87–89.
[50]Hardy, *L'enseignement au Sénégal*, pp. 13–15, 23–24; and Gaucher, *Les débuts*, pp. 54–55, 79–85, 102–111.
[51]Hardy, *L'enseignement au Sénégal*, pp. 26–27, 31–32.
[52]Ibid., pp. 37, 39ff.; Gaucher, *Les débuts*, pp. 94ff., 104.

From the late 1830s onward, official opinion turned decidedly against both serious scientific instruction and higher education more generally. Citing the failure of African students to learn anything beyond the speaking and writing of French, Governor Soret requested that the Frères de Ploërmel (a teaching order) assume responsibility for education in Senegal. As their choice of instructors indicates, Soret and his successors favored religious and moral education for peoples who were at an "infant" stage of development and who had no sense of progress and little need for scientific knowledge.[53] Thereafter, increasing emphasis was placed on teaching Africans the importance of manual labor in order to enhance their moral and material well-being. Education was viewed as a means for overcoming African indolence and disdain for hard work. As scientific instruction was downgraded, training in handicraft skills, both African and European, was promoted on a growing scale. Colonial officials and religious instructors agreed that the main aims of education for Africans, beyond moral improvement, ought to be the inculcation of the European work ethic and training in carpentry, weaving, masonry, and ironworking.

In the early 1840s this view was challenged briefly by the Abbé Boilat, a mulatto priest from Senegal. Boilat, a strong advocate of the assimilationist approach to French rule in Africa, worked for the establishment of a secondary school at St. Louis where Africans could pursue advanced studies in literature and the sciences. He was convinced—in part by the success of Anne-Marie Jahouvey in educating African schoolboys in France—that African youths, both male and female, possessed the intelligence to master European science and French culture and would do so if opportunities for education beyond the elementary level were open to them. He envisioned the founding of a college in St. Louis that would eventually produce African doctors, pharmacists, magistrates, and military officers, who could assist the French in opening up the interior of Senegal.[54] But Boilat's schemes prompted a strong reaction from the Frères de Ploërmel, who were angered by the loss of their best students and Boilat's challenges to their approach to African education. Boilat's vision did not long survive his departure from Senegal in the mid-1840s; his *lycée* was absorbed by the Ploërmels' institution in 1849.

After that date, despite the great impact of the policies of the dynamic Governor Louis Faidherbe, who was a strong partisan of racial equality and a believer in the perfectibility of the Africans,[55] industrial arts train-

[53]Gaucher, *Les débuts,* pp. 58ff.

[54]Abbé Boilat, *Esquisses sénégalaises* (Paris, 1853), pp. 9–13, 478; Hardy, *L'enseignement au Sénégal,* pp. 62–65; and André Villard, *Histoire du Sénégal* (Dakar, 1943), p. 98.

[55]Villard, *Histoire du Sénégal,* pp. 126–28; Georges Hardy, *Faidherbe* (Paris, 1947), pp. 84–87.

ing and preparation for manual labor dominated both official and missionary thinking on education in Senegal. Though small numbers of carefully selected African students continued to be sent to France for advanced study, views similar to those of the much-quoted traveler Anne Raffenel prevailed in the colony. Raffenel favored the elementary education of large numbers over higher education for a privileged elite, believing that well-run schools would instill in the colonized population an enthusiasm for hard work. She was also a strong advocate of the demonstration effect: if African youths were made aware of "all the marvels of [France's] clever industry" and shown the "beautiful and ingenious" machines that had produced them, they would readily acknowledge the Europeans' superiority and seek to emulate their colonial masters.[56]

From the 1840s on, officials responsible for "native" policy in Britain's colonial enclaves in West Africa more and more strongly insisted on educational policies similar to those of the French, in which industrial training was favored over literary learning. These demands were in part a response by British officials and missionaries to the perceived arrogance of English-educated Africans, but they were also fed by a growing acceptance of the belief that Africans were inherently incapable of mastering subjects suitable for European students. In the late 1840s a special committee, working from the assumption that African intelligence differed in kind from that of Europeans, endorsed an educational program aimed at impressing African students with the virtues of hard work and self-discipline and instructing them in Christian teachings, European agricultural techniques (for boys), and home economics (for girls).[57] Thus, in roughly the same time period both French and British missionaries and officials moved away from literary and all but the most elementary scientific and mathematical instruction for their African subjects. Though in the following decades English- and French-educated Africans successfully resisted measures aimed at substituting the teaching of "practical skills" for book learning, the new orientation and the racist assumptions on which it was increasingly based soon dominated educational policy-making in colonial Africa.

In view of the great stress in the existing literature on the roles of scientific racism in shaping European attitudes toward African and Asian peoples in the nineteenth century, it is remarkable how little the British and French colonial policies in the first half of the century were

---

[56]Anne Raffenel, *Nouveau voyage dans les pays des nègres* (Paris, 1856), vol. 2, p. 214.
[57]Curtin, *Image of Africa*, pp. 425–28.

influenced by the ideas associated with such writers as Edward Long, Charles White, Julien Virey, George Combe, and William Lawrence. In India from the time of Bentinck's governor-generalship to that of Dalhousie, initiatives taken by British officials in fields as varied as public works and education were premised on the assumption that the colonized peoples were capable of mastering even advanced Western learning and adopting Western technology. It is noteworthy that these policies were pursued in an era when British social exclusivism vis-à-vis the Indians was increasing dramatically. As numerous studies have shown, this exclusivism was frequently buttressed by notions of racial superiority which were rooted in perceptions of physical and cultural difference, as well as by British responses to the growing British political hegemony in the subcontinent, the shifting composition of the British community in India, and reactions to social changes in Britain itself.[58] These patterns suggest that there is no necessary correspondence between the development of ideological racism and racism at the level of social interaction. The fact that there is little evidence of a linkage between racist theories claiming scientific backing and the growth of social racism in India in this era also indicates that popular racism can arise with little or no validation from the writings of social theorists and other intellectuals.

Though they could do little to promote the introduction of Western technology or shape the curriculum of Chinese schools in this period, Europeans who wrote on policies toward China considered the Chinese equal to the Indians in their capacity for acquiring Western technology and scientific learning. Racist slurs about Chinese cowardice and entrenched conservatism, which were becoming more and more common in the popular press and European writings in the early nineteenth century, had little effect on the high regard of European missionaries and officials for the Chinese potential for scientific inquiry and mechanical innovation. At least until the 1840s, doubts that French or British officials expressed about the Africans' aptitudes for these pursuits resulted less from the influence of the writings of scientific racists than the low estimate of African abilities that centuries of enslavement and its concomitant humiliations had impressed upon European observers. An image of African culture in general, and particularly its material culture, as primitive and stationary reinforced these doubts and led European colonial officials and missionaries to stress rudimentary technical training,

---

[58]See Percival Spear, *The Nabobs* (London, 1963); Philip Mason, *Prospero's Magic* (London, 1962); Kenneth Ballhatchet, *Race, Sex, and Class under the Raj* (New York, 1980); and Francis G. Hutchins, *The Illusion of Permanence* (Princeton, N.J., 1967).

the inculcation of the work ethic, and (ironically) the revival of indigenous craft skills in their educational programs.

The Search for Scientific and Technological Proofs of Racial Inequality

In the eighteenth century the effort to demonstrate the scientific validity of the belief that there were innate differences in mental and moral
capacity between racial groups had been confined to small circles of
physicians and essayists. From the early 1800s, it grew into one of the
central preoccupations of European and North American intellectuals.
A substantial literature has developed on the decisive shift in this period
from skin color to skull configuration as the key determinant of racial
identity and the main gauge of racial potential.[59] But both science and
technology played other roles in nineteenth-century attempts to delineate racial boundaries and to provide empirical support for the belief that
European peoples were intrinsically superior to all others. Scientific and
technological achievements were increasingly cited as evidence of racial
abilities or racial ineptitude. These were used as key criteria by which to
rank different racial groups in the hierarchies of civilized, barbarian, and
savage peoples which nineteenth-century thinkers were so fond of constructing. The ability of African or Asian peoples to use Western tools
or firearms was also frequently cited as evidence of their potential as
races for improvement. Though the ways in which gauges based on
scientific or technological accomplishment were used remained fairly
constant, the frame of reference for arguments about racial capacity
changed significantly between the first half and the last decades of the
century. In the earlier period, a static view of racial divisions and attributes prevailed; from the 1860s onward a more fluid, evolutionary
view of racial development gained increasing acceptance both in intellectual circles in Europe and among colonial policymakers and
educators.

As the focus of racial thinking shifted from eighteenth-century speculation about the origins of racial differentiation to an extended debate

---

[59]Among the best recent works are Curtin, *Image of Africa*, chaps. 9 and 15; William B.
Cohen, *The French Encounter with Africans* (Bloomington, Ind., 1980), chap. 8; William
Stanton, *The Leopard's Spots: Scientific Attitudes towards Race in America, 1815–59* (Chicago, 1960); John S. Haller, *Outcasts from Evolution: Scientific Attitudes of Racial Inferiority,
1859–1900* (New York, 1971); George Fredrickson, *The Black Image in the White Mind*
(New York, 1971), chap. 3; Nancy Stepan, *The Idea of Race in Science: Great Britain,
1800–1960* (Hamden, Conn., 1982), chaps. 1–4; and Douglas Lorimer, *Colour, Class, and
the Victorians: English Attitudes to the Negro in the Mid-Nineteenth Century* (Birmingham,
Eng., 1978), chap. 4.

in the early nineteenth century over the consequences of racial differences, the efforts of such physicians as White and Soemmering to quantify physical distinctions between human groups came to dominate scientific thinking on the issue of race. Though measurements of arm length or genital size continued to be made,[60] the skull became the focus of investigation. Technology, in the guise of instruments from simple calipers to Gratton's craniometer, was enlisted in the search for an accurate technique of skull measurement and comparison.[61] Some investigators stressed the need to measure cranial capacity (Samuel Morton used white peppers and then lead gunshot; Friedrich Tiedemann used millet seed); others such as Anders Retzius and George Combe insisted that what mattered was the shape of the head or the proportions of different parts of the skull or brain. The usefulness of Camper's facial angle was debated, and it was gradually replaced by a bewildering variety of new phrenological gauges, including the cephalic index, the nose index, the vertical index, and the cephalo-orbital index. Thousands and then tens of thousands of skulls were measured in innumerable ways. The obsession with craniometry culminated at century's end with A. von Torok's thousands of measurements of a single skull.[62] As Stephen Gould has observed, few endeavors illustrate as well as phrenology and craniology the "allure of numbers" for nineteenth-century practitioners of the human sciences and their "faith that rigorous measurement could guarantee irrefutable precision, and might mark the transition between subjective speculation and a true science as worthy as Newtonian physics."[63]

Despite vigorous challenges throughout the century,[64] the enduring influence of phrenology and craniometry rested on the assumption, first fully elaborated in George Combe's *System of Phrenology*, that a correspondence between intelligence and skull shape or brain size could be

[60]For examples, see Charles Caldwell, *Thoughts on the Original Unity of the Human Race* (New York, 1830), pp. 74–91; and Cohen, *French Encounter*, pp. 231–32.

[61]For brief discussions of the nineteenth-century fascination with cranial measurements, see Curtin, *Image of Africa*, pp. 40, 46–47, 234–36, 366–69; Cohen, *French Encounter*, pp. 224–32; and Stephen Jay Gould, *The Mismeasure of Man* (New York, 1981), chaps. 2 and 3. For more extended considerations, see R. J. Cooter, "Phrenology: The Provocation of Progress," *History of Science* 14 (1976), 211–34; Francis Hedderly, *Phrenology: A Study of Mind* (London, 1970); and the contemporary work by Paul Topinard, *L'Anthropologie* (Paris, 1876), esp. pt. 2, chaps. 2 and 3.

[62]Earl W. Count, "The Revolution of the Race Idea in Modern Western Culture during the Period of the Pre-Darwinian Nineteenth Century," *Transactions of the New York Academy of Sciences*, 2d ser. 8 (1946), 151–52.

[63]Gould, *Mismeasure of Man*, pp. 73–74.

[64]In addition to the works by Firmin, Waitz, and Quatrefages (n. 4 above), see Thomas Winterbottom, *An Account of the Native Africans in the Neighborhood of Sierre Leone* (London, 1803), pp. 198–99; and "A System of Phrenology," *Edinburgh Review* 44 (1826), 253–318.

scientifically demonstrated for entire racial groups.[65] Some writers also sought to link cranial capacity to the temperament or moral development of various races. The American physician Samuel Morton, for example, one of the most prominent craniologists of the century, characterized the Hindus on the basis of skull measurements as "mild, sober and industrious" but "prone to fantastic religions," and the Turks as violent, passionate, cruel, and vindictive. Victor de l'Isle opined that the different races could be divided into conquerors and slaves on the basis of head shape and size, while Georges Cuvier suggested that there was a connection between the "beautiful form" of the Caucasians' heads and the dominion that they had historically exerted over most of the rest of humankind.[66] The fact that a priori conclusions had been reached about the most desirable sort of skull is made evident by descriptions of the Caucasian type, which are invariably exercises in European or American self-adulation. It is even more disturbingly apparent in the very unscientific efforts of craniologists such as Morton and Paul Broca to ignore or explain away anomalies in their data and outright contradictions between their data and conclusions.[67]

No aspect of what nineteenth-century Europeans considered the scientific study of human types had a greater impact on popular attitudes than phrenology. It left its mark on the fiction of this period, from the bump on Père Goriot's skull, indicating that he would make a good father (Balzac, 1834), to Dr. Mortimer's surprise at finding that Sherlock Holmes had a dolichocephalic skull with "such well-marked supraorbital development" (Arthur Conan Doyle, *The Hound of the Baskervilles*, 1902). As Walter Houghton has observed, phrenology was a subject on which English gentlemen were expected to have an opinion; at midcentury educated workers displayed an avid interest in George Combe's writings and his *Phrenological Journal*; and by the last decades of the century, phrenology had become a marketable object of mass culture as evidenced by the booths for "head reading" found at British seaside resorts.[68] In France an opponent of phrenology confessed in the

[65]George Combe, *A System of Phrenology* (Edinburgh, 1825). For sample expressions of this point of view at various points in the century, see B. G. E. Laville C. de Lacépède, *Histoire naturelle de l'homme* (Paris, 1827), p. 249; Charles Smith, *The Natural History of Man* (London, 1848), pp. 159–60; Robert Dunn, "Some Observations on the Psychological Differences Which Exist among the Typical Races of Man," *T.E.S.L.* 3 (1865), pp. 9–25; and Alfred Russel Wallace, *The Wonderful Century* (London, 1899), pp. 160, 193.

[66]Samuel Morton, *Crania Americana* (Philadelphia, 1839), pp. 32–33, 43; Victor Courtet de l'Isle, *Tableau ethnographique du genre humain* (Paris, 1849), p. 12; and Cuvier, *Animal Kingdom*, p. 97.

[67]For detailed examinations of these practices see Gould, *Mismeasure of Man*, chaps. 2 and 3.

[68]Walter Houghton, *The Victorian Frame of Mind* (New Haven, Conn. 1957), p. 103; G. M. Young, *Victorian England: Portrait of an Age* (New York, 1964), p. 36; Rowland Ryder, *Edith Cavell* (London, 1975), pp. 12–13.

1840s that its doctrines had "invaded everything," a claim given credence by Pierre Proudhon's dismissal of his rival François Fourier's ideas
in part because of the small size of Fourier's skull and his "mediocre
cerebral development."[69] Nor was the impact of phrenology restricted
to fiction and intellectual quarrels. It exerted some influence on the
course of events from the mid-nineteenth century onward, an influence
that has never been fully explored. Charles Darwin recounted, for example, that he was nearly denied a place on the *Beagle* because Captain
Fitzroy thought anyone with a nose like his could not possibly "possess
sufficient energy and determination for the voyage."[70] Nearly a century
later British Prime Minister David Lloyd George, a firm believer in
phrenology, was won over to the support of the plans of the French
General Robert Nivelle for the allied offensives in the spring of 1917 on
the Western Front because in addition to being a persuasive speaker and
fluent in English, Nivelle had the sort of bumps on his head that were
"deserving of every confidence."[71] But apparently it was not the sort
that ensured successful strategic planning: Nivelle's offensive led to the
disaster known as the second battle of the Aisne, which cost the French
nearly 190,000 casualties.

References to skull shape and type are abundant in the travel literature
on Africa and Asia in the late nineteenth century. Unlike the old doctor
in Conrad's *Heart of Darkness* who "in the interest of science" measured
the skulls of all Europeans going out to serve as company officials in the
Congo, the focus of the craniological investigations of most travelers
and missionaries, and occasionally colonial administrators, was "native"
heads. David Livingstone and Lord Lugard simply remarked on the
finely shaped skulls of some peoples they encountered and the degraded
crania of others.[72] Other observers collected skulls and devised elaborate
tables that ranked different African or Asian peoples on the basis of skull
size or shape.[73] J. D. Ball, a British civil servant in Hong Kong, based a
detailed discussion of Chinese character on phrenological charts worked
out in Europe. Decades earlier the journalist George Cooke had sought

[69]Cohen, *French Encounter*, p. 225.

[70]*The Life and Letters of Charles Darwin*, ed. Francis Darwin (London, 1887), vol. 1, p.
50.

[71]John Terraine, *The Western Front, 1914–1918* (London, 1964), p. 72. Perhaps Lloyd
George had heard of Gustave Le Bon's well-publicized measurements of the heads of the
Generals Wurmser and Jourdan: Le Bon found that the consistently victorious Jourdan
had the larger brain. See Robert Nye, *The Origins of Crowd Psychology* (Beverly Hills,
Calif., 1965), p. 33.

[72]David Livingstone, *Missionary Travels and Researches in South Africa* (London, 1858),
pp. 55, 109, 315, 486, 587; and F. D. Lugard, *The Rise of Our East African Empire*
(London, 1983), vol. 1, pp. 326–27.

[73]For examples, see Paul Du Chaillu, *A Journey to Ashango-Land* (New York, 1867),
pp. 285, 318, and App. I; and G. Schweinfurth, *The Heart of Africa: Travels, 1868–72*
(London, 1873), pp. 88–89.

to give the readers of *The Times* insights into the personality and motives of the Chinese leader Ye Ming-chen on the basis of a careful analysis of Ye's skull.[74] Phrenological and craniometric evidence both bolstered the Europeans' conviction that they were destined to be masters of humankind and served as a means of reminding bright or uppity African and Asian subordinates of their proper place in the larger scheme of things. Oginga Odinga recalls the remark of his British superior that though Odinga was "very intelligent" for an African, his brain was no better than the brain of the supervisor's six-year-old son.[75]

Phrenological findings were well suited to the static approach to the study of human types that dominated racial thinking in the first half of the nineteenth century. Emphasis on the structure of the skull or configuration of the brain lent a sense of permanence to discussions of racial differences. The tendency to regard racial attributes as fixed in time was partly the result of the transference of the eighteenth-century belief that species were "immutable prototypes fashioned 'in the beginning' by an all-wise Creator and perfectly adapted to their role in the divine economy of nature."[76] This position was quite consistent with the views of those who supported the polygenist theory that each race had originated in a separate creation and that races were equivalent to species. F. W. Farrar forcefully summarized this stance in an 1866 essay: the different races, he declared, "have always been as distinct as they now are, and . . . it is impossible for their limits to be confused either by degeneracy on the one hand or progress on the other."[77] Though present-day biological thinking would lead one to assume that those who adhered to the theory of monogenesis (and thus believed that all human races were the product of a common creation and represented varieties of the same species) would treat racial differences as fluid and changeable, this was often not the case. In this period many of those who adhered to the monogenist view also wrote of racial characteristics as if they had been fixed very early in time.[78]

[74]George W. Cooke, *China* (London, 1858), p. 397; and J. Dyer Ball, *Things Chinese* (London, 1892), pp. 73–75.

[75]Oginga Odinga, *Not Yet Uhuru* (New York, 1967), p. 58.

[76]John C. Greene, "Some Early Speculations on the Origin of Human Races," *American Anthropologist* 56 (1954), 31. See also Count, "Race Idea," pp. 139–41; and Gladys Bryson, *Man and Society: The Scottish Inquiry of the Eighteenth Century* (Princeton, N.J., 1945), pp. 58, 77.

[77]Fredric Farrar, "Aptitudes of Races," *T.E.S.L.* 5 (1867), p. 116. See also John Crawfurd, "On the Physical and Mental Characteristics of the European and Asiatic Races of Man," in ibid., p. 81; and J. C. Nott and George Gliddon, *Types of Mankind* (London, 1854), p. 411.

[78]See, e.g., the following discussion of the views of William Lawrence, one of the most respected advocates of monogenesis in the early nineteenth century; and Stepan, *Idea of Race,* pp. 38–40.

Both phrenology and the predominance of a static view of racial differentiation shaped the additional ways in which European scientific and technological superiority reinforced racist thinking in the first half of the nineteenth century. European writers frequently assumed a causal relationship between skull shape or brain size and the past achievements of non-Western peoples in science and invention. Speculation about the future potential of these peoples was in turn influenced by the assumption that there were innate limits to what different racial groups could be taught and to their capacity for invention and experiment.

As early as the second decade of the century, these connections were made in a series of lectures delivered by William Lawrence at the Royal College of Surgeons. Lawrence, a physician, was a prominent advocate of the monogenist viewpoint and an outspoken foe of slavery. He included in his lectures a series of vignettes of such "Negroes" as the meteorologist Lislet and the mathematician Fuller, whose lives he believed proved Africans capable of acquiring Western learning. He also commended the Africans for the "native ingenuity" they had shown in textile production and cited approvingly the explorer Barbot's comments on their "quick and accurate minds" and the "extraordinary strength" of their memories.[79] Yet though Lawrence even conceded that outstanding Africans might surpass average Europeans in talent and intellect, he stressed that he believed Africans as a race to be decidedly inferior to Europeans. He mentioned as proof of the Europeans' superiority a number of areas—including government, literature, and the treatment of women—in which they had surpassed all other peoples. But he returned repeatedly to achievements in technology and science as evidence of their superiority. "White men," he asserted at one point, had been responsible for the development of the arts and sciences in Asia as well as Europe.[80]

Lawrence attributed the greater attainments of the Europeans to their innately superior mental and moral endowments:

> If the nobler attributes of man reside in the cerebral hemispheres; if the prerogatives which lift him so much above the brute are satisfactorily accounted for by the superior development of those important parts; the various degrees and kinds of moral feeling and intellectual power may be consistently explained by the numerous and obvious differences of size in the various cerebral parts, besides

[79]Lawrence, *Lectures*, pp. 496–98.

[80]Ibid., pp. 364–65, 484–90, 498–99. Lawrence's contemporary Robert Hamilton viewed the capacity for invention as one of the surest signs of the "superiority" of an individual's understanding; see *The Progress of Society* (London, 1830), p. 19.

which there may be peculiarities of internal organisation, not appreciable by our means of inquiry.[81]

Lawrence suggested that the Africans' inferiority in government and invention was due to the fact that like animals they had highly developed sensory organs but stunted intellects in comparison to Europeans. He intimated that this imbalance helped to explain why Africans spent so much of their time in debauchery and lewd pleasures rather than the pursuit of knowledge or aesthetic appreciation.[82] He felt that some improvement of the Africans was possible but that they could never reach the exalted heights attained by the Europeans. The barriers were hereditary and physical: "The retreating forehead and the depressed vertex of the dark varieties of man make me strongly doubt whether they are susceptible of these high destinies [settled habits, the cultivation of the "useful arts," civilized life];—whether they are capable of fathoming the depths of science; of understanding and appreciating the doctrines and the mysteries of our religion."[83]

In the decades following Lawrence's lectures his arguments were elaborated by a succession of authors seeking to demonstrate the scientific validity of theories of racial inequality. Like Lawrence, most of these authors focused on Africans or black slaves in the Americas. Unlike Lawrence, the majority were polygenists who saw little in African history or culture worthy of praise and who were very pessimistic about the possiblity of educating even the brightest of Africans. In the 1826 edition of his *Histoire Naturelle*, Julien Virey went to great lengths to refute the claims of those whom he labeled "friends of the blacks" that Africans were equal to Europeans in innate intelligence and ability. After dismissing noble savages, African or otherwise, as "ignorant fools,"[84] Virey launched into a detailed comparison of the anatomical similarities between Africans and apes. Quoting Meiners, Camper, and other well-known authorities, Virey argued that scientific investigations had established beyond all doubt that the brain of the African was smaller than that of the European and that the latter was thus unquestionably superior in intelligence. The lower level of the Africans' intelligence, Virey asserted, was conclusively demonstrated by their lack of achievement in any field. Though he mentioned in passing the idolatrous re-

[81]Lawrence, *Lectures*, pp. 499–500.

[82]Ibid., pp. 363, 476.

[83]Ibid., p. 501.

[84]Julien Virey, *Histoire naturelle des races humaines* (Paris, 1826), pp. viii–ix; this view was shared by his more famous contemporary, Saint-Simon. See George W. Stocking, Jr., *Race, Culture, and Evolution* (Chicago, 1982), p. 38.

ligions of the Africans, their failure to devise laws or stable governments, and their lack of artistic creativity, Virey stressed the complete
absence of scientific inquiry and technological innovation in African
culture as proof of black inferiority. Reviving an evaluative standard
from the early centuries of expansion, Virey commented on the Africans' failure to develop monumental architecture and build great cities.
He implied that their contentment with huts of the "most primitive
kind" was symptomatic of their low level of intelligence. He asserted
that they were unable to manufacture even light cotton textiles, and he
characterized them as indolent, sensual, and without discipline. They
were, he concluded, *grands enfants* who lacked the capacity for abstact
thought and were able to conceptualize only material objects. He dismissed examples of intelligent "Negroes" as mere exceptions and concluded that the Africans were incapable of advancing toward a higher
level of culture without European tutelage. And even with the best
instruction imaginable, he clearly believed, the Africans would never
attain a level of civilized development comparable to that of the Europeans.[85]

In the middle decades of the century, prominent advocates of theories
of racial inequality highlighted a number of the arguments that centered
on African and Asian scientific and technological deficiencies. Like
Josiah Nott and a number of other leading exponents of what was styled
"scientific racism," both Samuel Morton and Robert Knox were physicians by profession, and both employed phrenological and craniometric evidence extensively. Morton was sensitive to the great diversity
of the peoples who were categorized as Africans but concluded on the
basis of his widely cited skull measurements that Negroes as a race had
no capacity for invention. He conceded, however, that they had "great
powers of imitation" and consequently readily acquired the mechanical
techniques of the Europeans.[86]

The Scottish anatomist Robert Knox disagreed, contending that like
the other "dark races" the Africans were markedly deficient in "the
generalizing powers of pure reason—the love of perfectibility—the desire to know the unknown—and, last and greatest, the ability to observe
new phenomena and new relations."[87] Clearly, Knox believed that it

[85]See Virey, *Histoire naturelle,* esp. vol. 1, pp. 366–67; vol. 2, pp. 29–60, 98–121. See
also Caldwell, *Original Unity,* p. 136.

[86]Morton, *Crania Americana,* p. 88.

[87]Robert Knox, *The Races of Men: A Philosophical Enquiry into the Influence of Race over
the Destinies of Nations* (London, 1862), p. 287. For a discussion of Knox's background
and the controversies that his work aroused, see Michael D. Biddiss, "The Politics of
Anatomy: Dr. Robert Knox and Victorian Racism," *Proceedings of the Royal Society of
Medicine* 69 (1976), 245–50.

would be difficult for Africans to excel in scientific inquiry or invention. He felt that the same was true for the Asian races, including the Chinese. He adhered to the view that China's early advances in technology were the results of borrowing from another, unspecified, race. The fact that the Chinese had not improved upon these initial gains for millennia was, in Knox's view, proof that they had not been their originators. In his writings the stationary or stagnant quality of Chinese civilization, stressed by European writers since the mid-eighteenth century, became a racial attribute rather than a product of environmental or social conditions.[88]

Ironically, assertions that rigid and permanent differences in mental and physical capacity between the races could be scientifically demonstrated peaked in frequency and certitude in the decade after Darwin published *On the Origin of Species* (1859). This trend is consistent with George Stocking's caution that the belief in polygenesis did not collapse with the publication of Darwin's theory of evolution.[89] In fact, a polygenist stance informed the arguments of James Hunt and John Crawfurd, who stressed the widely varying levels of scientific and technological development attained by different races as a reflection of innate differences in intelligence. Hunt was perhaps the most influential (or notorious) exponent of these views. He had become a disciple of Robert Knox in the 1850s and was one of the leaders of the revolt against the more moderate position on racial issues fostered by the Ethnological Society of London. The revolt led to the the founding of the Anthropological Society of London in 1863, which Hunt and his followers dominated for the remainder of the decade.[90] Hunt focused his racist pronouncements on blacks, citing examples from societies in both the New World and Africa. His views were developed most fully in "On the Negro's Place in Nature," a paper he presented to the Anthropological Society in 1863. Little in the essay was original; it was in fact a compilation of the preceding half-century's main arguments for the inherent inferiority of the "dark races." But Hunt sought to give his arguments scientific respectablity through comparisons of skull measurements and

[88]Knox, *Races of Man*, pp. 283–84. For examples of the wide acceptance of this view, see Nott and Gliddon, *Types of Mankind*, pp. 52–53, 185–89, 423–30; C. Smith, *Natural History of Man*, p. 196; and Haller, *Outcasts from Evolution*, p. 58.

[89]Stocking, *Race, Culture, and Evolution*, pp. 46–47. See also Stepan, *Idea of Race*, pp. 79, 88–93, 104–10.

[90]J. W. Burrow, "Evolution and Anthropology in the 1860s: The Anthropological Society of London, 1863–71," *Victorian Studies* 7 (1963), 137–54; and George W. Stocking, Jr., "What's in a Name? The Origins of the Royal Anthropological Institute (1837–71)," *Man* 6/3 (1971), 369–90.

by frequent references to the works of Morton, Broca, and the other masters of craniometry.

Hunt had disparaging things to say about virtually all aspects of African history and culture and particularly the failure of black peoples to produce either scientists or inventors. He challenged those who argued that the Negro was "equal in intellect" to the European to name one "pure" Negro who had distinguished himself as a "man of science" (listed first) or as a statesman, warrior, or poet. He quoted at length a Colonel Hamilton Smith who charged that the Africans had never devised an alphabet, developed a grammatical language, or "made the least step in science or art." In a footnote, Hunt pointed out that the only scientific achievement ever attributed to the Africans was the creation of an alphabet by a member of the "Vei tribe," but even that had been derived from another race. He acknowledged that some African peoples knew how to work metals and produce fairly serviceable tools and weapons. But he was quick to remind those who might be tempted to interpret this as a sign of civilized accomplishment and African inventiveness that there was strong evidence that in all cases African metalworkers were simply imitating European techniques. He went so far as to argue that African spears, often cited to illustrate African proficiency in metallurgy, were fashioned on old Anglo-Saxon lines. Even the use of cowrie shells for trade, Hunt averred, had its origins in European precedents.[91]

Several years later J. C. Nott, who had done much to popularize theories of Negro inferiority in the United States, echoed Hunt's categorical dismissal of African material achievement. The Africans, Nott wrote, had "no art, no ruined temples, no relic of science or literature." For four thousand years, he asserted, the Negro's intellect had remained "as dark as his skin." The black African peoples had gained nothing from their long contact with Egypt, which had done so much to set other races "on the road to civilization."[92] Similar sentiments were expressed in an article published by F. W. Farrar in England in the same year. The "savage races," Farrar declared, "have not added one iota to the knowledge, the arts, the sciences, the manufactures, the morals of the world, nor out of all their teeming myriads have they produced one single man whose name is of the slightest importance in

[91]James Hunt, "On the Negro's Place in Nature," *Memoirs Read before the Anthropological Society of London* I (1863–64), 30–31, 37–38. The "name a single scientist" challenge had been made as early as 1851; see John Campbell, *Negro-Mania* (Philadelphia, 1851), pp. 8–9.

[92]J. C. Nott, "The Negro Race," *Popular Magazine of Anthropology* 3 (1866), 107–9.

the history of our race."[93] Like Nott, Farrar lamented the inability of the "pure blooded negro" to learn from superior peoples such as the Egyptians, and he went on to discuss their resistance to the improving influences of the Phoenicians, the Dutch, the French, Spaniards, Americans, and Anglo-Saxons. Despite millennia of contact with advanced races, Farrar observed, the Africans had developed "no proficience even in the mechanical arts."[94] In the same year Charles McKay also sought to silence those who opposed his polygenist views by defying them to name a single African who was an engineer, architect, or mathematician.[95]

Other writers in this period expanded the polygenist approach to include Asians. Unlike Knox, Farrar believed the Chinese had themselves created the technology that had allowed them to fashion a civilization in ancient times, but he agreed with Knox that China's development had been "arrested" because of shortcomings inherent in the race. The innate tendency of the Chinese toward conservatism and stagnation was evidenced by, among other things, the backwardness of their technology and their scientific thinking.[96] The famed traveler John Crawfurd also emphasized deficiencies in science and technology to illustrate the contrast between progressive European and decadent Asian civilizations. He traced the divergence in their patterns of development to fundamental differences in the "intellectual and moral qualities of the European and Asiatic races." Crawfurd believed that these differences were far more significant than variations in physical appearance. He saw apathy and despotism, which had been stressed by eighteenth-century critics of India and China, as effects rather than causes. For Crawfurd, a tendency toward despotism, like indifference or hostility to innovation, was inherent in the racial composition of the Asiatic peoples.[97]

In supplying the craniometric evidence that was central to his case for the racial origins of African inferiority, Hunt developed a number of arguments that had been raised in a general way by earlier writers but were elaborated and increasingly emphasized by later proponents of scientific racism. He contended that the Africans' backwardness in sci-

[93]Farrar, "Aptitudes of Races," p. 120.
[94]Ibid., pp. 121–22.
[95]Charles McKay, "The Negro and the Negrophilists" (1866), in Michael D. Biddiss, ed., Images of Race (New York, 1979), p. 104.
[96]Farrar, "Aptitudes of Races," pp. 123–24. For earlier dismissals of Asian achievement, see C. H. Saint-Simon, "Lettres d'un habitant de Genève à ses contemporains" (1803), in Oeuvres (Paris, 1868), vol. 1, pp. 56–57; and Caldwell, Original Unity, p. 136.
[97]John Crawfurd, "On the Conditions Which Favor, Retard, or Obstruct the Early Civilization of Man," T.E.S.L. 1 (1861), 159–61, 165, 169–70; and "Physical and Mental Characteristics," pp. 60–64, 73.

ence and technology, as well as their overall failure to develop civilizations, could be explained not only by the smaller size and unfortunate shape of their skulls but also by the deficient configuration of their brains. Citing authorities on the structure and workings of the brain such as the traveler Paul Du Chaillu, Hunt asserted that the reflective faculties of Africans were decidedly underdeveloped; as a result they had "little power of forethought" and "a total lack of generalization." He conceded that Africans had good memories but reminded his readers that memory was "one of the lowest mental powers." He also admitted that many observers had commented on the quick intelligence displayed by black children. The failure of black adults to display a similar intelligence was due, he suggested, to the "arrested development" of the Negro's brain resulting from the "premature union of the bones of the skull" which accompanied the onset of puberty. Hunt believed that these physical deficiencies were the root cause of African racial inferiority. They accounted for African improvidence, indolence, and immorality. They explained why, despite thousands of years of contact with advanced peoples, African cultures had not advanced beyond the savage state. They made it impossible for the Africans to reach a level of culture comparable to that of the Europeans, even with extensive education and technical training.[98]

The precocious youth–arrested adolescent sequence of African mental development was already an issue for debate by the end of the eighteenth century.[99] The closing-sutures explanation for the stunting of African intelligence at puberty appears to have been first advanced in the 1850s by the French anthropologist Louis Gratiolet.[100] An exchange at the Anthropological Society during a discussion of Hunt's paper indicates that some nineteenth-century observers were skeptical about the notion of arrested development:

> *Mr. [Winwood] Reade*: Mrs. Walker [the wife of an American missionary who had taught African children for twelve years] told me that Negro children were more precocious than the American children, but had not such retentive memories, and that, generally speaking, they came to a state of *in statu quo* about sixteen, and after that forgot all they had learnt.

[98]Hunt, "Negro's Place," pp. 10–11, 27, 30, 36–37, 42–47, 52–53.

[99]As Thomas Winterbottom's vehement dismissal of these arguments suggests; see his 1803 *Sierre Leone*, pp. 216–18.

[100]Stocking, *Race, Culture, and Evolution*, p. 55; and Gould, *Mismeasure of Man*, pp. 83, 98.

Mr. Dingle: I could give you plenty of such instances among our own countrymen.

Mr. Reade: I am only speaking of Negroes.[101]

But despite either sarcasm or more scholarly reservations,[102] the idea of arrested development steadily gained in popularity in the last decades of the nineteenth century and was expounded by physicians, social theorists, and explorers well into the twentieth.[103] Though it was usually applied to Africans, some authors enlisted the precocity–stunted mental growth syndrome in their efforts to explain the stagnation of Asian civilizations. Just as Asians had emerged from savagery and barbarism sooner than Europeans, John Crawfurd and a number of other authors observed, Asian children did as well as or better than European pupils in schools where they were educated together. After the age of puberty, however, the European student quickly surpassed his Asian peer, whose learning abilities either (depending on the author in question) remained stationary or degenerated.[104]

Though racial arguments and visions of closing sutures were, of course, set aside in discussions of the differences in aptitudes between the sexes, a number of nineteenth-century writers on gender issues employed the ideas about variations in brain organization and rational versus sensory development that were common fare in racial comparisons. In 1839, for example, Gustave d'Eichthal, then the secretary of the Société Ethnologique, published a series of exchanges with Ismail Urbain in which he noted the similarities between the mental makeup of Africans and women in contrast to white males. Like blacks, d'Eichthal asserted, women liked dancing and jewelry, and they lacked the sort of intelligence required for politics or scientific investigation. Neither Africans nor women had ever distinguished themselves as astronomers,

---

[101]*Journal of the Anthropological Society of London* 2 (1864), xl.

[102]For one of the earliest critiques, see "Ethnology of the Science of the Races," *Edinburgh Review* 88 (1848), 469. See also Thomas F. Gossett, *Race: The History of an Idea* (New York, 1965), pp. 76–80.

[103]For examples, see Richard Burton, *A Mission to Galele, King of Dahomey* (London, 1864), vol. 2, pp. 178–79 (Burton explicitly cites Gratiolet for scientific validation); Samuel Baker, *The Albert N'yanza, Great Basin of the Nile* (London, 1866), vol. 1, p. 288; Abel Hovelacque, *Les nègres de l'Afrique sus-équatoriale* (Paris, 1889), pp. 425, 458–59; Albert Fouillée, "Le caractère des races humaines et l'avenir de la race blanche," *Revue des Deux Mondes* 124 (1894), 93; A. H. Keane, *Ethnology* (Cambridge, Eng., 1896), p. 266; and A. Cureau, *Les sociétés primitives de l'Afrique équatoriale* (Paris, 1912), pp. 68–72. Léon Fanoudh-Siefer has pointed out that African mental stagnation was also blamed on sexual excess; see *Le mythe du nègre* (Paris, 1968), pp. 178–79.

[104]Crawfurd, "Physical and Mental Characteristics," p. 60. See also G. O. Trevelyan, *The Competition Wallah* (London, 1864), pp. 55–56.

mathematicians, or naturalists, nor had either produced even a competent mechanic. Decades later these judgments continued to be made by even more influential figures such as the Count de Gobineau, Gustave Le Bon, and James Hunt. Their notions about innate differences in mental makeup between the sexes were used to justify the expectation that women, at least middle- and upper-class women, would tend to their domestic responsibilities and frivolous pleasures, leaving politics, science, and invention to those biologically suited to such endeavors—white males.[105]

The notion of stunted intellectual development in non-Western races was paralleled by the less well articulated but far more pervasive nineteenth-century belief that Africans and Asians thought and behaved like children. As the casual references of Virey in 1826 and Edmond Ferry nearly a century later illustrate, it was commonplace to regard the Africans as *grands enfants*; even as sympathetic an observer as the Abbé Boilat felt they had to be "dominated by the force of [the Europeans'] character."[106] In European writings of this period and well into the twentieth century, Africans were depicted as credulous, emotional, impulsive, and lacking in foresight—all qualities that were associated with childhood. A number of writers compared African reactions to those of "cruel schoolboys," and Richard Burton pointed out that African bearers or servants insisted on being treated like the children of white explorers or missionaries.[107] Similar though perhaps less frequent characterizations of Asian peoples as children were scattered through late nineteenth-century works. Indian religious beliefs were deemed mere childish superstitions; Chinese responses to European demands were seen to be as capricious as those of children; Asian potentates were caricatured as overgrown children delighting in pomp, splendor, and parades. Perhaps most significant, because of its impact on political dynamics in colonized areas, late nineteenth-century French and British writers depicted the behavior of Western-educated Africans and Asians as petty and childlike.[108]

---

[105]Gustave d'Eichthal and Ismail Urbain, *Lettres sur la race noire et la race blanche* (Paris, 1839), p. 22; and Cohen, *French Encounter*, pp. 236–37. For later examples, see Gobineau, *Essai sur l'inégalité*, vol. 1, pp. 150–52; Hunt, "Negro's Place," p. 10; Gustave Le Bon, *The Psychology of Peoples* (London, 1899), pp. 35–36; and Courtet de l'Isle, *Tableau ethnographique*, p. 28.

[106]Virey, *Histoire naturelle*, p. 4; Edmond Ferry, *France en Afrique* (Paris, 1905), p. 216; Boilat, *Esquisses Sénégalaises*, p. 124.

[107]Burton, *Mission to Galele*, vol. 1, p. 18, and vol. 2, p. 201; Winwood Reade, *Savage Africa* (London, 1863), p. 552; Hovelacque, *Les nègres de l'Afrique*, pp. 424–25.

[108]For examples, see Louis Figuier, *The Human Race* (London, 1872), p. 342; J. E. Bingham, *Narrative of the Expedition to China* (London, 1842), vol. 1, p. 345; Bolt, *Victorian Attitudes*, pp. 179–80; "Race in History," *Anthropological Review* 3 (1865), 246–

Very often, remarks about the childlike qualities of Africans or Asians were linked to their perceived deficiencies in science and technology. Gustave Le Bon, for example, categorically dismissed Indian scientific learning as "mere childish speculations"; Joseph Conrad wrote of the "pathetically childish" appearance of grass-walled African huts.[109] Peoples as diverse as the Yoruba and the Chinese were ridiculed for their childlike fear of European devices, from pipe organs to matches. The belief that African and Asian rulers regarded European inventions and scientific instruments as mere playthings provided a particularly telling manifestation of their childlike natures. Equally noteworthy were African and Asian attempts to offset with magical talismans the decisive European advantages in the art of killing. The Europeans regarded this resort to magic as much a product of childlike, irrational thinking as the feeble attempts at deception discussed in Chapter 3.[110] And James Bryce observed that such "backward races" as the "kaffirs" of South Africa refused like stubborn children to learn the lessons that repeated trouncings at the hands of the Europeans ought to have taught them: "Though they feared the firearms of the whites, whom they called wizards, it was a long time before they realized their hopeless inferiority. Their minds were mostly too childish to recollect and draw the necessary inferences from previous defeats and they never realized that the whites possessed beyond the sea an inexhaustible reservoir of men and weapons."[111] Richard Burton and others thought that these failings were racially ingrained, which meant that the Africans, at least, could not "improve beyond a certain point," could not advance intellectually beyond the level of a child.[112]

As H. A. C. Cairns has pointed out, technological superiority and the dominance that resulted gave credibility to European notions of African and Asian childishness which they might otherwise have lacked. Individuals from any background, Cairns notes, are bound to behave in ways that seem childlike when judged by the standards of cultures other than their own or when they are attempting to acculturate to other

47, and "Race in Legislation and Political Economy," 8 (1866), p. 120; S. J. Thomson, *The Real Indian People* (Edinburgh, 1914), pp. 304–6; and Martine Loutfi, *Littérature et colonialisme* (Paris, 1971), pp. 61–62.

[109]Le Bon, *Psychology of Peoples*, p. 68; and Joseph Conrad, *Heart of Darkness* (New York, 1950), p. 85.

[110]On European responses to these patterns of resistance, see Michael Adas, *Prophets of Rebellion: Millenarian Protest against the European Colonial Order* (New York, 1987), esp. chaps. 1 and 6.

[111]James Bryce, *Impressions of South Africa* (London, 1897), pp. 116–17.

[112]Burton, *Mission to Galele*, vol. 2, p. 203.

societies.[113] In the late nineteenth century, however, only the standards of the ruling Europeans mattered, and whatever acculturation was occurring was almost invariably on the part of Africans and Asians to European ideas, languages, norms and behavior. The Sister Niveditas and Annie Besants who embraced non-Western religions and adopted non-Western life styles were rare and regarded as eccentrics or aberrations by other Europeans.

The attribution of childlike qualities to Africans and Asians served to bolster the civilizing-mission ideology by which the Europeans justified their dominance over colonized peoples. If the "natives" were racially incapable of ruling and developing their societies, it was the duty of the Europeans, as a mature or fully adult race, to take charge. When the Africans or Asians quarreled among themselves, which like peevish children they always seemed to be doing, it was up to their European superiors to keep them from doing injury to one another and to try to arbitrate their differences. When they became angry and responded with violence, Europeans had to repress and punish them. When they were lazy, it was the responsibility of the Europeans to put them to work. Because they were ignorant, Europeans must try to educate them. But the fact that African and Asian immaturity was increasingly seen as racially determined meant, for racist thinkers at least, that there were limits to what the colonized could be taught and how far they could advance in the direction of civilized life. Some writers saw a problem even in teaching them the rudiments of Christian theology, and the majority came to see Western technological innovation and scientific inquiry as endeavors that were simply beyond the capacity of virtually all African and most Asian peoples. This conclusion, of course, served to justify European global dominance. If the progress of humankind depended on science and technology, and the Europeans and North Americans were the only "race" capable of mastering test tubes and machines, then Europeans and North Americans must decide the investment strategies, formulate the social policies, and make the political decisions that would shape the course of global development in the twentieth century.

In the last decades of the nineteenth century, efforts to construct and defend racial hierarchies became increasingly dominated by the evolutionary approach to human development, which the works of Charles Darwin and Alfred Wallace had elevated to a major focus of scientific

[113]H. A. C. Cairns, *Prelude to Imperialism: British Reactions to Central African Society, 1840–1890* (London, 1965), p. 95.

inquiry and the writings of Robert Chambers and Herbert Spencer did much to popularize. Though Darwin's responsibility for "social Darwinism," particularly in its racist, determinist guise, has been a matter of considerable controversy,[114] he appears to have subscribed to the prevailing view that innate differences in mental capacity did exist between different races. As John Greene has shown through a careful study of Darwin's unpublished papers and correspondence,[115] he also acknowledged that interracial competition and struggle had played a role in the evolution of the human species. In the sections on the early development of human societies in his *Descent of Man*, Darwin linked this competition and social advance directly to technological innovation. He saw technology as a key determinant of the potential of human groups to adapt to and survive in different environments, and at one point he suggested that advances in technology improved the intellectual endowment of the groups that devised them. In his view, technological innovations also explained why some groups of humans advanced more rapidly than others. In the primitive stage of human evolution, he argued, those "tribes" that excelled in toolmaking and weapons manufacture were the ones that grew in numbers and extended their dominion over others. At later stages of development, civilized societies, whose superiority was explicitly linked to their skills in "the arts," steadily supplanted savage peoples who had failed to keep pace in invention.[116]

The connections that Darwin sketched between technological attainments and overall social development were by no means original. As we have seen, late eighteenth-century European writers had begun to equate civilization with a high level of technological development. Many of Darwin's contemporaries were equally convinced that scientific inquiry and technological innovation had been essential to the ascent of humankind from naked cave dweller to Victorian gentleman. In his 1871 compendium of assorted ethnological lore, *Primitive Culture*, Edward Tylor identified the extent of scientific knowledge and the degree of development of the "industrial arts" as two of the four main gauges by which all human societies could be ranked on a continuum

[114]For statements of varying positions in this often heated debate, see the special issue of *Current Anthropology* 15 (September 1974), which is devoted to the controversy.

[115]John C. Greene, "Darwin as a Social Evolutionist," *Journal of the History of Biology* 10 (1977), 1–27, reprinted in his *Science, Ideology, and World View* (Berkeley, Calif., 1981).

[116]Charles Darwin, *The Descent of Man* (New York, 1936), esp. pp. 431, 496–98. See also Greene, *Science, Ideology, and World View*, pp. 114–16. Darwin's linkage of technological advance and mental growth was made more explicitly and elaborately in David Page, *Man: Where, Whence, and Whither* (Edinburgh, 1867), pp. 77–78, 88.

ranging from savagery to civilization.[117] Armand de Quatrefages, who like Tylor is often acknowledged as one of the founders of the discipline of anthropology, also stressed the importance of technological and artistic accomplishments as gauges of the degree of advancement achieved by different peoples. Even though he later used his considerable influence as the anthropologist of the French Museum of Natural History to challenge Darwin's theory of evolution, Quatrefages strongly disputed the findings of Morton and other phrenologists who sought to identify fixed, biological differences between human groups. In his *Histoire générale des races humaines*, Quatrefages gave great attention to toolmaking, pottery, stoneworking, metallurgy, and other evidence of material mastery in discussing different cultures and their level of development.[118]

Darwin's overall approach to evolution envisioned advance as local and episodic rather than part of a progressive teleology. But the ambiguity of some of his remarks on human competition and the piecemeal adoption of his ideas by those who claimed to be his followers resulted in theories of human development that equated technological advance with progress and both in turn with superior intelligence.[119] John Lubbock, for example, espoused a version of human evolution that was decidedly progressivist. His works on prehistory, which won a wide reading audience in this period, stressed the importance of toolmaking skills and other forms of material accomplishment as key indicators of social development.[120] Lubbock, like Tylor and Quatrefages, explored the connections between technology and evolution primarily with reference to "tribes" and "peoples" or "primitives" and "savages." Other authors sought to link technology and progressive advance to the innate capacities of various racial groups, a connection that Darwin had only hinted at in his published works.

Similar links were made by writers with a static approach to racial

[117]Edward Tylor, *Primitive Culture* (London, 1871), pp. 23–24. See also Stocking, *Race, Culture, and Evolution*, pp. 77, 82. For a general discussion of the Victorian reliance on this criterion, see Gérard Leclerc, *Anthropologie et colonialisme* (Paris, 1972), pp. 26–33.

[118]Quatrefages, *Histoire générale*, esp. pp. 238–46. It is interesting to note the great shift in Quatrefages's views on the fixity of races between the 1840s—when he wrote such essays as "La Floride," *Revue des Deux Mondes* 1 (1843), 732–73—and 1870.

[119]On the ambiguities in *The Descent of Man*, see Greene, "Darwin as a Social Evolutionist," and Greene's response to Freeman in *Current Anthropology* 15 (1974), 224. On Darwin's rejection of a concept of evolution that was inherently progressive, see Stephen Jay Gould, *Ever Since Darwin* (New York, 1977), esp. pp. 34–38; and Derek Freeman, "The Evolutionary Theories of Charles Darwin and Herbert Spencer," in *Current Anthropology* 15 (1974), 218–19.

[120]John Lubbock, *Prehistoric Times* (London, 1865), and *The Origin of Civilisation and the Primitive Condition of Man* (London, 1870). On Lubbock's progressivism, see Stocking, *Race, Culture, and Evolution*, pp. 77–78.

divisions, such as Crawfurd and Farrar.[121] But beginning in the last decades of the nineteenth century, evolutionary biology gave the construction of racial hierarchies a fluidity and diachronic dimension they had previously lacked. In England, where the influence of evolutionary ideas dominated the thinking of social theorists earlier than on the Continent,[122] elements of this approach can be found in the works of popularists such as Herbert Spencer, who argued for the superiority of "industrial" over "military" societies, and the writings of more stridently racist supremacists such as Charles Pearson, who saw the capacity for invention and the successful application of technology to war and production as keys to survival in an ongoing struggle between the races.[123] Benjamin Kidd noted in the mid-1890s that most of the thinkers who argued for an insurmountable gap in intellect between "high" and "low" races supported that position with illustrations of European achievements in the "arts of life." Kidd listed devices for communication over long distances, astronomical knowledge, the phonograph, and the astounding complexity of European machines as commonly cited "proofs" of the Europeans' superior evolutionary advance.[124] By the last decades of the nineteenth century, British colonizers—whether missionaries, explorers, or government officials—tended to measure "evolutionary distance" in terms of technological development. That gauge impressed them with the "immense distance in time"[125] that separated Europeans from all other peoples but particularly from the "primitive" or "savage races" of Africa, Australia, and the Pacific islands.

Alfred Wallace was one of the more prominent authors who explored the links between race, technology, and social advance. Working from the theory of natural selection that he and Darwin had independently developed, Wallace argued that at some point in their evolution humans had developed a strong immunity to the environmental forces that were so critical in the selection process within lower species. Wallace saw the

[121]See esp. John Crawfurd, "On the Effects of the Commixture of Locality, Climate, and Food on the Races of Man," *T.E.S.L.* I (1861), p. 77; and "Early Civilisation of Man," p. 159. For a survey of the persistence and transmutations of the static view, see Stepan, *Idea of Race*, esp. pp. 104–10.

[122]Two of the best studies of the widespread impact of these ideas are John Burrow, *Evolution and Society: A Study in Victorian Social Theory* (Cambridge, Eng., 1966); and Greta Jones, *Social Darwinism and English Thought: The Interaction between Biological and Social Theory* (Sussex, Eng., 1980).

[123]See, e.g., Herbert Spencer, *Principles of Sociology* (London, 1882), esp. vol. 2, pp. 240–42; and the detailed discussion of Spencer's thinking on this point in J. D. Y. Peel, *Herbert Spencer: The Evolution of a Sociologist* (New York, 1971), chap. 8. See also Charles Pearson, *National Life and Character* (London, 1893).

[124]Benjamin Kidd, *Social Evolution* (London, 1894), pp. 265–66.

[125]Cairns, *Prelude to Imperialism*, pp. 76–77.

superior intellectual capacity of humans as the key to this immunity and to their limitless potential for upward evolution. Superior intelligence gave humans the capacity to develop both the technology and the social cooperation that made them "in some degree superior to nature, inasmuch as [they] knew how to control and regulate her action." The continued improvement of the species from this point onward occurred not by "a change in body [like that essential to survival in lower species], but by an advance of mind." Within the human species, the more morally and intellectually advanced races—those most technically advanced and in control of their environment—dominated and displaced the "lower and more degraded." The latter, Wallace concluded, were destined to perish or be blended into a "single nearly homogeneous race" dominated by racial stocks that had demonstrated superior intelligence and material mastery.[126]

Perhaps more than any other English writer, C. S. Wake singled out scientific aptitudes and inventiveness as qualities by which to measure levels of racial evolution. Wake, the author of a work on human development titled *Chapters on Man* and long a member of the Anthropological Society of London, distinguished five main stages of social development: infancy, willful, emotional, empirical, and rational. During the first two stages, in which Wake grouped such "races" as the Australian aborigines and the North American Indians, there was "little mental activity of any sort" and "not the smallest taste" for the mechanical arts or the sciences. Races at this level of evolution were incapable of sustained thought or abstract reasoning. Their material culture was crude, and their inventive and imitative capacities were "very humble." Surprisingly, Wake had little to say about the emotional stage of development, despite the fact that he judged the Negro race to have reached this level. Citing Hunt's essays, he repeated the standard view that Africans were childlike, sensuous, and morally deficient. He left the reader to assume that they had done little in the sciences or technology, and he had praise only for their ability to imitate more advanced peoples. Wake deemed the Asiatic races, who had risen to the empirical level of development, to be moderately intelligent and "full of sagacity" for the "useful arts." They were, however, innately unable to generalize from specific discoveries, to think abstractly, or to recognize general truths.

[126]Alfred Russel Wallace, "The Development of Human Races under the Law of Natural Selection" (originally published in the *Anthropological Review* in 1864), in his *Natural Selection and Tropical Nature* (London, 1895), pp. 166–85 (quoted portions pp. 182, 185). For similar views on the progress of human evolution, see Page, *Man*, pp. 80–91; and Walter Bagehot, "Physics and Politics," *Fortnightly Review*, n.s. 3 (1868), esp. 452–56.

These failings hampered their capacity to improve on early advances and doomed them to the stagnation that Wake believed characteristic of empirical societies.

As the designation of Wake's highest stage of evolution suggests, the rational faculties were dominant among the most advanced or European races. Wake argued that the Europeans had gone through all the earlier stages of development and, alone of the races, had avoided stagnating at a lower stage of evolution. They were also the only race that had advanced far enough in science and technology to be able to force nature to divulge "her" secrets and yield to the needs of human societies. Though he felt that the aptitudes and proclivities of races tended to become fixed with the passage of long periods of time, Wake did allow that the "negroes" and Asiatics, who had begun to advance, were capable of further progress if assisted by the Europeans.[127]

The approach to human evolution that envisioned technology as a prime agent of progressive social advance found perhaps its strongest and certainly its most lasting support among American anthropologists. Though the bulk of Lewis Henry Morgan's 1877 study *Ancient Society* was devoted to the "growth" of political organization, the family, and the idea of property, he saw technological breakthroughs as essential preconditions of human development from the "older period of savagery" to civilization. Tools and modes of sustenance defined each stage in the progression, and invention was the key to a society's capacity to move from a lower to a higher stage of evolution: "The most advanced portions of the human race were halted, so to express it, at certain stages of progress, until some great invention or discovery, such as the domestication of animals or the smelting of iron ore, gave a new and powerful impulse forward."[128]

Morgan, an American Indian specialist writing of tribes and primitives rather than races, argued that human mental capacity increased as the result of successive inventions and the growth of institutional complexity. Thus, the most technologically proficient and organizationally sophisticated peoples were also the most intelligent.[129] Morgan's work was not well received in England, and his evolutionary approach was repudiated by Franz Boas and his disciples at the end of the nineteenth

[127]C. S. Wake, "The Psychological Unity of Mankind," *Memoirs Read before the Anthropological Society of London* 3 (1867–69), 134–47, and *The Classification of the Races of Mankind* (London, 1880). In the later work Wake's categories were reduced to four, with the "industrial races" the highest. For a schema similar to Wake's but aimed at a popular rather than a scholarly audience, see A. H. Keane, *The World's Peoples* (London, 1908).

[128]Lewis Henry Morgan, *Ancient Society* (Cambridge, Mass., 1964), p. 40 (quoted portion), and chaps. 1–3.

[129]Ibid., pp. 38–39.

century. But the influence of Morgan's ideas on technology and human evolution can be traced through the writings of later American anthropologists from John Wesley Powell to Leslie White.[130]

As Linda Clark and others have shown, French social thinkers were slower than their British counterparts to embrace an evolutionary frame of analysis.[131] From the 1890s, however, many prominent French authorities on colonial affairs, including Jules Harmand, Léopold de Saussure, and Georges Hardy, based their prescriptions for colonial administration and education on both evolutionary and racist assumptions.[132] Numerous writers on colonial issues, especially Saussure, relied heavily on the works of Gustave Le Bon to buttress their assertions about varying racial aptitudes and differing levels of evolutionary ascent. Le Bon's arguments provide a superb example of the uneasy combination of social evolutionism and a fixed or static view of human racial divisions which, as Nancy Stepan has shown, was accepted by a surprising number of late nineteenth-century authors.[133] Le Bon grouped different races in four categories from primitive to superior and sketched their characteristic features. In the early sections of his *Psychology of Peoples* he treated the groups in each category as clearly demarcated and permanent entities. But his general approach and specific observations—such as his remark that Negroes (whom he classified in the second or "inferior" category) had rudiments of civilization but had been unable to advance beyond barbarism—suggested that the categories were in fact stages of development through which various races evolved over long periods of time. This is certainly the way in which Le Bon's schema was interpreted by others and may explain his influence on such authors as Saussure, who quoted extensively from passages suggesting differential evolutionary advance as the cause of racial disparities.

Like Wake and A. H. Keane, Le Bon regarded evidence of scientific discovery and invention as a key means of determining which category each race was allotted to. Though his comments on the lower races in this regard were a good deal more vague than Wake's, Le Bon's superior races (exclusively Indo-Aryan peoples) were accorded that status chiefly on the basis of their achievements in science and technology. He argued

[130]For discussions of Morgan's work and evidence of his long-term impact, see Leslie White's introduction to the 1964 edition of *Ancient Society*; Stocking, *Race, Culture, and Evolution*, pp. 119–20, 128–29; and White, "Energy and the Evolution of Culture," in *The Science of Culture* (New York, 1949), pp. 363–93.

[131]Linda Clark, *Social Darwinism in France* (Birmingham, Ala., 1984), esp. pp. 50–55, 62; and Yvette Conry, *L'introduction du Darwinisme en France au XIXe siècle* (Paris, 1974), esp. chap. 2.

[132]Their views are discussed in the following section of this chapter.

[133]Le Bon, *Psychology of Peoples*, pp. 4–5, 10, 17–19; Stepan, *Idea of Race*, esp. pp. 88–93.

that in both ancient and modern times Indo-Aryans had been responsible for all significant innovations in the arts, science, and industry. He added that the European branch of the Indo-Aryans had been responsible for steam power and electricity; the "Hindus," whom he considered the "least developed" people of the Indo-Aryan family, had excelled only in the fine arts and philosophy.[134]

Le Bon's seemingly contradictory pronouncements on racial development were further muddled by his insistence that it was not the average members of a racial group but a "creative elite" who determined a race's achievements and potential. He conceded (rather alarmingly, one imagines, from the viewpoint of his fellow white supremacists) that the average Indian was equal in intelligence and ability to the average Englishman. The superiority of the English as a race, he surmised, could be traced to a small number of brilliant and inventive individuals, whom the Indians had proved incapable of producing—at least in the sciences and technology. Le Bon believed the Indians quite able to operate a telegraph or run a train but inherently incapable of inventing these marvels. He also doubted that Indians possessed the kind of intelligence that would allow them even to direct railway lines, let alone carry out scientific research and make great scientific discoveries. For Le Bon, these deficiencies were not a matter of cultural preference; they were racially ingrained. The English and other Europeans excelled at science and technology because they possessed a creative elite that was inherently more capable of rational thinking, precise measurement, self-discipline, and invention. However much Le Bon might admire Indian art and philosophy and deplore the sorry state of these fields in industrial Europe, he concluded that the low level of technological development and scientific investigation in Indian society when contrasted with the remarkable achievements of the Europeans left no doubt that the latter were racially superior.[135]

Le Bon's theory that racial achievement depended on a gifted elite found few adherents,[136] but his belief that innate intellectual deficiencies made it difficult or impossible for non-Western races to advance in the sciences and technology was even more widely held by his contemporaries than it had been by earlier racists with a static view of human differentiation. Such qualities as the capacity for abstraction and patient

[134]Le Bon, Psychology of Peoples, pp. 26–30.

[135]Gustave Le Bon, Les civilisations de l'Inde (Paris, 1887), pp. 188ff.; and Psychology of Peoples, pp. 40–43.

[136]He may have influenced Gabriel Tarde's thinking on elite-mass interaction, and both in turn clearly shaped Albert Fouillée's ideas. See Clark, Social Darwinism, pp. 124–25; and Fouillée, "Le caractère des races," pp. 88–91.

testing, once considered culturally nurtured, were now seen as the products of millennia of racial adaptation and selection. Races that had evolved the least, and were thus closest to the apes and other animals, had the most acute sensory perceptions but poorly developed rational faculties. The higher one ascended on the evolutionary scale, the greater the powers of reasoning and thus the greater the capacity for scientific investigation and invention. Though this gradation from predominantly sensual to primarily rational was not always explicitly stated,[137] it underlay assertions that the Negroes were incapable of inductive reasoning, the Indians illogical, and the Chinese unable to think abstractly. Popular expressions of these arguments can be found in contemporary novels that stressed the sensual, unpredictable, and deceitful nature and the highly emotional or irrational behavior of African and Indian characters.[138] In the best-selling stories of the Australian mystery writer Arthur Upfield, which were published in the decades after the Great War, the differentiation of races on the basis of their propensities for sensual perception or rationality is embodied in a single character. Upfield's hero Napoleon Bonaparte is both a half-caste and a master detective. In the *Winds of Evil*, "Bony" confides to the reader the advantages of his mixed aboriginal and Caucasian parentage: "Added to my inherited maternal gifts are those inherited from my white father. I see with the eyes of a black man and reason with the mind of a white man, and in the bush I am supreme."[139]

Most but by no means all late nineteenth-century observers agreed that Africans and other races that had not evolved beyond the primitive or savage stage possessed virtually none of the mental aptitudes necessary to develop scientific thinking or devise any but the most rudimentary tools. Explorers and missionary schoolteachers, colonial bureaucrats, and social theorists judged African peoples incapable of reflection, concentration, sustained or systematic observation, generalization, abstraction, and inductive reasoning.[140] Some authors conceded

---

[137]Though often it was. For examples, see Henry Guppy, "Notes on the Capabilities of the Negro for Civilization," *Journal of the Anthropological Society* 2 (1864), ccix; Cureau, *Les sociétés primitives*, p. 83; Jerome Dowd, *The Negro Races: A Sociological Study* (New York, 1907), pp. 360–61, 382; Stocking, *Race, Culture, and Evolution*, pp. 117–18, 122–26; and Gould, *Mismeasure of Man*, pp. 97–98.

[138]See Fanoudh-Siefer, *Le mythe du nègre*, pp. 86–91, 94–97; Lewis D. Wurgaft, *The Imperial Imagination: Magic and Myth in Kipling's India* (Middletown, Conn., 1983), esp. pp. 130–44; and Benita Parry, *Delusions and Discoveries: Studies on India in the British Imagination* (Berkeley, Calif., 1972), esp. pp. 59–69, 70–88, 106–21.

[139]Arthur Upfield, *Winds of Evil* (Sydney, 1972) p. 65.

[140]For examples, see Fouillée, "Le caractère des races," pp. 80–83, 88–93; Bryce, *South Africa*, p. 113; Kidd, *Social Evolution*, p. 269; Hovelacque, *Les nègres de l'Afrique*, pp. 425–26, 456; Dowd, *Negro Races*, pp. 358, 361–70; C. H. Phillips, "The Lower Congo: A

that the Africans had "prodigious memories" and a great capacity for imitation, but they frequently offset this praise by adding that like all primitives they were utterly lacking in curiosity and invention.[141] These deficiencies were seen as manifestations of fundamental differences in mentality between Africans and more developed races. The belief, for example, that savage races perceived causal relationships in ways entirely different from the perceptions of civilized peoples contributed to the formulation by anthropologists such as Lucien Lévy-Bruhl of elaborate theories exploring the "prelogical" modes of thinking and the primitive mentalities of groups thought to be low on the evolutionary ladder.[142]

Paradoxically, one of the most emphatic proponents of the notion that African thinking differed from European in kind rather than degree was Mary Kingsley, one of the most knowledgeable and outspoken defenders of Africans and African culture in this era. In the midst of a characteristic assault on missionary strategies for civilizing the "natives," Kingsley rather abruptly remarks:

> You cannot associate with them long before you must recognise that these Africans have often a remarkable mental acuteness, and a large share of common sense; that there is nothing "childlike" in their form of mind at all. Observe them further and you will find that they are not a flighty-minded, mystical set of people in the least. They are not dreamers; or poets, and you will observe, and I hope observe closely—for to my mind this is the most important difference between their make of mind and our own—that they are notably deficient in all mechanical arts; they have never made, unless under white direction and instruction, a single fourteenth-rate piece of cloth, pottery, a tool or machine, house, bridge, picture or statue; that a written language of their own construction, they none of them possess. A careful study of the things a man, black or white, fails to do, whether for good or evil, usually gives a truer knowledge of the man than the things he succeeds in

Sociological Study," *Journal of the Royal Anthropological Institute* 17 (1888), 220–21; Stocking, *Race, Culture, and Evolution*, p. 23; and Haller, *Outcasts from Evolution*, pp. 124–27.

[141]Hovelacque, *Les nègres de l'Afrique*, p. 426; Dowd, *Negro Races*, pp. 370–71; Fouillée, "Le caractère des races," p. 82; and H. H. Johnston, "British West Africa and the Trade of the Interior," *Proceedings of the Royal Colonial Institute* 20 (1888–89), p. 82.

[142]For examples, see Phillips, "Lower Congo," p. 220; Bernard Struck, "African Ideas on the Subject of Earthquakes," *Journal of the Anthropological Society* 8 (1908–9), 398–41; Paul Giran, *De l'éducation des races* (Paris, 1913), pp. 25–37; and Hardy, *Une conquête morale*, pp. 228–30. Of Lévy-Bruhl's numerous works, see esp. *Les fonctions mentales dans les sociétés inférieures* (Paris, 1910), and *La mentalité primitive* (Paris, 1922), both of which have been translated into English by Lilian Clare; see also Jean Cazeneuve, *Lucien Lévy-Bruhl* (New York, 1972); and Fanoudh-Siefer, *Le mythe du nègre*, pp. 34, 45, 175–79.

doing. When you fully realize this acuteness on the one hand and this mechanical incapacity on the other which exist in the people you are studying, you can go ahead.[143]

Though they lacked Kingsley's remarkably relativistic appreciation for African beliefs and behavior, numerous observers at the turn of the century shared her conviction that Africans were by nature woefully inept at creating machines and using those devised by other races.[144]

Because it was believed that the Indians and Chinese had advanced to a higher stage in the evolutionary progression, most late nineteenth-century observers considered the differences in thinking that separated them from the Europeans to be less extreme. Nonetheless, many writers in this period treated Chinese conservatism, passivity, and alleged inability to think abstractly, to generalize or to reason by analogy, as if they were racially derived. Features of the Chinese language that were said to render it unfit for scientific discourse were seen as symptoms of racial shortcomings. Rather than despotism or China's isolation, "Oriental" racial characteristics were seen by many observers as the key to China's stagnation at a middle stage of evolution.[145] Some writers saw differences in the perception of time and space and in attitudes toward work and discipline as inherited rather than cultural traits. Indian specialists contrasted the emotional and irrational races of the subcontinent with their analytical and logical British masters, pitting the vaunted empiricism of the British against Indian proclivities for flights of fantasy and imagination. Despite considerable contemporary evidence to the contrary, commentators on India shared Le Bon's conclusion that the Indians' mental makeup rendered them incapable of great scientific discoveries or technological breakthroughs.[146] Like those directed against the Africans, most of these allegations about Chinese or Indian deficiencies were not new. But the racial and evolutionist assumptions fre-

---

[143]Mary Kingsley, *Travels in West Africa* (London, 1897), pp. 439–40.

[144]Those among the observers discussed in the second section of Chapter 3, who believed the Africans *racially* unsuited for technological endeavors, included Henry Drummond, *Tropical Africa* (London, 1888), pp. 104–5; Hermann von Wissman, *My Second Journey through Equatorial Africa* (London, 1891), p. 291; and Dowd, *Negro Races*, pp. 370–71.

[145]Giran, *De l'éducation des races*, pp. 19–20, 29, 219–21; L. J. Beale, "The Brain and the Skull in Some of the Families of Man," *T.E.S.L.* 1 (1865), 50; Arthur H. Smith, *Chinese Characteristics* (Shanghai, 1890), pp. 42ff., 83, 133–34, 151–60; Emile Burnouf, *La science des religions* (Paris, 1885), pp. 229–30; and Samuel W. Williams, *The Middle Kingdom* (New York, 1879), vol. 1, pp. 578–79, and vol. 2, pp. 143–45, 178–79.

[146]Figuier, *Human Race*, p. 340; Le Bon, *Les civilisations de l'Inde*, pp. 192–94, 547–48; Trevelyan, *Competition Wallah*, p. 250; and George W. Knox, *The Spirit of the Orient* (New York, 1905), pp. 123–24.

quently associated with them greatly enhanced their impact on European educators and administrators whose power to shape the future of non-Western peoples peaked in this era.

By the early 1900s the eighteenth-century belief in the fundamental unity of humankind found few adherents among European intellectuals and politicians. Though missionaries continued to insist on the equality of all men and women before God, even an overwhelming majority of them readily conceded that there were profound and innate differences in physical aptitudes and especially intelligence between racial groups. A beleaguered minority of European social thinkers and policymakers challenged the validity of racial stereotyping and at times the racial categories themselves. But they could do little more than fight to preserve existing avenues for education and upward mobility available to colonized peoples. Social evolutionists had completed the work begun by anatomists and phrenologists. The racist premise that hereditary, biological differences distinguished human groups had not only won widespread scientific acceptance; it had been disseminated at all class levels by adventure novels, travel accounts, and the penny press.

The forces that led to the predominance of racist thinking in the decades before the Great War were diverse and complex. The arrogance engendered by European global hegemony, reactions to "native" revolts in India and elsewhere, accounts of mighty China humbled by British gunboats, and the ferocity of slave risings in the Caribbean all contributed to the growth of racist attitudes.[147] Historical events and popular sentiment helped shape intellectual views. But these were equally, in many cases more, affected by what were regarded as scientific proofs and a growing awareness of the gap that the scientific and industrial revolutions had opened between western Europeans and all other peoples. The civilized state that eighteenth-century thinkers had regarded as a goal to which all peoples might aspire was now reserved largely for the white races.[148] Depending on the author, this exalted condition might be defined by political institutions, marriage patterns, or ethical codes, but almost invariably the capacity for invention and scientific thinking were seen as essential attributes of those races that had evolved to the highest levels of human development.

Qualifying the Civilizing Mission: Racists versus Improvers at the Turn of the Century

Race was an omnipresent issue in the often heated debates over colonial policy that occupied British and French officials in the decades

[147]See the works by Curtin, Spear, Lorimer, Bolt, and Cohen cited above, and Bernard Semmel, *The Governor Eyre Controversy* (London, 1962).

[148]On this shift, see esp. Stocking, *Race, Culture, and Evolution*, pp. 26–36.

before the Great War. Armed with the findings of phrenologists and the pronouncements of social evolutionists, French essayists and overseas administrators launched a sustained assault on the cherished but ill-defined tenets of assimilation which had long been considered the core of France's *mission civilisatrice*. The changes envisioned by those who sought to reform British colonial policy were far less sweeping and categorical than those demanded by the French critics of assimilation. But numerous British administrators and social thinkers questioned the wisdom of advanced Western education for Indians and other colonized groups, and the policy of indirect rule through "native" leaders and institutions received widespread and enthusiatic support. Those who defended assimilation and the potential of non-Western peoples to master European learning and technology were confronted by a large body of purportedly scientific evidence that innate intellectual and moral differences made cultural exchanges between "superior" and "inferior" races impossible. Like writers throughout the century, those on both sides of these debates saw the potential of colonized peoples for mastering mathematics and the sciences and adapting to Western technology as the most reliable gauges of their overall racial capabilities. At the same time, the degree to which different writers and officials were committed to theories of racial inequality was often indicated by the level of scientific and technical education they advocated for colonized peoples. This barometer of racist sentiment also revealed important distinctions that Europeans made between different non-Western peoples: an author who recommended only the most rudimentary training in hygiene and simple tools for "primitive" or "savage" races might propose fairly advanced instruction in calculus or chemistry for the Indians or Chinese. By the end of the nineteenth century, questions of race and scientific and technological measures of human capability had become integral both to debates over colonial policy and to the future of the peoples of much of the non-Western world.

Though scientific and technological gauges and racist arguments were increasingly important for both British and French colonial policymakers, there were major differences in emphasis between them regarding education and more generally the acculturation of subjugated peoples to Western thinking and values. For even the strongest advocates of assimilation among the French, training in the sciences never assumed the importance that it had in the campaigns of the British improvers for the expansion of English-language education in India. Instruction in the French language and French literature and philosophy, and the inculcation of French manners, taste, and ethical sensibilities remained central to the assimilationists' sense of what the acculturation of colonized peoples entailed. Most assimilationists also considered instruction in mathematics and the sciences essential, especially at the secondary level. But in contrast to the improvers, few thought these subjects as vital as the inculcation of French culture, and only a handful defended French-language education because it was necessary for scientific instruction.

Because assimilation had been a vague concept that took on different meanings in different eras and colonial locales, the turn-of-the-century debate over whether or not it should continue to provide the guiding principles for French colonial policy ranged over a wide variety of issues.[149] The advocates of alternative approaches to colonial rule, including the equally amorphous policy of association, drew attention to the political dangers inherent in attempting to make full French citizens of the tens of millions of colonized peoples. They also pointed out that approaches to colonial administration which emphasized the preservation of indigenous leaders and institutions were more economical and less disruptive for the subject peoples. But whatever their position on these other issues, all who were involved in controversies over colonial policy in this era were forced to address the underlying issue of the extent to which it was possible to transmit key elements of French culture in a meaningful way from "civilized" races to less evolved or "savage" ones. Though earlier writers had raised doubts about the capacity of African and Asian peoples to be genuinely assimilated, the writings of Gustave Le Bon mark the beginning of the sustained assault on the assimilationist approach. His opinions were frequently cited at length by such leading critics of assimilation as Saussure and Harmand, and his basic arguments provided the framework for alternative approaches to French policy. Of special importance were his conviction that higher education in India had done little more than produce a disgruntled and "uprooted" class among the colonized and his insistence on the innate incapacity of non-European peoples for invention and scientific discovery. These arguments were frequently repeated, at times quoted verbatim, in virtually all subsequent discussions of French colonial educational policy.

The rather abrupt shift in policy toward the colonized that occurred in the 1890s is strikingly illustrated by the contrast between the openly hostile reception to the paper Le Bon presented to Congrès Colonial

---

[149]These controversies are the focus of Raymond Bett's *Assimilation and Association in French Colonial Theory, 1890–1914* (New York, 1961). Also useful is Hubert Deschamps, *Les méthodes et les doctrines coloniales de la France du XVIe siècle à nos jours* (Paris, 1953), esp. pp. 121–60; and Martin D. Lewis, "One Hundred Million Frenchmen: The 'Assimilation' Theory in French Colonial Policy," *Comparative Studies in Society and History* 4/2 (1962), 129–53.

Internationale de Paris in 1889 and the vindication of his positions a decade later by such prominent authorities on colonial affairs as H. H. von Kol, Paul Bernard, and especially Léopold de Saussure at the Congrès International de Sociologie Coloniale.[150] A year before the second gathering, Saussure had published a diatribe against assimilation that was based heavily on evidence from Le Bon's works on India and racial psychology. Before the publication of his *Psychologie de la colonisation française*, Saussure was an obscure naval officer who had served for a time in China. His relentless polemic against the assimilationist position and his energetic participation in the 1900 congress transformed him into a major "expert" on French colonial policy. He dismissed the assimilationist approach as one based on eighteenth-century philosophical abstractions concerning the fundamental unity of humankind, which, he claimed, ignored the reality of racial differences and the demands of diverse colonial situations.[151] Different races, Saussure argued, had evolved at varying rates and thus had attained very different levels of development. Differential evolution produced great discrepancies in the mental and physical capacity of each racial group, and this accounted for the insurmountable gap that separated the highest (white) from the lower (black or brown) races. Saussure scorned those who believed that a few years of Western education could compensate for mental and moral deficiencies that were the product of millennia of evolutionary lag. Though he thought the lower races might someday reach a level of civilization comparable to that of Europe, he was convinced that this process would take thousands of years. Despite this slight concession to the possibility of change, Saussure held to a rather fixed view of racial differences, similar to that espoused by Le Bon.[152]

Calling repeatedly for an "empirical" and "scientific" approach to colonial policy, Saussure pushed for educational policies that were geared to the level of evolutionary development of each culture in question. He made clear his opinion that the advanced sciences were beyond the learning abilities of the colonized peoples. Like Le Bon, he lamented the lack of precision displayed by inferior races and suggested that though the Africans, for example, might be able to memorize mathematical formulas or passages from French literature, they merely "parroted" this information without understanding it. A Negro, he declared, might with Western education be made into a barrister, but he

---

[150]Lewis, "One Hundred Million Frenchman," esp. pp. 137–49.

[151]These arguments are trenchantly developed in Saussure's paper to the 1900 international congress; see Arthur Rousseau, ed., *Congrès international de sociologie coloniale* (Paris, 1901), vol. 1, esp. pp. 145–49.

[152]Léopold de Saussure, *Psychologie de la colonisation française* (Paris, 1899), esp. pp. 12–14, 33, 132–33, 233–37; and Rousseau, *Congrès international,* pp. 153–56.

could not be taught to reason like a European. Thus, both colonizer and colonized would profit most if education emphasized literacy, elementary arithmetic, and training in the *metiers manuals* and agricultural sciences.[153]

Saussure's rapid rise to prominence owed a good deal to the fact that he appeared to provide empirical validation for fears and policy positions that had more and more frequently been voiced in French colonial circles. Many of the delegates at the 1900 congress were colonial officials who applauded his pronouncements and provided case evidence to support his antiassimilationist recommendations. Colonial administrators such as Jules Harmand, who had long shared many of Saussure's views, grew increasingly vociferous in their attacks on the assimilationist approach. Harmand, whose disciples included such giants of French colonial administration as Étienne, Galliéni, and Lyautey, had championed indirect rule and association for decades before his highly influential *Domination et colonisation* appeared in 1910.[154]

Though much of what he had to say on race and against assimilation had already been said by Le Bon, Saussure, and others, Harmand's recommendations for French colonial administration, particularly those on education, were a good deal more specific and detailed than those of earlier writers. Quoting extensively from both Le Bon and John Strachey, a British colonial official whose book on India he had translated into French, Harmand charged that Western education in India had done little more than encourage sedition, agitation, and violent rebellion by disoriented English-educated Indian youths.[155] Advanced scientific learning, he continued, was unsuited to the Indian, the Vietnamese, the Filipino, and most certainly the African mind. Given their inferior evolutionary development these less advanced peoples lacked the precision, "moral discipline," and related values that European school children instinctively possessed, because the latter had been reared in highly mechanized societies and came from more highly developed racial stock. Nonetheless, Harmand recommended that in the interests of both the colonizer and the colonized, technical training ought to be a high priority in French efforts to educate subjugated peoples. The French needed mechanics and railway engineers to run the empire, he reasoned, and the peoples of Africa and Asia were mainly interested in what the Europeans had to teach them about science and technology. But it was

[153]Saussure, *Psychologie*, pp. 35, 50–51, 66, 108, 127–29.

[154]See Deschamps, *Méthodes et doctrines*, pp. 147–58. A fine discussion of Harmand's colonial career and broader policy views can be found in Betts, *Assimilation and Association*, pp. 51–55.

[155]Jules Harmand, *Domination et colonisation* (Paris, 1910), esp. pp. 259–65.

essential for French officials to realize that what the colonized were capable of mastering was limited. Book learning, theorizing, and "*mnemonique* cramming" were to be eschewed. Secondary and higher education should be geared to the practical and the concrete; scientific instruction should be concentrated on everyday work and problems encountered in the workshop and on the plantation. Physics and chemistry and scientific discovery should be left to Europeans, who could handle them. Colonial education, Harmand concluded, should be aimed at producing mechanics, craftsmen, and agricultural experts, not savants or professors of literature.[156]

Harmand made little effort to distinguish one colonized "race" from another. But differences in capacity were implicit in the evolutionary progression of human development that undergirded his arguments. He also acknowledged that colonized peoples had the capacity to acquire technical skills that were denied to them by writers who saw colonial education as little more than the means of teaching better hygiene and assisting the colonized in acquiring a "taste for work."[157]

Paul Giran repeated Harmand's arguments almost verbatim in a work on colonial education that appeared just three years after *Domination et colonisation*. Giran, who had served for years as an administrator in Indochina, also included recommendations for educational changes running counter to the long-standing notion that the more evolved colonized races were capable of advanced learning in the sciences and mathematics. Giran observed that at the secondary level and above, French-sponsored education for the Vietnamese had increasingly emphasized the teaching of the French language and the physical and natural sciences. He believed this trend to be both wrongheaded and dangerous. It was wrongheaded because it ignored the "psychological abyss" that separated Frenchmen from Vietnamese. The "irreducible" intellectual differences between the two races[158] rendered the Vietnamese incapable, despite their acknowledged centuries of civilized development, of truly mastering scientific ideas or mathematical concepts. In adapting the "higher learning" of the West to their own mentality, Giran asserted, the Vietnamese fundamentally transformed and disfigured it.

Exposing peoples like the Vietnamese and Indians to this learning was dangerous because it led them to aspire to positions they could not handle (and indeed, although Giran did not say so, positions they had

---

[156]Ibid., pp. 266–77.

[157]See, e.g., the selections in Rousseau, *Congrès international,* vol. 1, p. 11; vol. 2, pp. 394–95.

[158]And, Giran stressed, between the sexes; he believed that the females of all races were inferior in intelligence to their male counterparts. See *De l'éducation des races*, p. 249.

little chance of occupying in colonial society). Flaunting their garbled versions of European ideas and manners but frustrated at not receiving what they considered their just rewards, Western-educated students were easily won over by religious fanatics and xenophobes who sought to fabricate nationalist loyalties and arouse anticolonial hostility. But, Giran argued, if the French devised educational programs geared to the level of evolutionary development of each of the races in their empire, all such problems could be avoided. Like Harmand, Giran appeared to suggest that these programs ought to differ from one colonial context to another. But his actual recommendations were uniform and familiar: play down theory and abstract speculation (which must remain the province of the higher races); push technical and vocational training and practical problem-solving; and strive to instill in colonized peoples a sense of the dignity and satisfaction to be found in manual labor.[159]

Before and after World War I, the views and recommendations of Saussure, Harmand, and Giran had a major impact on the ways in which the French actually administered their numerous and diverse colonial territories.[160] Though assimilationist rhetoric did not disappear from French colonial documents, and the new policies of association were as ill defined and varied as those they supplanted, in most colonies there was a clear shift away from creating black and brown citizens of France to a stress on the preservation of indigenous institutions, languages, and customs. French approaches to colonial education both reflected and were essential to this shift.[161] Africans were by far the most affected, and perhaps no individual was more instrumental than Georges Hardy in shaping the structure and determining the content of education in France's African possessions. After his appointment as Inspector General of Education for French West Africa in 1914, Hardy was able to oversee the implementation of an educational program that stressed technical and vocational training and ensured that instruction in the sciences remained at the "most modest" level. As his 1917 summary of his views made clear, Hardy was a loyal ally of Saussure and Harmand, though he was less fearful than they that French-sponsored schools would become seedbeds for revolt. His policies were premised on the

[159]Ibid., pp. 30–37, 218–24, 296–309, 237–38.

[160]See, e.g., the views on education of the revered William Ponty, as quoted in Denise Bouche, "Autrefois, notre pays s'appelait la Gaule . . . Remarques sur l'adaptation de l'enseignement au Sénégal de 1817 à 1960," *Cahiers D'études Africaines* 8/1 (1908), 111.

[161]For a convincing critique of the long-held view that the shift to association had little impact on French educational policy in the colonies, see Prosser Gifford and Timothy Weiskel, "African Education in a Colonial Context: British and French Styles," in Gifford and W. R. Louis, eds., *France and Britain in Africa* (New Haven, Conn., 1971), esp. pp. 667–77.

assumption that the Africans were millennia behind the Europeans in evolutionary development, that their less evolved intellect had little capacity for abstraction and the other modes of thought necessary for serious work in the sciences. African students ought to be instructed only in the "notions simples de technologie," and prepared to become mechanics, carpenters, blacksmiths, and railway engineers rather than inventors and experimental chemists or physicists. Hardy concluded that for their own good (he pointed out that Africans greatly enhanced their status by working with Western machines) and the good of France's vast empire, Africans should be taught to run and repair the machines that made the *mission civilisatrice* possible.[162]

Although most authors acknowledged differences in evolutionary development and intellectual capacity between colonized peoples but then treated educational issues in a monolithic fashion, the actual content and structure of French education in its overseas dependencies continued to vary significantly from one area to the next. There was, for example, little secondary or college education anywhere in West Africa before World War I, while hundreds of Vietnamese attended several universities in French Indochina. At the elementary level the curriculum in Vietnamese schools included serious instruction in mathematics and the natural sciences, but African students were fortunate to be taught basic arithmetic and "practical sciences" such as hygiene and plant classification. Vietnamese colleges turned out genuine medical doctors, while the medical school at Dakar graduated only medical auxiliaries.[163] But even in Africa, where little science was taught, some French educators challenged the assumption that school children were inherently incapable of complex mathematical calculations or scientific instruction. Despite his advocacy of policies that relegated the best African students to vocational instruction, Georges Hardy rejected the view that Africans could never master the sciences or mathematics of the West. He argued that their level of social development had prepared few Africans of his day for instruction in these subjects, but he suggested that French tutelage could in time (duration unspecified) advance them to this level.[164]

Paul Bernard, director of the Ecole Normale d'Alger, was even more certain than Hardy of the capacity for improvement of the students in

[162]Hardy, *Une conquête morale*. The discussion of Hardy's views is based heavily on pp. 89–97, 205–7, 228–230, 339–48, supplemented by the background information in Gifford and Weiskel, "African Education," pp. 690–93; Gaucher, *Les débuts*, pp. 12–13; and Bouche, "Autrefois," pp. 116–19. For a stress on vocational education similar to Hardy's, see Ferry, *France en Afrique*, pp. 242–43.

[163]Gail P. Kelly, "The Presentation of Indigenous Society in the Schools of French West Africa and Indochina," *Comparative Studies in Society and History* 26 (1984), 524–29.

[164]Hardy, *Une conquête morale*, pp. 6–7, 223–26.

his charge. Bernard's lengthy *mémoire* on education in Algeria, presented to the 1900 colonial congress, was for the most part well suited to the anti-assimilationist mood that dominated the gathering. He outlined a curriculum that was geared to practice rather than theory, to the teaching of vocational skills and the inculcation of the work ethic. He gave some credence to the notion that Arab schoolboys' capacity for learning slowed after they reached puberty, though he confessed that he was not sure whether social or physiological factors were responsible. But he disagreed strongly with the dominant view that the Arabs and other colonized peoples could gain little from education, especially after the age of puberty, when their minds were absolutely closed to scientific thought and new ideas. Bernard pointed out that many educated Arabs and Berbers had gone on to distinguished careers in the professions. He even intimated that the nature of the educational system, rather than arrested development at puberty, might be the root cause of the failure of the majority of Arab students to excel in the sciences and liberal arts. The education that was available to all but a small fraction of the colonized, he observed, was so elementary and practical that it was difficult to know whether Arab thinking did in fact stagnate in adolescence and just how much most Arab students were capable of learning.[165]

Though confidence in the ability of the colonized to rejuvenate their societies through the adoption of Western science and technology was decidedly on the wane in French official and intellectual circles by the end of the nineteenth century, outspoken advocates of the possibility for improvement and advance on the part of non-Western peoples could still be found. Perhaps none of these was more contemptuous of arguments of racial incapacity and the notion that science and technology were a white man's monopoly than Jacques Novicov. The epitome of the cosmopolitan intellectual, Novicov, son of an Odessa industrialist, had adopted France as a second home.[166] In his early writings he espoused a view of social development that owed much to Spencer's pronouncements on the primacy of the struggle for existence between human groups. But by the late 1890s, Novicov had become a convinced internationalist, and he produced a succession of works attacking protectionism, the subjugation of women, and predictions of an imminent

[165]Paul Bernard in Rousseau, *Congrès international,* vol. 2, esp. pp. 385, 392–96, 400, 404–5. At the 1889–90 congress, Etienne Aymonier buttressed a lengthy plea for the use of the French language in colonial education with the argument that it was essential if the sciences were to be taught effectively to the Vietnamese; see "L'enseignement en Indo-Chine," in *Recueil des délibérations du Congrès colonial national* (Paris, 1890), vol. 2, pp. 325–44.

[166]Biographical details have been taken from Novicov's obituary by René Worms in *Revue Internationale de Sociologie* 20/7 (1912), 481–83.

race war on a global scale. In his 1897 work, *L'avenir de la race blanche*, he sought both to allay widespread fears of the yellow and black "perils" and to challenge the racist assumptions that informed contemporary writings on white supremacy and the fate of the white race. Novicov unequivocally dismissed the belief that the "so-called inferior races" were incapable of progress. He cited the achievements of educated blacks in the United States and Jamaica, of the Maoris in New Zealand, and of the Chinese and Japanese to refute the arguments of those who contended that non-Western peoples were doomed to wallow forever in ignorance and backwardness. He not only drew on contemporary evidence but pointed to the past civilizations created by the Chinese and Africans as proof of their capacity for improvement.[167]

In his eagerness to refute allegations of innate incapacity or inferiority, Novicov embraced the technological and scientific measures of human worth so often enlisted by his adversaries. In contending that social conditions, not racial attributes, determine achievement, he speculated on why the black and yellow races had not produced a scientist the caliber of Laplace. To demonstrate that even the Chinese, whom his contemporaries regarded as the race most resistant to change, were capable of improvement, he pointed out that they had set up a cotton mill in Shanghai with their own capital and directors.[168] For Novicov, as for British and French improvers earlier in the century, the key to improvement was education. In his works, science and technology persist as the ultimate standards by which to judge advance. Racial arrogance is rejected, but cultural chauvinism remains.

In prewar discussions of colonial educational policies, the British were characteristically less concerned than the French with general principles and attempts to devise approaches that could be applied uniformly throughout the empire. Surprisingly, given their English origins, social evolutionist ideas played a much less important role in policy formulation in this period than one might expect. The emphasis of key French writers—Le Bon, Saussure, Harmand—on assumptions of racial differentiation rooted in varying degrees of evolutionary progress was greatly muted in most British works and altogether absent in many. But racist convictions of a more diffuse and often less explicit sort were very much in evidence. They informed much of the opposition to colonial policies designed to promote acculturation among colonized peoples and there-

---

[167]Jacques Novicov, *L'avenir de la race blanche* (Paris, 1897), pp. 74–99, 105–13, 119ff., 138–40.
[168]Ibid., pp. 108–11.

by to continue the work of the improvers in India in the first half of the century. Racist assumptions also frequently motivated those who advocated indirect rule through "native" leaders and institutions. This shift from an emphasis on Europeanization to the preservation of indigenous non-Western cultures was the British counterpart to the French move from assimilation to association in this era.[169]

As was true for the French, political considerations were often as important as racist convictions in accounting for growing British opposition to Europeanization and support for policies of indirect rule. It was, for example, the nationalist threat posed by the Western-educated Indians, especially the Bengalis, whose disorientation and discontent Le Bon had so vividly caricatured, that impressed most British authors.[170] They tended to ignore or mention only in passing Le Bon's more fundamental argument (which had so strongly influenced French colonial commentators) that learning must be adapted to the level of development attained by the race for which it is intended. Nonetheless, the question of the compatibility of race and curricular content did figure in British thinking. In a passage quoted approvingly by Jules Harmand, John Strachey challenged the long-cherished belief that "opening the gates of the temple of Western science" to Hindu students would prove instrumental in the regeneration of Indian civilization. On the contrary, Strachey observed, it had simply created a "new caste" of disgruntled pseudo-Europeans, infatuated with themselves and utterly lacking in concern for others.[171] Decades later, Arthur Mayhew sought to account for what had gone "wrong" with Western-educated Indians (that is, why they had become so nationalistic) by pointing out the fallacies in the improvers' approach to higher education. Macaulay and his allies had assumed that all humans were alike in mental makeup, but the subsequent development of "evolutionary philosophy" had demonstrated the fundamental differences in kinds of intelligence between races. In contrast to the Europeans, Mayhew argued, the Indians had emphasized the emotional over the "cognitive" side of their intelligence. They could develop an argument as effectively as their European counterparts, but they could not criticize it or relate it to other aspects of their learning or to general laws. Thus, the sort of education suitable for the Indians was necessarily very different from that best for British students.[172]

---

[169]For sample discussions of the links between racism and these shifts, see Hutchins, *Illusion of Permanence,* esp. chaps. 5 and 6; Cairns, *Prelude to Imperialism,* pp. 204–20, 238–39; and Gifford and Weiskel, "African Education," esp. pp. 685–90.

[170]For the classic statement of this position, see Valentine Chirol, *Indian Unrest* (London, 1910), chaps. 17–21.

[171]Harmand, *Domination et colonisation,* pp. 263–4.

[172]Mayhew, *Education of India,* pp. 55–59.

Arguments that British authors made rather tentatively for Indians were broached with far greater assurance for Africans. In asserting that different kinds of education were appropriate for different races, Sidney Shippard strongly disavowed any suggestion that he equated the "highly intellectual races of India" with the "natives" of South Africa. Shippard lamented the fact that English education had bred discontent in India by arousing ambitions that could not be satisfied and had caused its Indian recipients to despise "those useful manual occupations" that had been pursued by their forefathers. But he considered the very introduction of this sort of education into South Africa "absurd and impossible." Citing a conversation with a missionary teacher, he contended that the two thousand years of development separating the white from the black race rendered the latter utterly unable to comprehend such forms of learning as Greek or mathematics. Differential rates of racial development, he concluded, destined the "vast majority of the natives of South Africa" to be "hewers of wood and drawers of water to the end of time."[173]

The ubiquitous Harry Johnston sought to present the social evolutionist position in the most scientific of guises by interspersing his remarks on African education with references to natural laws, orchids and primulas, dogs and pigeons. Though he reversed Le Bon in concluding that some exceptional Africans were superior to the average European but most blacks were "far below" whites in intelligence, Johnston strongly concurred with Le Bon's criticisms of educators who believed that a few years of instruction could bridge differences that it had "taken Nature herself a thousand years" to develop. The attempt to transport "our backward fellow-man" from the Stone Age to the Age of Steel, he charged, could not possibly succeed, because it sought to abridge the evolutionary process. Savages whom Johnston judged to be at a level of development comparable to that of the peoples inhabiting Britain in the early postglacial epoch would not be able "to grasp or assimilate one-third of the wonders presented [for their] consideration, and [would] probably even suffer from the shock." Though Johnston made some distinctions between peoples, he doubted that the blacks of West Africa as a whole could be rendered even "normally intelligent" or self-governing within as many as three generations.[174]

For a small number of European observers in the prewar era, opposition to the Europeanization of Africans and Asians arose from a genuine

----

[173]Sidney Shippard, reply to A. P. Hillier's paper "The Native Races of South Africa," *Proceedings of the Royal Colonial Institute* 30 (1898–99), p. 53. For similar views by another South African educator, who quotes Walter Bagehot for support, see William Grenswell, "The Education of the South African Tribes," ibid. 15 (1883–84), pp. 82–84.

[174]Johnston, "British West Africa," pp. 91–98.

appreciation of the cultures of colonized peoples and a concern for their preservation. Perhaps the most noteworthy of these was Mary Kingsley, who in her admiration for Africans and African culture was matched by only a handful of anthropologists, such as the remarkable Frenchman Maurice de la Fosse. Kingsley contrasted the physically robust, bright, and self-assured peoples of the interior and Islamicized areas with the coastal peoples; the latter, she argued, had been deprived by Christianity of their own customs and identity and left with a culture that was "second-hand, rubbishy [and] white." By the 1890s the Christianized Africans themselves had come to accept this highly unfavorable assessment of the results of their long struggle to learn European ways.[175] As we have seen, Kingsley believed that African intelligence differed fundamentally from European, and this led her to conclude that Africans could not genuinely adopt European values or views—nor should they. Because she held that it was possible for backward peoples to be uplifted only by more advanced societies of their own race, she saw Islam, not Christianity, as the hope of the African continent.[176]

The impact of racist and social evolutionist views on the actual formulation of colonial policy varied as widely from one area to another in the British as in the French empire. In China, which with the exception of trading enclaves remained free of direct colonial rule, racist attitudes and social exclusivism were readily apparent in everyday interaction between Europeans and Chinese. But resistance to assumptions of innate inferiority was strong among the missionaries and lay teachers who took on the responsibility of educating ever greater numbers of Chinese. The resolve of British educators in this regard was bolstered by their American counterparts, whose influence was vastly greater in China than in areas formally colonized by the European imperial powers. Though there were both British and Americans who doubted that the Chinese would be able to master fully the sciences and machines of the West, the majority view was succinctly stated by W. A. P. Martin, an American missionary and professor of international law at a succession of Chinese universities: "It would be superfluous to vindicate the Chinese from the charge of mental inferiority in the presence of that immense social and political organization which has held together so many millions of people for so many thousands of years, and *especially* of [indus-

---

[175]E. A. Ayandele, *The Missionary Impact on Modern Nigeria, 1842–1914* (New York, 1967), pp. 249–50.

[176]Kingsley, *West African Studies*, (London, 1901), pp. 326–31, and *Travels in West Africa*, pp. 20, 205, 403. For an early expression of the Muslims-as-civilizers view, see d'Eichthal and Urbain, *Lettres*, pp. 17–18.

trial] arts, now dropping their golden fruits into the lap of our own civilization, whose roots can be traced to the soil of that ancient empire." Martin did feel that "servility to antiquity" and a language that he dismissed as "so imperfect a vehicle of abstract thought" might impede Chinese acquisition of Western knowledge and techniques. But he insisted that these were cultural, not biological, barriers and that with Christian education as a catalyst a "stupendous intellectual revolution" was in progress.[177]

Other missionary educators also grounded their faith in the potential of the Chinese on the abundant evidence of past achievements among a people they pictured as patient, hardworking, and clever. Alexander Williamson, the Chinese representative of the National Bible Society of Scotland, reminded his readers that the early Chinese discovery of numerous technologies, including gunpowder and printing, had provided the basis for the West's transformation in "modern times." He saw these breakthroughs as proof of the "inventive genius" of the Chinese and judged that the "ease with which they are acquiring the ability to utilize European science . . . prognosticates a grand future." In contrast to observers earlier in the century, who conceded the capacity of the Chinese to imitate the techniques of the West but doubted their ability to innovate on their own,[178] Williamson believed that competition with Westerners would push the Chinese to discover new facts, machines, and production processes that would benefit all humankind. John Nevius deplored the tendency of Westerners in China to view their own present superiority in science and technology as proof of the innate inferiority of the Chinese intellect and to forget the great contributions the Chinese had once made in both these fields. He pointed out that if they were to go back only a few hundred years, these boastful Europeans would find the situation completely reversed. What folly then, he reasoned, for them to conclude that a people's backwardness at one point in time doomed them to perpetual inferiority—especially when the culture in question had produced the original inventors of so much of man's basic technology. B. C. Henry, who taught for ten years in Chinese mission schools, was confident that in both the practical and the theoretical sciences the Chinese had a "glorious future." He believed it inevitable that the intellectually gifted Chinese would soon regain in the

---

[177]W. A. P. Martin, *The Chinese* (New York, 1898), pp. 147–49 (my italics).

[178]E.g., Williams, *Middle Kingdom*, vol. 1, pp. 143–45, and vol. 2, pp. 178–79; Huc, *Chinese Empire*, p. 302; Charles Gutzlaff, *China Opened* (London, 1838), vol. 1, p. 507, and vol. 2, p. 159; and "De Guignes' 'Voyage à Pékin,'" *Edinburgh Review* 14 (1809), 425–26.

sciences and engineering the position among the "front rank of nations" they had so long occupied in the past.[179]

The confidence of missionaries and lay educators in the Chinese capacity for instruction in Western sciences and technology was reflected in the curricula of British- and American-sponsored schools in China. As we have seen, as early as the 1830s schools had been established to train Chinese medical doctors. From the 1880s major efforts were made to upgrade the teaching of the sciences and mathematics in both missionary schools and institutions funded by Western philanthropists. Western instructors of some distinction also taught in Chinese institutions, such as a school specializing in mechanical engineeering established at Shanghai in 1867 and the Wujang College of Mining and Engineering.[180] The goal of those who went out to teach physics and chemistry to Chinese students or instruct Chinese engineers in the latest Western techniques of bridge-building was nothing less than a fundamental reorientation of Chinese education that in turn would generate profound changes in Chinese culture and society. Few of them challenged American engineer James Wilson's conviction that the Chinese possessed a "natural intelligence" equal to that of any other race, an intelligence ensuring that they would soon join the "march of progress."[181] By the last decades of the nineteenth century it was clear to such men as Martin and Wilson that if the Chinese people wished to prosper, or perhaps even to survive, they must adopt and improve upon Western science and technology.

Even more than the French critics of higher education in Indochina, critics in India found it difficult, however strong the racist sentiment in both civilian and official circles, to undo the work of the early nineteenth-century improvers. Fed by a seemingly insatiable Indian demand for education in English-language schools, the number of university students grew rapidly, despite the warnings of such observers as Le Bon and Strachey. The doubts of educators regarding the potential of the "quite underdeveloped minds" of Indian students to understand scientific instruction when they could not even speak or write "correct English"[182] may have slowed but did not stop the establishment and

179W. A. P. Martin The Lore of Cathay or the Intellect of China (Edinburgh, 1901), pp. 8–9, 23–29; Alexander Williamson, Journeys in North China (London, 1870), vol. 1, pp. 38–39; John Nevius, China and the Chinese (New York, 1869), pp. 279–80; and B. C. Henry, The Cross and the Dragon (New York, 1885), pp. 428–49.

180Jessie Lutz, China and the Christian Colleges (Ithaca, 1971), pp. 21, 29, 68, 111–12; Buck, American Science and China, pp. 33–36, 47–48, 82–83; L. G. Morgan, The Teaching of Science to the Chinese (Hong Kong, 1933), pp. 63–69; and Norman Goodall, A History of the London Missionary Society, 1895–1945 (Oxford, 1954), pp. 488–90.

181James H. Wilson, China: Travels and Investigations in the Middle Kingdom (New York, 1887), pp. 84, 311–12.

182S. Lobb, "Physical Science in the Calcutta University," Calcutta Review 106 (1871), 326–27.

expansion of schools of medicine and engineering, or stem the flow of Indians receiving higher degrees in science and mathematics.[183] Though the numbers of Indians receiving bachelor of science degrees remained miniscule even within the small minority who received a higher education, and though technical training was neglected, racist convictions of Indian incapacity were less responsible for these trends than Indian preferences and British shortcomings. An overwhelming number of Indian students opted for degrees in literature and the humanities because these programs were reputed to be easier than the sciences; because they carried more prestige than engineering or technical training; and because they were more likely to lead to relatively high-paying positions in colonial administration or in business and the professions. As John Strachey pointed out, the rather anemic state of scientific instruction and technical training in India was not at all surprising, given the neglect of these subjects in England itself. Britain was losing its long-held industrial preeminence to such rivals as Germany and the United States in this era; thus, its reluctance to equip and educate further competitors, especially colonized ones, is not surprising.[184]

Despite these trends, some of the most prominent British experts on India, including J. S. Mill, Henry Maine, and George N. Curzon, strongly advocated scientific and technical instruction for Indians and expressed confidence in their ability to do well in these fields.[185] Their faith was well rewarded: India produced several scientists of world renown in the last half of the nineteenth century. J. C. Bose, for example, did pioneering work on the polarization of electric rays by crystals; P. C. Ray discovered the compound mercurous nitrite.[186] In India, these and other Indian scientists had to fight prejudice against their being given government appointments commensurate with their attainments, as well as a general tendency in Indian institutions to stress applied at the

[183]The best estimates of the state of scientific education in India in this period can be found in Government of India, *Review of Education in India in 1886* (Calcutta, 1886), esp. pp. 27–28, 145–51; and Government of India, *Progress of Education in India* (annual reports, beginning 1893).

[184]John Strachey, *India: Its Administration and Progress* (London, 1888), p. 269. For a detailed survey of the frustrations encountered by those who were seeking to upgrade scientific and technical education in Britain in this era, see D. S. L. Cardwell, *The Organisation of Science in England* (London, 1972), chaps. 5 and 6.

[185]J. S. Mill, *Memorandum of the Improvements in the Administration of India during the Last Thirty Years* (London, 1858), pp. 66–67, 76–77, 82; Henry Maine, *Short Essays and Reviews on the Educational Policy of the Government of India from the "Englishman"* (Calcutta, 1866), pp. 8–9; Thomas Raleigh, *Lord Curzon in India* (London, 1906), pp. 312, 332–34, 360. For more general opinions, see F. W. Thomas, *The History and Prospects of British Education in India* (Cambridge, Eng., 1891), pp. 48–49, 54–55, 141–42, 146.

[186]P. C. Ray, *Life and Experiences of a Bengali Chemist* (Calcutta, 1932), includes a discussion of J. C. Bose's discoveries and their impact on other Indian students. See also Shiv Visvanathan, *Organizing for Science* (New Delhi, 1985), pp. 28–38.

expense of theoretical research. But in Europe their scientific peers readily acknowledged the research talents of Indian scientists and offered them funding and facilities.[187] The discoveries of such innovators belied the assumption of Lord Landsdowne, Viceroy in the late 1880s, that "original scientific research demands mental and physical qualifications which are not apparently found in races bred in tropical climates."[188] The achievements of Indian researchers did not silence the critics of scientific and technical education for Indians, but they stiffened the resolve of those British officials and educators who were arguing against racial determinism and for the advancement of Indians in these fields. In winning British support and laying the basis for Indian-run institutions, scientists such as Ray and Bose helped to sustain and build upon the initiatives taken by the improvers in the middle of the century.

As was the case for the French, British educators who wished to fashion policies on the basis of racist and social evolutionist assumptions found their most abundant opportunities in Britain's growing empire in Africa. Policies and institutional arrangements favored earlier, when the improvers were more dominant, were less well established in Africa, where most colonies were acquired only in the last decades of the century. In addition, in Sierre Leone and such trading enclaves as those at Cape Coast and Lagos, which the British had controlled formally or indirectly for decades, African converts to Christianity—not British missionaries or officials—had been responsible for most educational initiatives.[189] Ironically, in the early 1870s the efforts of African pastors affiliated with the Church Missionary Society to win greater control over both affairs within their own congregations and Christian churches in Africa more generally provided ammunition for those who argued that racial differences between Africans and Europeans made special educational programs in Africa imperative. The resistance to Europeanization that was central to the "Native Pastorate Controversy," plus claims by African leaders such as Edward Blyden that there were innate differences between races and that educational systems should take these into account, meshed neatly (though unintentionally) with rationales

[187]On restrictions in India, see Ray, *Bengali Chemist*, pp. 78–81; Deepak Kumar, "Patterns of Colonial Science in India," *Indian Journal of the History of Science* 15/1 (1980), 108–11, and "Racial Discrimination and Science in Nineteenth Century India," *Indian Social and Economic History Review* 19/1 (1982), esp. pp. 68–73. On better conditions in Europe, see Ray, *Bengali Chemist*, pp. 59, 68, 131–32; and J. C. Bagal, *Pramatha Nath Bose* (New Delhi, 1955), pp. 26–29.

[188]Kumar, "Racial Discrimination," p. 74. For discussions of these attitudes in official circles generally, see Patrick Geddes, *The Life and Work of Sir Jagadis C. Bose* (London, 1920), pp. 33–34, and Mayhew, *Education of India*, pp. 54–59. The latter work develops the evolutionary line of reasoning that French administrators favored.

[189]Gifford and Weiskel, "African Education," pp. 678–84.

offered by those who advocated education geared to the Africans' allegedly lower level of evolutionary development.[190] African vulnerability to social evolutionist arguments was also increased by the fact that Negroes were invariably singled out as members of a "race" that had not evolved beyond the savage state. Most of the missionaries and officials who shaped educational policy in British Africa from the 1880s onward saw little purpose in instructing peoples at this level of development in the sciences and advanced mathematics.

The move to a system which, like its counterpart in French West Africa, emphasized elementary at the expense of secondary and university education, and stressed the need for vocational training rather than instruction in the liberal arts and sciences, can be linked to the British missionary offensives of the last decades of the nineteenth century. Whether the conviction that the African was a racially inferior being, "half-devil, half-child," arose mainly from shifts in missionary education or from exposure to James Hunt's brand of anthropology, the effects for Africans were the same.[191] A strong consensus emerged among British missionaries and colonial officials, who wrested control over education away from the African pastors in this period, that teaching in African schools should concentrate on religious and moral instruction, hygiene, and elementary reading and arithmetic skills. Many writers emphasized the need to teach Africans the "dignity of labor" and "habits of industry," and in East Africa, where Indian migration was concentrated, to prepare them to compete with "more advanced" peoples such as the Gujaratis in the rapidly expanding market economy. Most educators agreed that the highest education Africans were likely to need was vocational training, which would give them the practical skills in carpentry, ironworking, and machine repair which were assumed to be in demand in the colonial economy. Some observers found even vocational education too advanced for peoples as "backward" as the Zulu, whom Henry Moore reckoned were not even up to classes in improved agricultural techniques.[192] Instruction in science and mathe-

---

[190]Hollis R. Lynch, "The Native Pastorate Controversy and Cultural Ethnocentrism in Sierre Leone, 1871–1874," *Journal of African History* 5/3 (1964), esp. 395, 402–5. The reemergence of these views in the late 1890s may have further buttressed policies that had already been put into effect; see Ayandele, *Missionary Impact*, pp. 246–51.

[191]For an emphasis on shifts in theological training, see Andrew Porter, "Cambridge, Keswich, and Late Nineteenth-Century Attitudes to Africa," *Journal of Imperial and Commonwealth History* 5/1 (1976), esp. 27–28. For a stress on Hunt and social evolutionism, see J. F. A. Ajayi, *Christian Missions in Nigeria, 1841–1891* (Evanston, Ill., 1965), pp. 260–64. Ayandele (*Missionary Impact,* pp. 213–17) takes a middle position between Porter and Ajayi.

[192]For examples from the period, see C. W. Orr, *The Making of Modern Nigeria* (London, 1911), pp. 262, 269; F. D. Lugard, *The Dual Mandate in Tropical Africa* (London,

matics was meager even in teacher-training institutions. Those who recommended science classes thought they should be taught at the "most elementary" level. The idea, advanced by educated Africans in the 1860s, that West Africans ought to have a medical school of their own had been summarily dismissed by British officials and doctors decades before the missionary offensive.[193] Had it been revived in the 1880s or 1890s, most British educators would have regarded it as the wildest of fantasies.

Despite the apparent rout of the nonracist improvers in British Africa, however, long before World War I there were significant doubts about the educational policies being pursued there and about the assumptions of innate differences between races on which those policies were based. Many missionaries continued to share David Livingstone's view that though the Africans as a whole belonged to an inferior race, some African peoples were "not naturally" inferior in intelligence and spiritual endowments to Europeans and were quite capable of improvement.[194] Though equally selective, other missionaries were even more forceful in their challenges to those who believed that the African capacity for improvement was limited. The Reverend Milum, who challenged Harry Johnston's assertion that the Africans were all savages using only the crudest of tools, went on to dismiss the notion that they were twenty thousand or even two thousand years behind the Europeans. Milum found that "generally speaking" West Africans made "apt scholars" and were the equals of any other people in learning new skills and "becoming clever and learned men." Milum met the social evolutionists head on: "It may, indeed, take thousands of years for the outward man to alter, especially if there be no intermixture of races. But the mind, which Providence has given all men, is on another footing. It may be developed rapidly under favorable conditions, and it is the business and duty of England to bring this about as speedily as possible."[195]

1922), pp. 431, 442–50; T. F. Victor-Buxton, "Missionaries and Education in East Africa," *Journal of the African Society* 8 (1908–9), 279–86; and the essays by W. G. Bennie (esp. pp. 253–59), S. de Lendestrey (pp. 261–64), and the Director of Education, Nigeria (pp. 265–70) in *Report of the Imperial Education Conference* (London, 1923). The best historical treatments of these themes are included in Arthur Mayhew, *Education in the Empire* (London, 1938), pp. 112–14, 136–39; and Goodall, *London Missionary Society*, pp. 479, 490–91.

[193]Ayandele, *Missionary Impact*, p. 287; H. A. Junod, "The Native Language and Native Education," *Journal of the African Society* 5 (1905–6), 7–14; and Gifford and Weiskel, "African Education," p. 685. Junod's essay provides a superb example of the "work at their own level" approach.

[194]David and Charles Livingstone, *Narrative of an Expedition to the Zambesi and Its Tributaries* (New York, 1866), pp. 624–25; and William Monk, ed., *Livingstone's Cambridge Lectures* (London, 1858), pp. 83–85, 124–25.

[195]Reply to H. H. Johnston, in *Proceedings of the Royal Colonial Institute* 20 (1888–89), pp. 121–22.

Even Sir Frederick Lugard, one of the most influential proponents of indirect rule and a champion of morality and manual skills as the core of the African curriculum, admitted the great need for African doctors and civil servants, as well as surveyors, telegraph operators, and mechanics. He had high praise for African service in technical posts and administration more generally, and he challenged the view that "literary educations" were wasted on Africans. At one point in his influential study on administration in Africa, Lugard suggested that exceptional Africans were qualified to be trained in law and medicine—but in English, not African, universities.[196] The educator W. G. Bennie was as confident as Lugard that selected Africans could handle the work in British universities and a good deal more forceful in supporting measures to open professional education to them. In a paper read before the Imperial Education Conference in 1923, Bennie cited the findings of an examiner in logic at the South African Native College at Fort Hare who had concluded that the brightest African students were on a par with their white peers. Bennie pointed out that African students who had qualified for medical or legal training in England had done well and gone on to professional practice.[197] Such recommendations, though limited to exceptional African students, struck at the very foundations of the social evolutionists' approach. Nothing has proved more devastating than exceptions to racist rules.

In the differences between colonial educational policies for the Africans on one hand and the Indians and Chinese on the other, the continuing importance of material measures of human potential are apparent. Those who argued for higher education and instruction in the advanced sciences and mathematics for Indians and Chinese almost invariably prefaced their recommendations with praise of the high achievements of these peoples in ancient times. References were made to the Indians' great scientific treatises; the Chinese were lauded for their inventiveness and technological contributions to all peoples. Though nineteenth-century European writers agreed that neither the Indians nor the Chinese had lived up to their full potential, that the work of both had stagnated beyond a certain point in their evolution, most observers were convinced that past attainments were proof of present abilities. Given their undeniable accomplishments in the material realm, it was hard to argue convincingly that either the Indians or the Chinese were innately incapable of mastering the machines and equations of the West. The Africans, by contrast, continued to be lumped with the Polynesians and

[196]Lugard, *Dual Mandate*, pp. 442–43, 451, 458.
[197]W. G. Bennie, "Special Means of Educating the Different Non-European Races within the Empire," in *Report of the Imperial Education Conference*, pp. 260–61.

Amerindians, who were thought to have advanced barely beyond the most primitive stages of human evolution. Their apparent lack of science and the alleged crudeness of their tools suggested that efforts to teach European scientific ideas to the Africans and introduce complex European machines into Africa (without whites to run them) would be wasted. Africans simply did not have the sort of mind or level of intelligence required to comprehend, much less themselves create, these wonders of Western imagination and research. Thus, the burden of the material measure of human worth, now concentrated in a combination of scientific and technological accomplishments, continued to weigh much more heavily on the Africans than on the Indians or Chinese.

Missing the Main Point: Science and Technology in Nineteenth-Century Racist Thought

The understandable post-Hitlerian obsession with racism has pervaded virtually all aspects of recent research on the interaction between peoples of European descent and non-European peoples in the nineteenth century. Studies of attitudes and contacts with respect both to intellectual discourse and social relationships have almost invariably been reduced to the analysis of perceptions of racial differences and the assumptions, policies, reactions, and social tensions that these engendered. Most recent work conveys the impression that in the age of high imperialism virtually all Europeans at every social level were convinced of their innate superiority over other peoples and expressed that conviction in racist terms. Nineteenth-century European authors who were critical of aspects of African or Asian societies or suggested ways in which Western societies had advanced beyond all others are routinely, and sometimes incorrectly, labeled racists. Colonial policies, whether advocating the preservation or the transformation of non-Western cultures, are evaluated according to considered judgments about the extent to which they derived from or reflected racial prejudice. Social discrimination, which was usually based mainly on rather crude and unthinking perceptions of physical-*cum*-cultural differences, is assumed to have been rooted in the findings of intellectual inquiries into the nature and varieties of humankind.

It would be absurd to deny that racism strongly influenced European attitudes toward non-Western peoples in the nineteenth century. But given the tendency toward racist reductionism that has been dominant in recent decades, it is important to stress that many European thinkers, both major and "middle range," were nonracists or anti-racists.[198]

198The views of such prominent thinkers as J. S. Mill, Theodor Waitz, and Jacques Novicov, discussed above, give some indication of the strength of anti-racist thinking in this era.

Equally important is the fact that the majority of nineteenth-century Europeans used the term "race" interchangeably with "nation," "people," or "ethnolinguistic group." Thus, its frequent occurrence in nineteenth-century speech and writing should not be seen as an accurate gauge of European adherence to racism in its narrower and more significant sense: the belief that humankind can be divided into subgroups that differ innately and fundamentally in physical and mental attributes.

As we have seen, for many nineteenth-century colonial observers, policymakers, and educators, race was not the main issue; for some, no issue at all. Racial differences were not a matter of concern for the British improvers who sought to introduce Western learning in the English language into India in the first half of the century. However much they deprecated Chinese culture, many of the missionaries and educators who throughout the nineteenth century pushed for China's regeneration through Western learning and technology dismissed or simply ignored race as a major barrier to Chinese borrowing. In their essays on the impact of European imperialism in Asia, Marx and Engels avoided racial terminology and eschewed racial categories of analysis.

This indifference to racial questions in the first half of the century is all the more remarkable because both improvers and critics of colonial expansion were active in an era when the ideas of those professing to offer scientific proofs of racial inequality were widely propagated and when perhaps the most pernicious variant of racist thinking, polygenism, was ascendant. Despite rapidly increasing racial discrimination and friction at the level of social interaction in the colonies, racial thinking appears to have played a major role in policy-making only in Africa, and even there its influence cannot be meaningfully separated from other considerations. By the last decades of the century it *was* difficult to deal with colonial issues without taking racist ideas into account. But even as European global power and arrogance peaked, nonracist educational policies continued to be pursued in China and India, and the racist, social evolutionist foundations on which British and French policy rested in Africa had begun to be eroded by the doubts and counterevidence of travelers, officials, and missionary educators.

It was not just that race was often a peripheral issue or that racist assumptions were strongly contested. There were other criteria by which Europeans measured the achievements and potential of non-Western peoples, criteria that numerous authors applied without any reference to race. These included the notions discussed in earlier chapters about what it meant to be civilized as well as judgments concerning such specific cultural mores as those governing the treatment of women, legal procedures, and marriage customs. But no standards were more frequently invoked in this era than attainments in science and technology. They became central to the European sense of what it meant to be civilized. They were by far the aspects of non-Western cultures most frequently scrutinized by the plethora of nineteenth-century authors who sought to construct hierarchies ranking human groups. In fact, though scientific and technological proficiency was often invoked by those who argued for European racial superiority, in some ways it was more fundamental than racial categories in European thinking in this era. Race in its broad and vague sense certainly crops up more in the literature than comparisons of tools and concepts of the cosmos. But the superiority of the Europeans in the latter areas was virtually universally accepted, whereas many European thinkers challenged the validity of racist suppositions in the narrower sense of innate and permanent differences. Indeed, many authors challenged the racist categories themselves. In addition, scientific and technological achievements had provided key standards by which most non-Western peoples were judged long before racist arguments were applied to them. For thinkers and policymakers concerned with the Chinese and Indians, racism in the narrower sense and at the level of intellectual exchange—as opposed to social interaction—was rarely an issue of central importance in this or any era. By contrast, questions relating to the diffusion of Western science and technology had been debated by travelers and missionaries as early as the sixteenth century, and they were prominent in colonial policy debates from the late eighteenth century to the decades of decolonization after World War II.

These arguments about the relative importance of scientific-technological versus racial standards apply less well to Africa, where European judgments were strongly influenced by the legacy of centuries of enslavement and the situation of black peoples in New World societies. My reading of the secondary evidence also suggests that they do not work for the United States, South Africa, and similar settler societies where large numbers or a majority of non-Europeans were present in the population and where slavery was long a prominent feature of social and economic life. As George Frederickson and others have shown, ideologies upholding white supremacy can differ significantly by time and place.[199] Studies focused on the pre–Civil War era, for example, suggest that the capacity for scientific thought, mathematical computation, or invention of the black slaves in the American South or their African forebears was not a critical issue for those who defended the institution of slavery on racist grounds.[200] Similarly, though English

[199]See George Frederickson, *White Supremacy: A Comparative Study in American and South African History* (Oxford, 1981).

[200]See William S. Jenkins, *Pro-Slavery Thought in the Old South* (Chapel Hill, N.C., 1935); and Fredrickson, *Black Image,* esp. chaps. 2 and 3.

travelers and missionaries had much to say about the tools and attitudes toward work of the peoples of South Africa, the racist doctrines of the long-resident Afrikaners were rooted primarily in perceptions of physical and moral differences and the Afrikaners' interpretation of the Bible. Having lived for centuries in isolation, first in the Karoos and later deep in the interior, the Boers knew little of the scientific or industrial revolutions in Europe until the last decades of the nineteenth century. Their initial education in these matters was painful and meted out by British missionaries, entrepreneurs, and officials whom they viewed as oppressers. Thus, it was unlikely that they would build their case for supremacy over the black majority on the values and achievements of industrial Britain.[201]

The realization that by reducing nineteenth-century European perceptions of and reactions to colonized peoples to manifestations of racism we may be missing the main point—focusing on subordinate themes and missing more dominant ones—suggests that we may need to modify some of our basic assumptions about European intellectual history in this era. At the very least, the ideas and positions of such thinkers as Thomas Babington Macaulay, who are rather frequently and often unthinkingly lumped with such champions of white racial superiority as Robert Knox, must be reassessed.[202] Cultural chauvinism, which Macaulay possessed in abundance, ought not to be confused with racist convictions of innate superiority or inferiority. Conversely, sympathy for subjugated peoples, and even considerable understanding of their cultures, cannot necessarily be taken as proof that an individual was free of racial prejudice. However much Mary Kingsley may have defended African culture from missionary charges of savagery and degradation, she clearly held racist views. Though most recent writers have tended to play down this side of Kingsley's thought, her conviction that the African mind was fundamentally different from the European was singled out by Western-educated African contemporaries as false and deeply harmful to the cause of African advancement.[203]

A move away from racist reductionism would also help to focus attention on much-neglected but significant nineteenth-century authors who ignored or questioned the validity of categories of analysis based on

[201]For studies of the sources of white supremacist ideology in South Africa, see Fredrickson, *White Supremacy*; T. Dunbar Moodie, *The Rise of Afrikanerdom* (Berkeley, Calif., 1975); and Leonard Thompson, *The Political Mythology of Apartheid* (New Haven, Conn., 1985).

[202]As John Clive's fine biography *Macaulay* illustrates, this process has begun.

[203]John Flint's "Mary Kingsley: A Reassessment," *Journal of African History* 4/1 (1963), 95–104, provides an exception to this rule in its frank exploration of Kingsley's racist sentiments. For a contemporary critique of her positions by an African writer, see Mark Hayford, *Mary H. Kingsley from an African Standpoint* (London, 1901), esp. pp. 5–6.

racial distinctions. It would prod us to give serious consideration to other standards by which non-Western peoples were judged and alternative views of the essential qualities that differentiated them from Europeans. Thus, though the vast majority of Europeans may indeed have considered themselves superior to Africans or Asians, significant numbers did not see or express this superiority in racist terms. For many of these the conviction that they possessed vastly better tools and weapons and attitudes toward work and discipline, or that they knew better how to treat women and to write legal codes, was sufficient to justify European conquest, commercial expansion, and efforts to educate and uplift the "benighted" peoples of the non-Western world.

PART THREE

THE TWENTIETH CENTURY

Mistah Kurtz—he dead.

Joseph Conrad, *Heart of Darkness* (1899)
T. S. Eliot, "The Hollow Men" (1925)

In this etching, *Stormtroops Advance under Gas*, Otto Dix captures the dehumanizing effects of scientific-industrial warfare, which had much to do with the crisis of European confidence in the civilizing mission in the years after World War I. (Reproduced by the courtesy of the Trustees of the British Museum)

CHAPTER 6

The Great War and the Assault
on Scientific and Technological
Measures of Human Worth

A s EARLY as the eighteenth century, some European thinkers were troubled by the increasing tendency to decide which peoples were civilized and which savage on the basis of evidence of scientific advance and technological accomplishment. Some observers implicitly questioned the centrality of these gauges by subordinating them to alternative standards such as religion; others mentioned them only in passing or ignored them altogether. A handful of early authors explicitly questioned their validity. In 1744, for example, the English traveler William Smith doubted that the Europeans' technological proficiency provided them with a quality of life superior to that experienced by the Africans. Early in his journey Smith had repeatedly dismissed the Africans as lazy and ignorant savages, but after visiting the "most potent city" of Benin and its environs, he concluded that the "natural and pleasant" life of the Africans was in many ways preferable to that of the Europeans. Noting that bestiality and sodomy, which he claimed were "common" among Europeans, were unheard of among the Africans, Smith wondered whether the Europeans who had "sought so many Inventions" and "put so many Restrictions upon Nature" could possibly be truly happy. That the restrictions Smith had in mind were more social than scientific is indicated by his tirade against celibacy, chastity, and the excessive modesty of priests and nuns. But he returned to England convinced that the Europeans had at least as many "Idle and Ridiculous Customs and Notions" as the Africans he had encountered.[1]

Several decades later the French traveler Cossigny Charpentier

---

[1]William Smith, *A New Voyage to Guinea* (London, 1744), pp. 233–34, 245, 267.

reached similar conclusions after a visit to China. He noted that in ancient times the Chinese had been far more inventive than the Europeans, but in recent centuries the Europeans had forged far ahead in the arts and sciences. He wondered, however, whether these advances had made them any happier—whether their quest to excel, to make grand discoveries, to be the best, had made them any wiser or better governed. A Dutch envoy to China in the years just before Charpentier's visit had reached a similar conclusion, though he restricted his reservations about European development to technology and its impact on work patterns. A. E. Braam van Houckgeest remarked on several occasions, particularly during the early stages of his stay in China, on the "simple and singular" construction of various Chinese machines from water mills to wheelbarrows. Though he found these and other examples of Chinese technology much less complicated and imposing than their European counterparts, he stressed that they were just as efficient and a good deal less expensive to build and maintain. His assessments were clearly designed to counter the widespread assumption of his contemporaries that better necessarily meant bigger, newer, and more complex.[2]

The themes of the noble savage and unspoiled tropical paradise alluded to by William Smith were more fully developed in other works on Oceania, North America, and Africa published in the eighteenth and well into the nineteenth century. Even though the main targets of the authors of these accounts were the social conventions and institutional restrictions of the "civilized" Europeans, they implicitly raised questions about the Europeans' obsession with work, scientific analysis, and continuous material innovation through their depiction of the unhurried and seemingly carefree lives of "savage" peoples. Ironically, some of the most explicit statements of the deleterious effects of excessive mechanization on the quality of human life are contained in accounts of Africa written in the late nineteenth century, when the noble savage was no longer in fashion and the Europeans' edge in matters technological and scientific had reached awesome proportions. The Scottish missionary Duff Macdonald, for example, contrasted the Africans' "stillness and repose that is in beautiful harmony with the scenery around" with the "crowds of pale-faced men and girls rushing along almost mechanically in response to some factory bell."[3] He derided the notion that inventive-

---

[2]Cossigny Charpentier, *Voyage à Canton* (Paris, 1799), pp. 95–96; A. E. Braam van Houckgeest, *An Authentic Account of the Embassy of the Dutch East-India Company to the Court of the Emperor of China* (London, 1798), vol. 1, pp. 72–75, 96–97. It is probable that Braam's views reflected the widespread preference of eighteenth-century French innovators and political economists for the small machine. Traian Stoianovich is presently completing a study on patterns of production in prerevolutionary France that will include a detailed discussion of the advocacy in this era of the small machine.

[3]Unless otherwise noted, these examples are taken from H. A. C. Cairns, *Prelude to*

ness and superior organization guaranteed progress and happiness, pointing out that the Africans were well off not because they had an abundance of material possessions but because they had fewer wants than the Europeans.

This sentiment was echoed by H. L. Duff, a British official who was generally very critical of the Africans. Duff confessed that as he trudged about in the bush laden with the "accoutrements" of civilized society, he was embarrassed by the extent of his needs and wants in comparison with the modest means by which the "savage" Africans were able to support themselves.[4] Like Macdonald, the traveler Walter Kerr stressed the contrast between the condition of carefree African children, "plump and round as distended bladders,"[5] with the working-class children of Britain, "penned in narrow slums" and "stunted in body and depraved in mind." These observers attributed the Africans' contentment to their preference for sharing over competition, their lack of excessive ambition, and their refusal to "work against time." The widely read essayist W. R. Greg conceded that the Europeans had "stern and persevering endowments" that suited them to invention and scientific inquiry, thus ensuring success in the struggle for existence in the "rough infancy of the world." But he predicted that the mild and humble Africans, whom he felt made much better Christians than the Europeans, would inherit the earth in the ages to come.[6]

These evocations of the noble savage and gentle Christian were largely throwbacks to themes more prominent in the preindustrial period. In the decades before World War I there were also challenges to the validity of scientific and technological measures of human worth that were more attuned to the broader intellectual preoccupations of the age. It is worthy of note, however, that scientific reassessments of understandings as fundamental as Newtonian celestial mechanics and the "whirl of infinite innovation"[7] in the arts, philosophy, and literature that preoccupied European intellectuals in the prewar era had surprisingly little impact on ideologies of imperialist dominance until after the Great War. In part this was due to the lag between scientific discovery and artistic innovation and an awareness of these breakthroughs on the part of the educated public. A vocal and growing minority sought to explore the unconscious or irrational, championed emotion and intuition at the expense of inductive reasoning, and experimented with new

---

Imperialism: British Reactions to Central African Society, 1840–1890 (London, 1965), pp. 103–5.

[4]H. L. Duff, *British Administration in Central Africa* (London, 1903), p. 304.

[5]It is possible that Kerr had mistaken symptoms of malnutrition for signs of plenty.

[6]W. R. Greg, "Dr. Arnold," *Westminster Review* 39/5 (1843), 8–9.

[7]Carl E. Schorske, *Fin-de-siècle Vienna: Politics and Culture* (New York, 1981), p. xix.

ways of viewing time and space. But most European thinkers, including an overwhelming preponderance of those concerned with overseas affairs, held to the faith in progress, in the primacy of rationality, and in the unbounded potential of scientific inquiry and technological invention for human improvement which had been dominant throughout the nineteenth century. In addition, challenges to the tenets of the civilizing mission and imperialism more generally diminished greatly in number and fervor in the decade or so before World War I. The rising tide of patriotism carried most European intellectuals as readily as the masses toward the catastrophic war that colonial tensions (including recurring crises in the Balkans) did so much to bring on.[8] Relative to the doubt and cynicism that clouded the civilizing mission in the decades after the Great War, questions about the ends of science and technological innovation were rather tentative in the prewar era. Suggestions that Africans or Asians might have something to teach the Europeans were invariably muted, yet there were warnings that the direction in which the European juggernaut was hurtling, and dragging the rest of humanity, was not toward the destination that the high priests of positivism and progress had prophesied.

In this period many of the criticisms of science and technology as gauges of European superiority and proof of the validity of its colonial mission were related to the broader anti-industrial sentiments that had won considerable support in England and France in the late nineteenth century.[9] Goldsworthy Lowes Dickinson's much-publicized reservations about life in industrial America and the dehumanizing effects of mechanization more generally were among the more prominent expressions of this view. Employing the literary device that Montesquieu had used so successfully in the eighteenth century, Dickinson published in 1901 a volume of letters that he attributed to a Chinese official who had visited the West. Though Dickinson, a Cambridge don and essayist on a wide range of literary and social issues, made no secret of the fact that he had written the letters himself, their Chinese authorship was widely accepted, as was most infamously evidenced by a spirited rebuttal from William Jennings Bryan.[10] Bryan's reply was prompted mainly by the unfavorable impression of America conveyed in the volume's prefatory comments. Dickinson's Chinese visitor contrasted the United States, where the "modern spirit" and "genius of industrialism" reigned

---

[8]See, e.g., Raoul Girardet's discussion of the weakening radical and socialist assault on French colonial activities in the years before the war, in *L'idée coloniale en France de 1871 à 1962* (Paris, 1972), esp. pp. 106–10.

[9]See the first section of Chapter 3, above.

[10]E. M. Forster, *Goldsworthy Lowes Dickinson* (New York, 1934), pp. 142–43.

"naked" and "unashamed," with Europe, where industrial society still had to compete with "ancient culture" for supremacy. He wondered whether the Americans' victories over matter and space had been gained at the cost of their souls, whether their achievements in the "practical arts" were matched by contributions in literature and the fine arts.[11] Believing that the future of the West would be determined by events in America, Dickinson was deeply disturbed by the obsession with gadgetry and material abundance that he felt had seized the entire nation. He feared that this mania would eradicate the "life of the spirit" and reduce American citizens to mere machines.[12]

Dickinson's reservations about life in America were mirrored in broader concerns for industrial Europe and the non-Western societies that the Europeans had come to dominate. Viewing Europe through the eyes of his imaginary Chinese traveler, he pointed to similar technological triumphs but lamented the same "failure in all that calls for spiritual insight" and the separation of man from nature that had resulted from those successes. Remarking on the dreary row houses of English cities and the "stream of fatuity" that pervaded literature and the daily press, he concluded that Western society was a "huge engine" that was out of gear. Speaking through his Chinese surrogate, Dickinson added that this soulless and uncontrolled machine had been unleashed on the world and had smashed its way into ancient societies, where it was causing untold havoc. China, once a rich and self-sufficient agricultural society, had been impoverished and destabilized by the aggressive competition of the industrial West. The Chinese, whom Dickinson depicted as more sensitive to the effects of mechanization than Europeans or Americans, had proved unable to defend themselves. As a result, they too were in danger of losing their souls to materialism and secularism. Was not all of this, he asked rhetorically, too high a price to pay for material progress?[13]

Similar questions were posed by other writers at the turn of the century. The French poet Paul Valéry, who would later be a powerful spokesman for those who insisted on the need for the moral and intellectual renewal of Europe after the Great War, was more explicit than Dickinson in affirming what he believed to be Chinese values as alternatives to those of the aggressive and materialistic West. In an impressionistic reverie entitled "The Yalu," written in 1895 during the Sino-Japanese war, Valéry recounted a conversation with a Chinese

---

[11]Goldsworthy Lowes Dickinson, *Letters from a Chinese Official: Being an Eastern View of Western Civilization* (New York, 1904), pp. viii–xiii.

[12]Forster, *Dickinson*, pp. 129–32; Dickinson, *Letters*, pp. xi–xiii.

[13]Dickinson, *Letters*, pp. 10, 14–16, 21–29, 32–34.

companion during a walk along the sea.[14] The deleterious effects of Western technology on China are suggested by a reference to the destruction caused by the "great white ships" of Europe's surrogates, the Japanese. "They are very strong," Valéry remarks. "They are imitating us." But the Chinese responds that the Europeans (and, by implication, the Japanese) are mere children. He suggests that his visits to the West have made its "crazy disorder" all too clear. Reversing many of the standards by which Europeans had criticized the Chinese and other non-Western peoples for decades, Valéry's Chinese friend declares: "You have neither the patience that weaves long lives, nor a feeling for the irregular, nor a sense of the fittest place for a thing, nor a knowledge of government. You exhaust yourselves by endlessly re-beginning the work of the first day."[15] Westerners, he continues, worship intelligence as if it were an "omnipotent beast" and place no limits on what they seek to know. The Chinese, by contrast, "do not wish to know too much" because they understand that "knowledge must not increase endlessly. If it continues to expand, it causes endless trouble, and despairs of itself. It halts, decadence sets in." Again reversing one of the stock nineteenth-century censures of Chinese civilization, he points out that though most of the Europeans' inventions were first devised in China, his people knew that they should not be developed so far that they "spoiled the slow grandeur of our existence by disturbing the simple regularity of its course." The Chinese invented gunpowder but used it only for firecrackers. Despite their present humiliations and temporary setbacks, his people are better off ignorant than stricken with the Europeans' "disease of invention" and their "debauchery of confused ideas."[16]

India was even more favored than China as a counterpoise for Western materialism at the end of the nineteenth century, partly because a good deal more was known at the time about the history and culture of India than of China. The Orientalist tradition that had flourished in Indian studies at the end of the eighteenth century had declined but did not die out in the nineteenth, and by the late 1800s it had begun to revive. Some of the impetus for its new life came from rather suspect sources, including spiritualists such as Madame Blavatsky, who regarded India as a fecund source of occult beliefs and practices.[17] More important, however, was the work of a new generation of Indologists, especially the

---

[14]The essay was not published until 1928. This discussion is based on the translation in Paul Valéry, *Collected Works*, trans. by Roger Shattuck and Frederick Brown (London, 1970), vol. 10, pp. 371–94.

[15]Ibid., pp. 372–73.

[16]Ibid., p. 374–76.

[17]One of the best exposés of Blavatsky and her circle is included in J. N. Farquahar's survey, *Modern Religious Movements in India* (Delhi, 1967), pp. 211–67.

German-born Oxford don Friedrich Max Müller. In the mid-1870s, Max Müller began to edit and publish the *Sacred Books of the East*, a series that would eventually run to more than fifty volumes. He also wrote numerous works on comparative religion and philology, which drew heavily on his own research on India and that of such contemporaries as Rudolf von Roth, Emile Burnouf, and H. H. Wilson. In both scholarly books and those aimed at a broader audience—for example, his *India: What Can It Teach Us?*—Max Müller argued that India had much to offer the rest of mankind in virtually all spheres of human endeavor.[18] The importance of India as a "spiritual" alternative to the materialistic West was also stressed by Indian thinkers in this period, most notably Swami Vivekananda, who had won international fame with a brilliant address at the 1898 Parliament of Religions in Chicago.[19]

Indian values and wisdom struck growing numbers of European writers as ideal antidotes for the "disease of invention" and obsession with inductive reasoning which they believed had distorted Europe's development and that of its colonial dependencies. The British colonial administrator and Orientalist Alfred Lyall conceded that British rule had contributed much to the "material progress" of India, but he had doubts about the "moral" consequences of colonial innovations. Trade, industry, and the area under cultivation had all greatly increased under the British, Lyall admitted, but these advances had undermined Indian religion and morality, destroyed its artisan traditions, and dulled the Indians' "spiritual instinct." Lyall feared that material gains had been achieved at the expense of India's ancient institutions and beliefs. The latter were in decline, and the British had provided nothing to take their place.[20]

The antithesis suggested by Lyall between the spiritual "East" and the material "West" was much more explicitly developed by Goldsworthy Dickinson in a volume describing his own travels in Asia. Dickinson made no secret of his preference for Chinese civilization over Indian, which he thought difficult for an Englishman to appreciate fully, but he felt that Europe could learn much from India. The lessons he outlined were standard in works touting Indian mysticism in this and the postwar era: overcoming strain through contemplation, for example, and becoming more attuned to nature's rhythms. But they are of particular

[18]Friedrich Max Müller, *India: What Can It Teach Us?* (London, 1883), p. 15.

[19]As V. S. Narvane has noted, there was considerable irony in the fact that Vivekananda's great successes came at meetings convened in conjunction with the Chicago World's Fair, which had "the express object of publicising the *material* progress achieved by the West"; see *Modern Indian Thought* (Bombay, 1964), p. 107 n.14 (original italics).

[20]Alfred Lyall, *Asiatic Studies* (London, 1899), vol. 2, pp. 4–6, 36, 44.

interest because they often directly contravene the work ethic and pride in applied technology which had for decades been central to the Europeans' explanations for their global dominance.[21]

In these same years, both the populist Benjamin Kidd and the novelist Pierre Mille directly contested the scientific and technological standards that they believed were pivotal to contemporary efforts to distinguish "higher" from "lower races." In his 1894 *Social Evolution*, Kidd argued that though such inventions as the telephone and phonograph indicated the social stability and creativity of the societies that produced them, they were not in themselves proof of intellectual superiority. In any case, he continued, intelligence should not be confused with "social efficiency," which more than any other quality ensured that a society would advance and prevail in ongoing evolutionary competition. Despite its technocratic connotations, social efficiency had more to do with moral values than inventive capacity. Though Kidd's use of the term was vague and mutable, he believed that "superior races" were those that exhibited a strong sense of cooperation and altruism, and fostered ethical systems stressing obligations to the group rather than individual gain. He held that societies could best be judged on the basis of the religions they nurtured, and because Christianity was the most other-oriented and social-minded of all religions, the European civilization in which it had flourished represented the apex of human development.[22]

In a 1905 essay in *La Revue de Paris*, Mille raised questions similar to those posed earlier by Kidd. He was skeptical of the widely held conviction of his contemporaries that whatever reverses Europeans might suffer at the hands of Japanese warriors and Asian factory workers, they could always be consoled by the "postulate" that the white race was the most "cerebral." This led to the reassuring but dubious conclusion, Mille argued, that Europeans would always be the most advanced people and the best organizers, and that their innate inventiveness would sustain them as the "demiurge indispensable de la terre." But Mille doubted that the potential for scientific discoveries and their practical application was unlimited and that the drive for material prosperity could be infinitely sustained. He asked whether it was not necessary for the Europeans to curb their desires if they were not to destroy themselves. He also wondered whether it would be possible for the Europeans to find other outlets once their obsession with the accumulation of worldly goods waned, whether they were capable of becoming more

[21] Goldsworthy Lowes Dickinson, *An Essay on the Civilisations of India, China, and Japan* (London, 1914), esp. pp. 12–14, 18, 82–85.

[22] Benjamin Kidd, *Social Evolution* (London, 1894), pp. 120–21, 245–46, 265–68, 312–13, 324–25.

just, moderate, and good. Mille saw as major barriers to this meta-
morphosis the Europeans' intense individualism and their propensity for
aggression, criticism, and the ceaseless destruction of "what is, in order
to see if they can find something else to put in its place." He feared that
the excessive passions and spiritual disarray characteristic of contemporary European society would severely constrict efforts to order "l'uni-
vers moral."[23]

As these examples illustrate, India and to a lesser extent China were
the focus of efforts by European intellectuals to arouse concern over the
dangers facing the powerful but, they believed, overly materialistic
West. Though the first serious and systematic studies of African societies were undertaken in the fin-de-siècle period, African beliefs and
institutions were rarely seen as serious alternatives to those of the West.
Frustrated young romantics such as Ernest Psichari found in Africa a
refuge from the "large stomachs and vain speeches of Paris"—an unspoiled, primeval world free from the decadence and corruption of civilized Europe.[24] But the vision of Africa conveyed by the two stories
that Joseph Conrad wrote in the late 1890s, *Heart of Darkness* and "An
Outpost of Progress," was much closer to the "dark continent" view
that prevailed in the prewar era. In these Conrad depicts Africa as an
accursed continent, a steaming pit of "rank grass," mud, and "matted
vegetation."[25] Though he abhors the sufferings inflicted upon the Africans by the greedy Europeans, who kill "natives" for sport and can talk
of nothing but ivory and profits, the Africans in his stories are savages.
They are cannibals, primitives, prehistoric beings, wild and naked men,
simple children. Marlowe, Conrad's protagonist in *Heart of Darkness,* is
perversely fascinated by the "black and incomprehensible frenzy" that is
unleashed in "a burst of yells, a whirl of black bodies, a mass of hands
clapping, of feet stamping, of bodies swaying, of eyes rolling, under the
droop of heavy and motionless foliage."[26] He insists that these creatures
are human, and he does what he can to prevent the Belgian Company's
"pilgrims" from wantonly slaughtering them. But it never occurs to
him that the Africans have qualities which Europeans might emulate. As
they appear in *Heart of Darkness* and "An Outpost of Progress," Africans
are without "law or social restraint,"[27] and they possess only the crudest
material culture.

[23]Pierre Mille, "La race supérieure," *La Revue de Paris* 1 (1905), 843–44.

[24]Martine Loutfi, *Littérature et colonialisme* (Paris, 1971), pp. 95–98; and Robert Wohl,
The Generation of 1914 (Cambridge, Mass., 1979), pp. 12–13 (quoted passage).

[25]Joseph Conrad, *Heart of Darkness* (New York, 1950), p. 94.

[26]Ibid., pp. 78, 84–85, 104, 105 (quoted portion), 111–13, 135, 139.

[27]Benita Parry, *Conrad and Imperialism* (London, 1983), p. 29.

Interestingly, the most human Africans in each story are those tied to white traders: Henry Price or Makola, a "Sierre Leone nigger" and the trading station bookkeeper in "An Outpost of Progress," and the fireman on Marlowe's steamboat in *Heart of Darkness*. The fireman has special significance: he not only serves the Europeans but is regarded by Marlowe as "an improved specimen" because he can "fire up a vertical boiler." The respect and affection Marlowe displays for the fireman—and only the fireman of the many Africans he encounters—is reminiscent of the praise for which the skillful foreman Peroo is singled out in Kipling's "Bridge-Builders." Admittedly, Marlowe's respect for the fireman (he compares him to "a dog in a parody of breeches and a feather hat, walking on his hind-legs") is much more qualified than that of the British engineers for Peroo—but the African's technical skills are elementary compared with those of the Indian foreman. The fireman is "useful" because he has been "instructed" in what he regards as a "strange witchcraft full of improving knowledge,"[28] while Peroo is considered capable of building bridges on his own.

Conrad's condemnation of European exploitation of the peoples of the Congo and his derisive critique of the civilizing-mission ideology have been extensively studied. The first became familiar to readers in this era when a flurry of exposés of European atrocities in the Congo Free State appeared. Both condemnation and critique resulted from Conrad's horrific experiences in the service of the Société Belge pour le Commerce du Haut-Congo.[29] But less has been made of Conrad's doubts about the Europeans' boundless faith in their technology and the assumption that it had elevated them above nature and the rest of humanity, a theme that runs through both his African stories and his tales of the South Seas.[30] The contrast between, on the one hand, "the insignificant cleared spot of the trading post"[31] and the tiny, dilapidated steamboats on which men like Marlowe make their way into central Africa and, on the other, the huge, seemingly endless river and the vast rainforest mocks the white man's pretensions to having mastered nature. These images parallel those of Chinese landscape painters who depict diminutive people and buildings against vast tree-covered moun-

[28]Conrad, *Heart of Darkness*, pp. 106–7, 124–25.

[29]Useful sources for biographical background include Gerard Jean-Aubry, *The Sea Dreamer: A Definitive Biography of Joseph Conrad* (New York, 1957), chap. 8; and Zdzislaw Najder, *Joseph Conrad: A Chronicle* (New Brunswick, N.J., 1983), chaps. 4 and 5.

[30]E.g., in *The Nigger of the "Narcissus,"* mentioned in Chapter 1, and other tales of men against the sea, such as *Youth: A Narrative* and *Typhoon*.

[31]Conrad, "An Outpost of Progress," in *Typhoon and Other Tales* (New York, 1963), p. 188.

tains, but Conrad substitutes the dense and brooding rainforest for the ethereal grandeur of the Chinese landscape.

As the ship that carries Marlowe to Africa makes its way along the coast, he sees a French man-of-war "shelling the bush": "In the empty immensity of earth, sky, and water, there she was, incomprehensible, firing into a continent. Pop, would go one of the six-inch guns; a small flame would dart and vanish, a little white smoke would disappear, a tiny projectile would give a feeble screech—and nothing happened. Nothing could happen. There was a touch of insanity in the proceeding."[32] On his journey into the interior, Marlowe comes upon a steam boiler "wallowing in the grass" and "an undersized railway-truck lying there on its back with its wheels in the air." Farther on, he discovers more "decaying machinery," "rusty rails," a vast hole that appeared to have been dug for no purpose, drainage pipes in a ditch, and a man looking after a road that does not exist.[33] In both stories the emissaries of progress and industrial civilization are overwhelmed by the continent's savagery, enveloped by its impenetrable forest. Kayerts and Carlier, the rather dim-witted managers of "An Outpost of Progress," reduce the station and its storehouse—which is called the fetish, "perhaps because of the spirit of civilization it contained"—to a shambles. Enervated by the tropical environment and increasingly mistrustful of each other, they quarrel over a packet of sugar; Kayerts murders Carlier and then commits suicide. Kurtz, whom Marlowe seeks to bring out of the interior, begins his work in the Congo convinced that the whites "from the point of development we had arrived at" should be worshiped by the "savages" as "supernatural beings." He is soon reduced to committing atrocities beyond anything the cannibalistic savages could imagine. He too dies in the jungle, physically drained and psychologically eviscerated by the realization of his own moral depravity.[34]

Conrad's vision of African savagery and social chaos was at odds with the findings of the first generation of field anthropologists, who began their work in the decade after his harrowing visit to the Congo in 1890. Administrators such as Colonel de Trentinian and F. J. Clozel, desiring to understand the institutions and customs of the diverse peoples who had come so rapidly under French colonial rule, instructed subordinates to gather information throughout West Africa. Local officials supplied mounds of raw data, and a few talented administrator-anthropologists —including Maurice Delafosse—began to publish studies based on years of field experience; these revealed for the first time the sophistica-

---

[32]Conrad, *Heart of Darkness*, p. 78.
[33]Ibid., pp. 78, 80, 86.
[34]Ibid., pp. 122–23, 138–45; "Outpost," pp. 193–205.

tion of African religions and the complexity of African social systems.[35] This research obviously had little impact on writers like Conrad, but information on the actual, as opposed to the imagined, practices and institutions of the Africans was beginning to shape the attitudes of some intellectuals in the years just before the war.

Of particular importance were the collections of African masks and carvings that had been housed in museums of ethnology in various European capitals from 1879, when the Trocadero was opened in Paris. Originally regarded as mere curiosities, the "barbarous fetishes" of primitive peoples, these works were "discovered" by avant-garde French artists such as Derain, Braque, Matisse, and, most important, Picasso in the early 1900s.[36] Picasso claimed that he had not been much affected by African art before beginning his stunning *Desmoiselles d'Avignon* in 1907. But the masklike faces of the bathers on the right side of the canvas leave little doubt that his emotional encounter with African sculpture at the Trocadero had profoundly reshaped his artistic vision by the time the painting was completed. As a friend remarked at the time, Picasso had found his true self and his own style in "the company of African soothsayers."[37]

The African contribution to the artistic revolution that began in the years before the outbreak of the war did not lead European intellectuals to advocate African beliefs and patterns of social organization as correctives or alternatives to those of industrial Europe. But it did provide inspiration for many artists and thinkers who had begun to experiment with alternatives to the sense of space and the approach to perspective that had dominated European painting since the Renaissance and had long been regarded as manifestations of Europe's scientific and technical superiority.[38] Like the revived interest in Indian religions, new revelations about African culture and society also strengthened the arguments of a growing number of thinkers who were calling for a relativistic approach to the thought and organizational patterns of different peoples

[35]Gérard Leclerc, *Anthropologie et colonialisme: Essai sur l'histoire de l'africainisme* (Paris, 1972), pp. 43–52; and Girardet, *L'idée coloniale*, pp. 158–64. A similar collection of information was occurring in British territories in this period, but (Mary Kingsley's revelations notwithstanding) the great age of British field anthropology came after World War I.

[36]Michael Leiris and Jacqueline Delange, *African Art* (London, 1968), pp. 7–10; Frank Willet, *African Art* (New York, 1971), pp. 35–36.

[37]H. R. Rookmaaker, *Modern Art and the Death of Civilization* (London, 1970), p. 117; Roland Penrose, *Picasso: His Life and Work* (Berkeley, Calif., 1981), pp. 134–37; Leiris and Delange, *African Art*, pp. 10, 13.

[38]Stephen Kern, *The Culture of Time and Space, 1880–1918* (Cambridge, Mass., 1983), esp. pp. 132, 140–48; Donald Lowe, *History of Bourgeois Perception* (Chicago, 1982), chap. 6.

and human creativity in general. The adoption of such an approach, which was even more vociferously advanced after the war, would of course have made a shambles of such narrow and culturebound gauges of human worth as scientific and technological achievement. Finally, African examples—though one suspects those drawn from the Conradian rather than the Delafossean vision of Africa—contributed to the growing interest in the irrational, in emotions, and in the unconscious which can be found in the works of thinkers of this period ranging from Le Bon and Sorel to Bergson, Freud, and Jung.[39] The full impact of these converging currents of experimentation, criticism, and doubt about ideologies of Western dominance based on evidence of material advancement would not be felt until after the Great War. But in combination with the emergence of a non-European industrial power, they had begun to erode the very foundations of the European imperial edifice.

The Specter of Asia Industrialized

The emergence of Japan as an industrial power in the period before World War I presented a very different sort of challenge to the scientific and technological proofs of European superiority from those posed by the revelations of field anthropologists and concerns about excessive materialism. The rapidity with which the Japanese acquired and then themselves manufactured and improved upon Western machines called into question the widespread assumption that "lower races" were inherently incapable of matching European inventiveness and material prowess. Japan's transformation also shattered the illusion that industrialization was a uniquely Western process. Growing fears of Japanese economic competition for markets that were already hotly contested by the European and North American industrial powers were compounded in the 1880s and 1890s by warnings that the Asian giants, India and China, were also beginning to industrialize. Visions of unlimited reservoirs of cheap and skilled laborers, lost colonial market outlets, and Asian manufactures flooding the markets of Europe began to haunt the already troubled dreams of British and French entrepreneurs and economic forecasters. By the 1890s the real "Yellow Peril" was increasingly

---

[39]On the importance of these preoccupations in prewar European thinking, see H. Stuart Hughes, *Consciousness and Society* (New York, 1976), pp. 4, 15–17, 35–65, and chap. 4; Kern, *Culture of Time and Space*, pp. 132, 137–39, 151–52, 179; G. D. Josipovici, "The Birth of the Modern: 1885–1914," in John Cruickshank, ed., *French Literature and Its Background* (Oxford, 1970), vol. 6, pp. 12–13; and Michael Biddiss, *The Age of the Masses* (Harmondsworth, Eng., 1977), pp. 55–61, 83–90.

seen as something very different from the invasion of Europe by Mongolian hordes that so obsessed such clamorous sentinels as the Count de Gobineau and the Kaiser Wilhelm.[40] As the Baron d'Estournelles de Constant, a prominent French diplomat, observed in the late 1890s, the economic threat to Europe posed by a mechanizing Asia was much more insidious than the remote possibility of a Japanese or Chinese invasion. A military challenge, he argued, would unify the Europeans, stiffen their moral resolve, and arouse their very considerable martial instincts. Asian economic competition, on the other hand, would divide the European powers and the classes within each nation. Undersold and eventually unemployed, the demoralized and impoverished workers of Europe would rise up in revolutions that would destroy Europe from within.[41]

From the time of their first contacts in the 1540s, European observers had found the Japanese one of the most remarkable peoples they had encountered overseas. Though many explorers and sea captains, mindful of the warlike disposition of the Japanese, might have questioned Francis Xavier's judgment that they were the "best [people] who have yet been discovered,"[42] there was clearly much in Japanese society to be admired as well as feared. The Japanese struck Jan Linschoten and other seasoned explorers as clean, industrious, and intelligent. Their architecture was much praised, as were their mining techniques and handicraft skills.[43] In the *shogun* overlords and their *daimyo* retainers and the disciplined and fierce armies of *samurai* warriors, the military-minded Europeans met their match. Numerous visitors remarked upon the great interest the Japanese displayed in European technology, and, above all, military hardware. But the Japanese were more than curious; they were astoundingly adaptive. Within a decade of their first encounters with European firearms in 1543, they were making Western-style muskets of considerable quality; within a generation, they had made major im-

[40]On Gobineau, see Jacques Barzun, *Race: A Study in Modern Superstition* (New York, 1937), pp. 81–83; on the Kaiser, Heinz Gollwitzer, *Die gelbe Gefahr: Geschichte eines Schlagworts, Studien zum imperialistichen Denken* (Göttingen, 1962), pp. 206–18. For a sense of the intensity and pervasiveness of "Yellow Peril" sentiments, see Jacques Novicov's *L'avenir de la race blanche* (Paris, 1897), in which the fears of invasion are discussed, and rebutted, at length (esp. pp. 44–45, 57, 60–71, 142–50).

[41]Baron d'Estournelles de Constant, "Le péril prochain: L'Europe et ses rivaux," *Revue des Deux Mondes* 134 (1896), 680–81.

[42]Quoted in C. R. Boxer, *The Christian Century in Japan, 1549–1650* (Berkeley, Calif., 1951), p. 37.

[43]Jan Linschoten, *Voyage to the East Indies* (London, 1885), vol. 2, pp. 151–53; Boxer, *Christian Century*, chap. 2; Donald Lach, *Asia in the Making of Europe* (Chicago, 1965), vol. 1, pt. 1, p. 444, and pt. 2, chap. 8.

provements on the matchlocks they had originally imported from the West.[44]

The decision of the early Tokugawa overlords to shut Japan off from Western influences and the gradual elimination of firearms from domestic conflicts in the early seventeenth century did not put an end to Japanese interest in the West and its impressive technology. From the 1640s the Japanese sought to keep abreast of developments in Europe through contacts with the Dutch merchants who were allowed to retain a trading station on the island of Deshima in Nagasaki Bay. Particularly after 1745, when Japanese scholars were permitted to learn Dutch, and translations of European books were allowed to circulate freely among the elite, great efforts were made to acquire works on the latest scientific advances in Europe and to obtain drawings, descriptions, or samples of new mechanical devices. As had been the case in earlier centuries, Japanese interest in European warfare and weaponry remained pronounced.[45] Thus, though the Deshima window to the West was small, by the end of the eighteenth century the Japanese were better informed about European civilization than any other non-Western people.[46]

Although European opinion was not a significant factor in the calculations that persuaded the Tokugawa *shoguns* to limit contacts with Westerners, Japan's long isolation did much to diminish the high regard in which early European explorers and missionaries had held the Japanese. By the nineteenth century few Europeans knew Japan firsthand, and even fewer Nipponophiles could be found in the ranks of European travelers, merchants, and missionaries. The "rough, brave, bold, honest" Japanese might be pitted against the "mild, timid, avaricious, cheating" Chinese by such writers as Hugh Murray,[47] but most observers in this period found little to recommend either. Japan, like China, was thought to be backward and stagnant, and the Japanese were classified as a branch of the stationary "Mongolian" race. At midcentury Nott and Gliddon ranked Japan with China as a "semi-civilization," while Samuel

[44] Noel Perrin, *Giving Up the Gun: Japan's Reversion to the Sword, 1543–1879* (Boston, 1979), pp. 5–23. My discussion of Japanese responses to Western military technology relies heavily on Perrin. Delmer Brown, "The Impact of Firearms on Japanese Warfare," *Far Eastern Quarterly* 7 (1947–48), 236–53, also remains useful.

[45] C. R. Boxer, *Jan Compagnie in Japan, 1600–1817* (Tokyo, 1968), esp. pp. 14–17, 42–43. The fullest account of the Dutch role as transmitters of Western learning and information to the Japanese can be found in Grant Goodman's recently reissued *Japan: The Dutch Experience* (London, 1986).

[46] Donald Keene, *The Japanese Discovery of Europe, 1720–1830* (New York, 1954), p. 123.

[47] Hugh Murray, *Enquiries Historical and Moral Respecting the Character of Nations and the Progress of Society* (Edinburgh, 1808), p. 150.

Morton deemed the Japanese "laborious artificers" like the Chinese, but less ingenious and less skilled in navigation.[48]

Japan's stunning success in adopting Western ways and machines after being forced open by Admiral Perry's warships in the mid-1850s soon compelled reassessments of these views. The revolution from above that overthrew the *shogun* and restored the emperor as at least the symbolic locus of power led to broad and profound changes in Japanese society and politics from the late 1860s onward.[49] The basis for a centralized bureaucracy, ironically, had been established in the "feudal" Tokugawa period. After the Meiji restoration this process was completed, and by the late 1880s the Japanese had adopted at least the outward trappings of Western parliamentary democracy. Under the leadership of young and progressive *samurai* factions in alliance with the improving landlords and great mercantile families, Japan rebuilt its army and navy along Western lines; thoroughly reformed its educational and land taxation systems; and, beginning with textiles in the 1880s, launched a full-scale effort to industrialize.

By the 1890s, and especially after its victory in the 1894–95 Sino-Japanese War, Japan was accorded a special status among non-Western societies. Some writers, most notably Lafcadio Hearn, savored its refined and ancient culture, but most singled out the Japanese because they had done what no other non-Western people had been able to do: remake their society in the image of industrial Europe. The achievements stressed by different writers varied from Japan's emergence as a major military power to its widespread application of science to industry.[50] But the general conclusions were clear. Progressive Japan was a society very different from stagnant China. Once pitied as the beleaguered

[48]William Lawrence, *Lectures on Physiology, Zoology, and the History of Man* (London, 1819), p. 483; Julien Virey, *Histoire naturelle du genre humain* (Paris, 1826), p. 416; J. C. Nott and George Gliddon, *Types of Mankind* (London, 1854), pp. 52–53; and Samuel Morton, *Crania Americana* (Philadelphia, 1839), p. 47.

[49]In recent years a sizable literature has developed on the Meiji transformation. For key works on various aspects, see H. D. Harootunian, *Toward Restoration: The Growth of Political Consciousness in Tokugawa Japan* (Berkeley, Calif., 1970); Carol Gluck, *Japan's Modern Myths: Ideology in the Late Meiji Period* (Princeton, N. J., 1985); Thomas C. Smith, *Agrarian Origins of Modern Japan* (Stanford, Calif., 1970), and *Political Change and Industrial Development in Japan: Government Enterprise, 1868–1880* (Stanford, Calif., 1955); W. G. Beasely, *The Meiji Restoration* (Stanford, Calif., 1973); and the essays in John W. Hall and Marius B. Jansen, eds., *Studies in the Institutional History of Japan* (Princeton, N.J., 1968), pt. 4.

[50]For samples of these differing reactions, see G. N. Curzon, *Problems of the Far East* (London, 1894), pp. 15–20, 45–60; Jean Dhasp, *Le Japon contemporain* (Paris, 1893), pp. 258, 270, 325–29; J. S. Lawrence, *A Wandering Student in the Far East* (London, 1908), pp. 17–21, 168–69; J. Griffith, *China: Her Claims and Call* (London, 1882), pp. 13–14; and Edward Clodd, *The Story of Primitive Man* (London, 1895), p. 195.

"hermit kingdom," Japan had transformed itself within a generation into a power to be reckoned with. The shock of Japan's victory over Russia in 1904–5 dispelled any lingering doubts.[51] A nonwhite, non-European people had not only mastered Europe's science and technology, but had proved more than a match for a mighty, if somewhat backward, European empire.

In most European accounts of Japan's accomplishments, admiration was tempered by foreboding that Japan's new-found power might increasingly be used to thwart European interests and imperial designs. The military dimensions of the Japanese threat were apparent by the 1890s, but many writers in this period gave equal or greater attention to the dangers of economic competition. The trends appeared to bear out the gloomiest of prognostications.[52] As the Japanese gained the capacity to produce their own machines and weapons, a lucrative market for European manufactures was steadily closing. Having acquired the industrial technology of the West and adopted Western organizational techniques, the workers of Japan were also producing a wide range of products for export. The efficiency of Japanese industrial firms and their significantly lower labor costs made it possible for them to compete successfully with European and American producers both for more accessible markets in Asia and Latin America and, increasingly, for outlets in European and North American centers of industrial production. The loss of markets in East Asia was of particular concern to the West, as indicated by fears that the Japanese would soon proclaim a "Monroe Doctrine" for the Orient.[53] Japanese textiles, both silk and cotton, posed the most obvious threats to European producers. But by the late 1890s, Japan watchers such as E. H. Norman and Baron d'Estournelles de Constant warned that the Japanese were competing globally with products ranging from matches and umbrellas to watches and iron girders. There were frequent references to the poor quality of Japanese goods, of course, and insinuations or outright charges that they were winning markets only because they were cheap. But superior quality mattered

[51]Martine Loutfi has stressed the importance of Japan's stunning victory in shaking the confidence of the French in their overseas mission and in continuing European dominance more generally; see Loutfi, *Littérature et colonialisme*, pp. 122–23, 129–30. For contemporary French and British assessments, see Mille, "La race supérieure," pp. 832–37; and George Knox, *Imperial Japan* (London, 1905), pp. 53–61.

[52]This summary of late nineteenth-century fear of economic competition is based chiefly on Auguste Moireau, "Le mouvement économique," *Revue des Deux Mondes* 130 (1895), 912–17; Baron d'Estournelles, "Le péril prochain," pp. 664–81; and Paul Leroy-Beaulieu, "Le Japon: L'éveil d'un peuple oriental à la civilization européenne," *Revue des Deux Mondes* 98 (1890), 633–68.

[53]Baron d'Estournelles, "Le péril prochain," p. 666.

little if there were no markets for Western manufactures, and pride in workmanship was equally meaningless if Europeans could not find jobs.

Anxiety about Japanese competition grew steadily in the prewar decades and soared during and after the war. As early as 1890, when Japanese industrialization was just getting under way, no less imposing a figure than the French colonial theorist Paul Leroy-Beaulieu called for a new "Berlin Conference" on world economic affairs. He saw as the main task of such a gathering, to which both China and Japan would be invited, the integration of these emerging powers into a new global economic order. If these measures were not taken, he warned, the Japanese and Chinese, having mastered European techniques and machines, would show the "soft" Europeans just how much peoples who had not lost the zest for hard work could achieve.[54]

As Leroy-Beaulieu's inclusion of China in his conference proposal suggests, European writers in this period saw dangers even greater than those already posed by Japan in the future spread of industrialization throughout Asia. Earlier complaints about the crudity of Chinese tools and the inefficiency of Chinese machines now took on new meaning. If modern tools were placed in the hands of China's hardworking and dexterous millions, what chance would European workers stand in the competition for market outlets? China's large surplus population meant an oversupply of labor and, in turn, minimal labor costs and low prices for manufactured goods.[55] Minimal living standards and labor demands also meant fewer strikes in China and India, at a time when work stoppages were a serious problem in industrialized Europe and the United States. Strikes by European workers would provide market opportunities that Asian producers were certain to exploit.[56] For many late nineteenth-century writers, the rapid growth of a modern textile industry in India was an ominous portent of Europe's future economic decline. India, after all, was a British colony. But by the last decades of the century it was also a major competitor for textile markets within the subcontinent, in Britain's other colonial territories, and in England itself.[57] These were clearly not the results the civilizing mission had been

[54]Leroy-Beaulieu, "Le Japon," p. 668.

[55]Charles H. Pearson, *National Life and Character* (London, 1893), pp. 124–30; Novicov, *L'avenir de la race blanche*, pp. 11–35, 106–9, 124–25; Baron d'Estournelles, "Le péril prochain," pp. 681–82; and Mille, "La race supérieure," pp. 838–42.

[56]Baron d'Estournelles, "Le péril prochain," p. 681–82, 686.

[57]Ibid., pp. 662–69; Moireau, "Le mouvement économique," pp. 909–12; Gustave Le Bon, *Les civilisations de l'Inde* (Paris, 1887), pp. 720–22; W. W. Hunter, "The New Industrial Era in India," *Proceedings of the Royal Colonial Institute* 19 (1887–88), pp. 260–63, 272–75.

intended to produce; nor was the related possibility that a China with railways and modern arms might reassert its position as a great power.[58]

There was not a little irony in the obsession of numerous European writers with competition from Asian laborers. Having for so long stressed the need to put the rest of the world to work, European colonizers now recoiled at the results of a mission's being fulfilled. As the Baron d'Estournelles de Constant's thoughtful 1896 essay "Le péril prochain" reveals, this was not the only contradiction in the civilizing mission that industrialization in Asia was bringing home to European thinkers. The acquisitive and energetic Japanese were relentlessly transforming inventions that had played central roles in Europe's rise to global hegemony, and had long been regarded as key symbols of European superiority, into weapons in an economic struggle that threatened to topple Europe from its position of global leadership. Cheap, mass-produced Japanese watches and clocks, he warned, were capturing the markets of the world, including those of Europe. Japanese businessmen and workers had not only mastered the new sense of time but had become more efficient and punctual than the Europeans themselves. Now they threatened to dominate production of the devices that were essential to the maintenance of the pace and discipline of industrial life. Even the railroads, which had been so vital to Europe's own industrial transformation and its overseas dominance, had become a double-edged sword. Building on a remark by the French essayist Joseph Renan, Baron d'Estournelles observed that the same railway lines that carried European goods and troops out to the Far East could carry Japanese products and even soldiers back to Europe. In an interesting reversal of present-day patterns, he reported that English and American manufacturers were relocating their plants in Japan, where labor was cheap and reliable, and government policies favorable. All these developments forced him to wonder whether Europeans would soon be left with anything other than ideas to export, whether the time had come when they must be content with the "noble role" of inventor of machines that would be used to enrich other peoples.[59]

As Pierre Mille saw so clearly in 1905, Japan's successful industrialization ought to have put an end to the illusion that the Europeans' scientific and technological accomplishments provided empirical and incontestable proof of their racial superiority.[60] A surprising number of Mille's contemporaries agreed. Jacques Novicov, who had little use for

[58]Pearson, *National Life*, pp. 46, 111–12.
[59]Baron d'Estournelles, "Le péril prochain," pp. 668–69, 673, 678, 680–81.
[60]Mille, "La race supérieure," pp. 837, 844.

racial classifications of any kind, concluded that the industrial achievements of the Japanese demonstrated that they were the Europeans' equals. At the end of the Great War, when Japan's industrial and military might had again been demonstrated in fulfilling the terms of its 1902 alliance with Britain, Benjamin Kidd pronounced the Japanese nation the equal of any Western power. Two years later H. H. Johnston, who was fond of ranking the human races, placed Japan, alone of non-Western societies, with Great Britain, France, and the United States at the top of a scale that rated national development on the basis of standard of living, educational and sanitary practices, government efficiency, and levels of agricultural and industrial production.[61]

Though the concessions of such strident white supremacists as Johnston were indicative of the extent to which Japanese achievements had shaken racial ideology insofar at it was based on scientific and technological measures of ability, the Japanese found that they remained a target of popular racial prejudice. This point was driven home by their failure to win approval of an antiracist clause at Versailles in 1919, by the barriers erected against Japanese emigration to the United States and the British dominions before and after the Great War, and by continued racist slurs against them in both official circles and popular thinking in the interwar years and through World War II.[62]

In addition, many writers who, like Mille, were considered authorities on colonial issues explicitly rejected his contention that Japan's industrialization demonstrated the unreliability of the technological and scientific standards of human worth. They dismissed the Japanese counterexample as an exception, pointing out that the Japanese were after all as different from other non-Western peoples as they were from the Europeans. They also argued that borrowing machines and imitating European manufacturing techniques were hardly proofs that the Japanese had the capacity to think scientifically or to invent new machines and methods of production. As Gustave Le Bon and numerous colonial theorists after him asserted, a Japanese student could garner every possible university degree, but he could never reason logically like a European, because that capacity came only through inheritance. The Japanese might have succeeded in adopting the outward forms of European civi-

---

[61]Novicov, *L'avenir de la race blanche*, p. 109; Benjamin Kidd, *The Science of Power* (London, 1918), pp. 109–10; and H. H. Johnston, *The Backward Peoples and Our Relations with Them* (Oxford, 1920), p. 7.

[62]John Dower, *War without Mercy: Race and Power in the Pacific War* (New York, 1986), chaps. 5–7. Abundant insights into official attitudes in this regard can be found in Christopher Thorne, *Allies of a Kind: The United States, Britain and the War against Japan, 1941–1945* (Oxford, 1978).

lization, but neither their basic beliefs nor their character had been transformed. Though the Europeans might be outproduced and undersold, they retained a unique capacity for invention and scientific discovery. They also remained morally superior to mere imitators like the Japanese—or so Le Bon and Saussure continued to argue.[63] Their pronouncements were a good deal less convincing after four years of slaughter on the Western Front.

Trench Warfare and the Crisis of Western Civilization

The generalized concerns expressed by a handful of European intellectuals at the turn of the century regarding the excessive materialism of industrialized societies and the more focused and widespread fears of non-Western economic competition were very soon overwhelmed by the shock waves of the catastrophic global war that science and industrial technology had done so much to make possible. Curiously, though pacifist movements and conferences abounded in the prewar decades, and war scares periodically threatened the long peace between the great powers, little serious discussion was devoted to the horrific potential of the new weapons that had been spawned by the union of science and technology in the ever changing industrial order. As I. F. Clarke has concluded from a study of pre-1914 books on wars of the future, "save for rare exceptions, they are distinguished by a complete failure to foresee the form a modern war would take." Writers engaged in "the great enterprise of predicting what was going to happen" in future wars, envisioned conflicts that turned on brief, decisive battles and heroic deeds, which had done so much to enhance the glory and valor associated with past wars.[64]

The failure of most Europeans to fathom the potential for devastation of the new weapons they were constantly devising owed much to their conviction that no matter how rapid the advances, *Western men* were in control of the machines they were creating. In an age dominated by plans for decisive offensives and insistence on élan as the key to military success, few civilian or military leaders paid much heed to the lessons that a handful of military specialists found in the bloody battles of the

[63]Le Bon is quoted at length and approvingly by Léopold de Saussure, *Psychologie de la colonisation française* (Paris, 1899), pp. 50–52, who adds a considerable discourse of his own (pp. 275–80, 287–92) on the failure of the Japanese to adopt Western values and thinking. See also Paul Giran, *De l'éducation des races* (Paris, 1913), pp. 201–13, 234–37; and Jules Harmand, *Domination et colonisation* (Paris, 1910), pp. 270–72.

[64]I. F. Clarke, *Voices Prophesying War, 1763–1984* (London, 1966), pp. 68–69 (quoted portion), 78–79, 89–90.

American Civil War and the Russo-Japanese conflict. In the American case, Grant's handling of the railways might be admired, but the appalling casualty rates at Shiloh and Antietam and the trench stalemates at Vicksburg and Petersburg could be chalked up to the lack of training and discipline among frontier bumpkin troops and the failings of their poorly schooled commanders. The military significance of the Russo-Japanese conflict was neglected in accounts of the war dominated by warnings about the growing power of the Japanese and by discussions of the meaning of the decline of the Tsarist empire.[65]

Aside from such visionaries as H. G. Wells and Albert Robida, military pundits in the prewar years chose to stress the ways in which the new weapons would shorten wars rather than make them more horrific. Prognosticators and the European public as a whole took comfort in the assurances of Norman Angell that long wars were economically unfeasible in the interdependent modern world and of Joseph Kohler that war between "civilized" nations had been rendered obsolete by the triumph of "modern reason." They ignored or dismissed out of hand the gloomy predictions of Ivan Bloch that, should it come, a war between the great powers would be a long and bloody stalemate.[66]

The industrial underpinnings of the nineteenth-century revolution in the organization and technology of war that had advanced European armies light-years ahead of African and Asian adversaries were obvious to contemporary specialists on military affairs. As the Baron von der Goltz observed in the 1880s, "All advances made by modern science and technical art are immediately applied to the abominable art of annihilating mankind."[67] Advances in metallurgy and machine-tooling made possible great increases in the size, range, accuracy, and rate of fire of both artillery and hand weapons. Rifled, breech-loading weapons had within decades increased the effective range of massed infantry from a

[65]On European responses to the Civil War, see Jay Luvaas, *The Military Legacy of the Civil War* (Chicago, 1959), esp. pp. 229–33. On the Russo-Japanese war, see William McElwee, *The Art of War: Waterloo to Mons* (Bloomington, Ind., 1974), pp. 241–55; and note 51 in this chapter.

[66]Norman Angell, *The Great Illusion*, was published in London in 1909. Joseph Kohler's arguments were made in his *Recht und Personlichkeit*, published in 1913 (cited in Peter Wust, *Crisis in the West* [London, 1931], pp. 41–42). Ivan Bloch's predictions were made in a six-volume study of economics and warfare published in French as *La Guerre Future* (Paris, 1898).

[67]Quoted in Clarke, *Voices Prophesying War*, p. 78. The following overview of the links between invention, scientific discoveries, and military innovation in the nineteenth century is based chiefly on McElwee, *Art of War*, esp. chap. 4; Michael Howard, *War in European History* (Oxford, 1976), chap. 6; Richard Preston and Sydney Wise, *Men in Arms* (New York, 1970), chap. 15; J. F. C. Fuller, *The Conduct of War, 1789–1961* (New Brunswick, N.J., 1961), chaps. 5–8; and Cyril Falls, *A Hundred Years of War* (New York, 1953), chaps. 1–11.

few hundred to thousands of yards. Field and heavy artillery, fed by the
same innovations, advanced apace. By 1914 the famed Krupp works
could produce massive 42–cm howitzers (the "Big Berthas") capable of
firing 1,800-pound shells as much as 10,000 yards in a trajectory that
reached three miles at its apex.[68] The development of machine guns,
beginning in the 1860s, provided infantry with a new source of firepower, which became (along with artillery) the great killer in World
War I. Factory production and the development of interchangeable parts
by American inventors such as Eli Whitney meant that millions of soldiers could be equipped with the new arms, and hundreds of thousands
of shells fed into the big guns that supported them in the field. One
gauge of the destructive potential of the new weapons is provided by
Michael Howard's estimate that by 1914 a single regiment of field guns
could deliver in one hour more firepower than had been unleashed by all
the adversary powers in the Napoleonic wars.[69]

Railways made it possible to move millions of soldiers into battle
within days and—equally critical once the trench stalemate set in—
reinforce points in the line where enemy forces threatened to break
through. Wireless communications allowed general staffs and division
commanders to coordinate the movements of tens or hundreds of thousands of troops over vast areas, and the smokeless explosives (lyddite,
cordite, and melinite) devised by chemists in the late nineteenth century
considerably improved visibility on the field of battle. New techniques
of food preservation and the invention of canning made it possible to
feed the huge national armies of recruits and conscripts over long periods of time, and mass production meant that they could be steadily
supplied with helmets, uniforms, boots, and trenching shovels—which
(ominously) the combatants of each of the great powers carried into
battle from the opening days of the war. Once the war of maneuver
sputtered to a halt after Schlieffen's grand design for the destruction of
the French army had been frustrated at the Marne, barbed wire (which
the Americans had invented to fence in cattle) and concrete and steel
(which the Germans used the most ingeniously) were combined to build
the massive field fortifications that would dominate the war in the west
until the spring offensives of 1918.

Extended warfare between the great powers intensified efforts to apply scientific knowledge and inventiveness to the business of destruction.[70] Georges Duhamel, upon observing a group of English long-

[68]Curt Johnson, *Artillery* (London, 1975), p. 38.

[69]Howard, *War in European History*, p. 120.

[70]The fullest account of technological changes during the war is provided in John
Terraine, *White Heat: The New Warfare, 1914–1918* (London, 1982). See also Howard,

distance artillerymen whose shirts and trousers were smeared with grease and oil, commented bitterly, "war has become an industry, a mechanical and methodical enterprise for killing."[71] Numerous writers lamented the extent to which scientific research, formerly seen as overwhelmingly beneficial to humanity, had been channeled into the search for ever more lethal weapons. Some of the most brilliant minds of a civilization "devoured by geometry" had labored for generations to ensure that death could be dealt on a mass scale "with exactitude, logarithmic, dial-timed, millesimal—calculated velocity." Many of those who watched their compatriots die cursed not the enemy, who was equally a victim, but the "mean chemist's contrivance" and the "stinking physicist's destroying toy."[72]

The succession of toxic gases devised by the chemists of the adversary powers were perhaps the most diabolical products of the mobilization of science—only grotesque masks prevented the inhalation of the lethal vapors. Flame throwers, improved explosives, and devices that made it possible to extend the war into the air and under the seas introduced other new ways to slaughter the youth of Europe. It was as if the dream of Jules Verne's gun club president had come true, C. E. Montague observed in 1917, and all of science had been reduced to ballistics.[73] But for Georges Duhamel, who served as a surgeon at the front, the ultimate perversion of scientific inquiry resulted from the efforts of physicians to recycle the wounded in order to sustain the war effort. The Ambulance Chirurgicale Automobile (ACA) he wrote in 1918, was:

> the most perfect thing in the line of an ambulance that has been invented. It's the last word in science; it follows the armies with motors, steam-engines, microscopes, laboratories—the whole lock, stock, and barrel of a modern hospital. It's the first great repair-shop the wounded man encounters after he leaves the workshop of trituration and destruction that operates at the front. Those parts of the military machine that are the worst destroyed are brought there. Skilful [sic] workmen fling themselves upon them, unwrap them at full speed, and examine them competently, for all

---

War in European History, chap. 7; Fuller, *Conduct of War*, chap. 9; and Falls, *Hundred Years of War*, chaps. 12–15.

[71]Georges Duhamel, *Civilization, 1914–1917* (New York, 1919), p. 35.

[72]The quotations are taken from Andre Malraux, *La tentation de l'Occident* (Paris, 1976), p. 20; and David Jones, *In Parenthesis* (London, 1982), p. 24. For additional examples, see Romain Rolland, *Above the Battlefield* (London, 1914), p. 8; H. G. Wells, *The Salvaging of Civilization* (New York, 1921), pp. 7–10; Georges Duhamel, *La possession du monde* (Paris, 1919), pp. 242–44, 255–56; and John Cruickshank, *Variations on Catastrophe: Some French Responses to the Great War* (Oxford, 1982), pp. 63, 66, 89–90, 159, 169.

[73]C. E. Montague, *The Western Front* (London, 1917), vol. 1, p. lxv.

the world as if with a hydro-pneumatic machine, a collimator. If the part is seriously out of order, they do what they can to set it right; if the human material is not absolutely worthless, they patch it up carefully, so as to get it back into service at the first opportunity. That is what they call "the conservation of the effective.". . . stretcher-bearers, with the clumsiness of drunken porters, would bring [the ACA] a few wounded men, who were immediately digested and eliminated. Then the factory would continue to rumble like a Moloch whose appetite has merely been awakened by the first fumes of the sacrifice.[74]

In this scene, which calls to mind Fritz Lang's 1927 film *Metropolis*, in which workers march zombielike into the mouthlike orifices of devouring machines, Duhamel grimly explores an outcome of the convergence of science and technology unimaginable to those who had earlier regarded very different products of this combination as undeniable proofs of European superiority.

The theme of humanity betrayed and consumed by the technology that Europeans had long considered the surest proof of their civilization's superiority runs throughout the accounts of those engaged in the trench madness. The enemy is usually hidden in fortresses of concrete, barbed wire, and earth. The battlefield is seen as a "huge, sleeping machine with innumerable eyes and ears and arms." Death is delivered by "impersonal shells" from distant machines; one is spared or obliterated by chance alone. The "engines of war" grind on relentlessly; the "massacre mecanique" knows no limits, gives no quarter. Men are reduced to "slaves of machines" or "wheels [or cogs] in the great machinery of war." Their bodies become machines; they respond to one's questions mechanically; they "sing the praises" of the machines that crush them. War has become "an industry of professionalized human slaughter," and technology is equated with tyranny. Western civilization is suffocating as a result of overproduction; it is being destroyed by the wheels of great machines or has been lost in a labyrinth of machines. Its very future is threatened by the machines it has created.[75] Like David

[74]Duhamel, *Civilization*, pp. 274–75.

[75]For expressions of these sentiments, see Henri Barbusse, *Under Fire: The Story of a Squad* (London, 1916), pp. 48–49; Ernst Jünger, *Storm of Steel* (London, 1929), pp. 80–81, 108–10, 118; Frederic Manning, *The Middle Parts of Fortune* (New York, 1979), pp. 181–82; Richard Aldington, *The Death of a Hero* (London, 1984), pp. 255, 264, 267; Georges Duhamel, *La pensée des âmes* (Paris, 1949), pp. 242–43, and *Civilization*, pp. 223–24; Sigmund Freud, "Reflections upon War and Death" (1915) in *Character and Culture* (New York, 1963), p. 108; Robert Wohl, *The Generation of 1914* (Cambridge, Mass., 1979), pp. 218–19; Cruickshank, *Variations on Catastrophe*, pp. 31–35, 61–65; Eric J. Leed, *No Man's Land: Combat and Identity in World War I* (Cambridge, Eng., 1979), pp. 29–31, 34, 55, 96,

Jones, many of those who fought on the Western Front and lived long enough to write about their encounter with war in the industrial age began to "doubt the decency of [their] own inventions, and [were] certainly in terror of their possibilities." To have any chance of survival, all who entered the battle zone were forced to "do gas-drill, be attuned to many newfangled technicalities, respond to increasingly exacting mechanical devices; some fascinating and compelling, others sinister in the extreme; all requiring a new and strange direction of the mind, a new sensitivity certainly, but at a considerable cost."[76]

In the face of the overwhelming destructiveness of industrial technology, the individual soldier was reduced to a "pigmy man [who] huddles in little holes and caves."[77] As thousands of combatant memoirs have testified, the soldier under fire was engulfed by a technological maelstrom. Frederic Manning recaptures this ordeal in the following passage describing his descent into this mechanized hell:

> The air was alive with the rush and flutter of wings; it was ripped by screaming shells, hissing like tons of molten metal plunging suddenly into water, there was the blast and concussion of their explosion, men smashed, obliterated in sudden eruptions of earth, rent and strewn in bloody fragments, shells that were like hell-cats humped and spitting, little sounds, unpleasantly close, like the plucking of tense strings, and something tangling his feet, tearing at his trousers and puttees as he stumbled over it, and then a face suddenly, an inconceivably distorted face, which raved and sobbed at him as he fell with it into a shell-hole.[78]

The numbers of shells expended reached fearsome levels, particularly during offensives. It has been estimated, for example, that between February and December 1916, an average of one thousand shells fell on *each square meter* of the contested area surrounding the French fortress of Verdun.[79] Those that exploded sent shards of metal ripping through human flesh. As a nurse at the front, Vera Brittain witnessed firsthand the effects of the handiwork of the engineer and the chemist. She treated "men without faces, without eyes, without limbs, men almost disembowelled, men with hideous truncated stumps of bodies." Shells that

107, 121–22, 155–56, 164–65, 192; and Hanna Hafkesbrink, *Unknown Germany: An Inner Chronicle of the First World War Based on Letters and Diaries* (New Haven, Conn., 1948), pp. 66–70.

[76]Jones, *In Parenthesis*, p. xiv.

[77]Lt. Edward Graham, quoted in Leed, *No Man's Land*, p. 133.

[78]Manning, *Middle Parts of Fortune*, p. 7; see also pp. 212–21.

[79]Terraine (*White Heat*, p. 208), citing an official notice posted in Fort Douaumont at Verdun.

released chlorine, phosgene, or mustard gas left their victims "burnt and blistered all over . . . with blind eyes . . . all sticky and stuck together, and always fighting for breath." Tens of thousands of soldiers, having escaped the explosions and fumes, were driven mad by the very sound of the great guns and the constant tension that the anticipation of incoming shells engendered in all who were on the front line. Brittain recalled how her friend Geoffrey "shuddered from the deathly cold that comes after shell-shock; his face was grey with a queer, unearthly pallor, from which his haunted eyes glowed like twin points of blue flame in their sunken sockets."[80]

The extent and structure of the industrial battlefield rendered the most logical responses to these horrors—flight, surrender, mutiny—very dangerous alternatives to enduring trench conditions with the help of one's comrades, waiting to be relieved, and very often hoping for a "blighty": a wound that was relatively superficial but serious enough to put one out of the war.[81] Overawed by the ponderous war machines and battlements of the adversary powers, infantrymen found some consolation in "*griping*," writing antiwar poetry, and satirizing their plight in trench newspapers such as the *Wipers Times*. Noteworthy among the "Letters to the Editor" that appeared in the latter were complaints by "a lover of fresh air" about gas leaks in the town of Ypres, and commentary on the effectiveness of the "friendly fire" that destroyed British trenches in reprisal for German shelling.[82]

As numerous writers have observed, one of the prime casualties of the trench debacle was the age-old notion of war as an honorable and valorous enterprise in which youths were initiated into manhood and nations proved their mettle in the struggle for survival of the fittest. Four years of slaughter which centered on men feeding shells and bullets into machines and which involved only a minimum of hand-to-hand combat made a mockery of the chivalrous lexicon ("foe," "peril," "gallant," "staunch" "warrior," "perish") which, as Paul Fussell has shown, dominated writing on war before 1914 and demonstrated a surprising staying power during the conflict.[83] Little that was glorious or noble could be found in cowering in ditches in the midst of a wasteland glutted with the bloated bodies of dead men and animals whose stench carried miles to the rear. "Chivalry here took a final farewell," Ernst Jünger declared.

[80]Vera Brittain, *Testament of Youth* (London, 1980), pp. 339, 395, 257. See also Duhamel, *Civilization*.

[81]For an incisive discussion of these issues, see John Keegan, *The Face of Battle* (New York, 1976), esp. pp. 269–79.

[82]Patrick Beaver, ed., *The Wipers Times* (London, 1973), pp. 20, 141.

[83]Paul Fussell, *The Great War and Modern Memory* (New York, 1975), pp. 18–35.

"All fine and personal feeling has to yield when machinery gets the upper hand."[84] Many who dwelt in this nightmare world saw clearly the responsibility that had to be borne by the celebrated scientists and technicians who had conceived and perfected so many instruments of death. No writer expressed this realization and its bearing on the illusion of the glory of war more powerfully than Wilfred Owen in a brief but deservedly often quoted poem that recounts the death throes of a single gas attack victim:

> If in some smothering dreams you too could pace
> Behind the wagon that we flung him in,
> And watch the white eyes writhing in his face,
> His hanging face, like a devil's sick of sin;
> If you could hear, at every jolt, the blood
> Come gargling from the froth-corrupted lungs,
> Obscene as cancer, bitter as the cud
> Of vile, incurable sores on innocent tongues,—
> My friend, you would not tell with such high zest
> To children ardent for some desperate glory,
> The old Lie: *Dulce et decorum est*
> *Pro patria mori.*[85]

But the nobility of war was only one of the casualties. The mechanized slaughter on the Western Front corrupted or undermined the credibility of most of the ideals and assumptions on which the Europeans had based their sense of superiority to all other peoples and from which they had fashioned that ideological testament to their unprecedented hubris, the civilizing mission. Years of suicidal devastation forced European intellectuals to question the very foundations upon which their thought and value systems had been built: the conviction that they were the most rational of all beings, in control of themselves, of other peoples, and of all creation. Their unswerving faith in reason was doomed by the central contradiction of the conflict: though the new technology required reasonable men to fight (if they must) a defensive war, the generals who dictated the terms of combat were firmly committed to an obsolescent war of offensives and maneuver in which *cran* (guts), élan, and individual acts of heroism would prove decisive. It was not merely that the casualty figures were so appalling (Theodore Ropp has reckoned that in 1917 the British lost 8,222 men for every square mile of Flanders they gained in the five-month offensive known as the Third

---

[84]Jünger, *Storm of Steel*, pp. 109–10.
[85]"Dulce et Decorum Est," in *The Collected Poems of Wilfred Owen* (New York, 1965), p. 55.

Battle of Ypres);[86] it was that so many men were killed and maimed to gain so little. In the offensives devised by Douglas Haig and his staff in 1917, to give one of the more sobering examples, the British and French suffered over 500,000 casualties to gain at the point of their farthest advance into the German lines a little over five miles of "worthless" and "muddy ground"—much of which was soon recovered by the counterattacking German forces. The futile year-long effort on the part of the Germans to take the fortress city of Verdun in 1916 resulted in combined German-French losses that may have reached as high as 420,000 dead and 800,000 gassed or wounded. By 1917, "routine" deaths in "lull" periods in the trenches (for which the British staff devised the term "wastage" which was as coldly calculating as the "kill ratios" of the Vietnam war) were averaging 7,000 a week for the British alone.[87]

In the face of this glaring disproportion between human losses and military or political gains, the war—as Basil Liddell Hart, one of its participants and great chroniclers, observed—ceased to have "reason or method."[88] But it was not just the war that was seen to be irrational; the civilization that had produced it appeared to have lost its bearings. The vaunted rationality of the Europeans had been unable to prevent the trench madness; indeed, advances in science and invention, which had been ranked among the premiere achievements of their reasoning powers, were now seen to have been preparing this catastrophe for generations. As Malraux, Valéry, and numerous other intellectuals warned, Europeans could no longer be sure that reason was in command, that the products of rational thought would necessarily lead to the improvement of humankind. The assumption that the control over nature achieved by science would bring "human happiness and well-being," Bertrand Russell pointed out in 1924, rested on the presupposition that men were rational, when in fact, as the recent war had so convincingly demonstrated, they were "bundles of passions and instincts."[89]

For those who struggled to survive in the trenches, the power of emotions or instinctual drives, which had long been associated mainly

[86]Theodore Ropp, *War in the Modern World* (London, 1959), p. 250.

[87]Alastair Horne, *The Price of Glory: Verdun 1916* (Harmondsworth, Eng., 1982), pp. 327–28; Leon Wolff, *In Flanders Fields* (New York, 1964), pp. 230–34.

[88]Quoted in Alan Lloyd, *The War in the Trenches* (New York, 1976), p. 165. See also Duhamel, *La possession*, p. 19.

[89]Malraux, *La tentation*, pp. 23–24, 49–51, 127–29; Paul Valéry, "Letters from France: I. The Spiritual Crisis," *Athenaeum*, April 11, 1919, pp. 182–84; Cruickshank, *Variations on Catastrophe*, pp. 63, 121, 169, 179; Duhamel, *La possession*, p. 19; L. P. Jacks, "Mechanism, Diabolism, and the War," *Hibbert Journal* 13/1 (1914), 29–32; Brittain, *Testament*, pp. 288, 376; Bertrand Russell, *Icarus, or the Future of Science* (New York, 1924), pp. 57–59.

with savage peoples, was all too obvious. As Frederic Manning's protagonist Bourne remarks after observing the stoical acceptance of the
war by French peasants whose fields were near the front line: "There is
nothing in war which is not in human nature; but the violence and
passions of men become, in the aggregate, an impersonal and in-
calculable force, a blind and irrational movement of the collective will,
which one cannot control, which one cannot understand, which one can
only endure as these peasants, in their bitterness and resignation, endured it."[90] The prevalence of instinct and the irrational in the battle
zone rendered book learning remote and vacuous for the many soldiers
at the front who had gone directly from the schoolroom to the trenches.
The charges of avant-garde, prewar intellectuals like André Massis and
Alfred de Tarde (who—cruel irony—wrote under the pseudonym of
"Agathon," the disciple of Socrates who was "good and brave in war")
that European education had degenerated into "empty science" and
"pedantic materialism" resounded in Erich Maria Remarque's impromptu epitaph for a compatriot named Leer in *All Quiet on the Western
Front*: "What use is it to him now that he was such a good mathematician at school."[91]

No thinker more thoroughly explored the ways in which the war had
shaken the European faith in the primacy of reason than Sigmund
Freud.[92] Both in essays written during the conflict and in his *Civilization
and Its Discontents*, published a decade later, Freud argued that the passions and destructive forces unleashed by the war confirmed the view of
human nature he had enunciated to a rather limited, often highly critical,
audience in the years before the war. After the years of carnage in the
very heartland of the civilization that Europeans had confidently proclaimed the highest ever achieved, Freud's theories regarding the primal
passions persisting in humans despite their repression by family, state,
and individual conscience found a wide and often sympathetic hearing.
Freud charged that the war had exposed the false confidence of the
dominant "white nations" that their mastery over nature and the rest of
humanity ensured their ability to find means other than war to resolve
their quarrels. The release of bound-up tensions and emotions that so

---

[90]Manning, *Middle Parts of Fortune*, pp. 108–9.

[91]The "Agathon" passage is quoted and discussed in Wohl, *Generation of 1914*, pp. 5–6;
Remarque's epitaph is on p. 246 of *All Quiet on the Western Front* (New York, 1975), and
related responses can be found on pp. 79–80, 152–53. See also "Repression of War
Experience," in *The War Poems of Siegfried Sassoon* (London, 1983), pp. 84–85.

[92]This discussion of Freud's views is based on "Reflections upon War and Death"
(1915) and "Why War?" (1932)—both of which are reprinted in *Culture and Civilization*—
and *Civilization and Its Discontents* (New York, 1961), esp. pp. 36–90.

many participants recounted feeling during the early weeks of the war[93] was seen by Freud as a predictable response of individuals whose innate drives had hitherto been thwarted and whose instinctual urges had been denied by the agencies of civilized society, which were internalized in their own superegos. "The horrors of the recent World War," he argued, demonstrated both the extent to which national rivalries and hostilities were rooted in powerful instinctual forces and the degree to which human intelligence was dependent upon human emotion. The war had forced Europeans to confront the fact that aggressive and destructive urges—which they had once associated with children, the insane, and primitive peoples—lurked in their own psyches as well; that they too possessed, as Conrad had seen decades before, "hearts of darkness." Freud pointed out that the war had forced the states of Europe to violate all the moral constraints that they had imposed on their citizens; it had corrupted the sciences, which had been seen as the pinnacle of Western rationality; and it had resulted in a mass regression to primitive states of thinking and behavior. The savagery that the war had released within Europe, Freud concluded, reminded Europeans that their "fellow-citizens" of the world had not "sunk so low as we feared, because [the Europeans] had never risen so high as we believed."

The theme of the reversion of European "man" to a primitive or savage state of existence as a result of the trench stalemate is pervasive in contemporary accounts. Combatants describe themselves as "savages," "wild beasts," "barbarians," "primitives," "bushmen," "mere brutes."[94] Those at the front compare their existence to that of prehistoric men who lived in caves or crude holes dug into the earth. For example, when Tulacque, one of the characters in Henri Barbusse's novel *Under Fire*, discovers a prehistoric flint and bone axe in the trench excavations, one of his compatriots remarks that Tulacque himself resembles an "ape-man, decked out with rags and lurking in the bowels of the earth."[95] In the trenches or behind the lines, the refinements of civilization receded. Decorum was associated with death; modesty became irrelevant to sol-

[93]The best discussion of this phenomenon is contained in Leed, *No Man's Land*, chap. 2. See also Wohl, *Generation of 1914*, pp. 216–17.

[94]Remarque, *Western Front*, pp. 103–4, 236–38; Barbusse, *Under Fire*, pp. 40, 44; Manning, *Middle Parts of Fortune*, p. 12; Aldington, *Death of a Hero*, pp. 279, 289, 298; Hafkesbrink, *Unknown Germany*, p. 71; Jünger, *Storm of Steel*, pp. 262–63; Brittain, *Testament*, p. 216; Roland Dorgelès, *Les croix de bois* (Paris, 1983), pp. 62, 113; Duhamel, *La possession*, p. 243; and Freud, "Reflections," p. 119, and *Civilization and Its Discontents*, pp. 65–66. Duhamel also compared French soldiers discussing their wounds to well-trained children, and the operating room to a "barbaric temple" or "animal's lair" (*Civilization*, pp. 280, 286).

[95]Barbusse, *Under Fire*, p. 10. See also Fritz Kreisler, *Four Weeks in the Trenches* (Boston, 1915), pp. 64–65

diers who used crudely fabricated latrines as places to congregate, gossip, and curse their leaders.[96] In battle, primal instincts—"the furtive cunning of a stoat or weasel"—were the key to survival. Europeans fought, as Frederic Manning observed, like peoples at a "more primitive stage in their development, and . . . [became] nocturnal beasts of prey, hunting each other in packs."[97] The constant shelling forced infantrymen at the front to listen for incoming shells, which they could not see until it was too late. Thus, an acute sense of hearing, which the Europeans had associated with primitive peoples since at least the eighteenth century, superseded sight, which had been regarded as the most developed sense of the civilized.[98]

The retrogression that was occurring on a massive scale in the heart of Europe also helps to explain the "terrible feeling of foreignness" that soldiers felt on returning to the home front and their inability to make friends, parents, and fiancées understand what it meant to fight in the trenches.[99] Though it has always been difficult for soldiers to communicate their battle experiences to civilians, those on leave or returning from the Great War found themselves trying to describe an alien, unbelievable world in which all that European civilization had exalted for centuries was smashed and defiled. The patriotic formulas and tales of heroism that had so often bridged the gap between soldier and civilian in the past had little to do with the reality of the trench experience. The language of polite discourse in civil societies could not begin to describe the obscene, troglodyte existence the combatants had been forced to endure.

The conditions in which the soldiers lived and the wasteland their combat made of northern France mocked the European conceit that their unprecedented mastery of nature was proof of their superiority over all contemporary peoples and past civilizations. The lice-ridden bodies of the youth of Europe, exposed for weeks on end to the cold and wet of winter in Flanders or the valley of the Somme, fighting with huge rats for their miserable rations or their very limbs, belied the conviction that science had given Western *man* dominion over nature.

[96]Manning, *Middle Parts of Fortune*, p. 12; Remarque, *Western Front*, pp. 12–13; Aldington, *Death of a Hero*, p. 362; Ludwig Renn, *War* (London, 1984), p. 110–11.

[97]Manning, *Middle Parts of Fortune*, pp. 8, 39–40. See also Remarque, *Western Front*, pp. 124–25, 236–37.

[98]Léon-François Hoffman, *Le nègre romantique* (Paris, 1973), p. 180; and Leed, *No Man's Land*, pp. 126–27.

[99]Remarque, *Western Front*, pp. 137–53 (quoted portion p. 152). See also Aldington, *Death of a Hero*, pp. 340–53; Cruickshank, *Variations on Catastrophe*, pp. 35–36; Hafkesbrink, *Unknown Germany*, pp. 60–65; Sassoon, "Blighters," in *War Poems*, p. 68; Wohl, *Generation of 1914*, p. 98; Leed, *No Man's Land*, p. 126; and Robert Graves, *Goodbye to All That* (Garden City, N.Y., 1957), esp. chap. 21.

The clouds of flies or "busy black beetles" that gorged themselves on the carcasses of men and domesticated animals,[100] the ravages of disease, and ailments such as trenchfoot revealed that even Europeans remained subject to "the blind forces of nature" which Frederic Manning compared to the war itself, conceding that neither could be controlled nor even comprehended.[101] Even in its most innocuous manifestations, nature could be a source of apprehension. In Celine's *Journey to the End of the Night*, Ferdinand confesses that he was "afraid of trees too, since [he] had known them to conceal an enemy. Every tree meant a dead man."[102] "Uprooted, smashed" trees, "pitted, rownsepyked out of nature, cut off in their sap rising," were also central images in participants' descriptions of the war zone wasteland. For many, the perils inherent in the European determination to master nature found their most sobering expression in these moonscapes where wounded horses (among the most faithful servants of "man") died "wild with anguish, filled with terror, and groaning"; where "not a bush or a tiniest blade of grass was to be seen"; where "mangolds, uprooted, pulped, congealed with chemical earth" and "the sap of vegetables slobbered the spotless breechblock of No. 3 gun"; and where the very soil lost "the usual clean pure smell of earth" and reeked as if it were "saturated with dead bodies— dead that had been dead a long, long time."[103] As Georges Duhamel observed in 1919, the Western obsession with inventing new tools and discovering new ways to force nature to support material advancement for its own sake had inevitably led to the trench wasteland in which "man had achieved this sad miracle of denaturing nature, of rendering it ignoble and criminal."[104]

The trench stalemate also deprived the Promethean European male of the possibilities for activity, aggression, and taking charge which had long been considered key sources of Western superiority over inert, passive, slavish African or Asian peoples. In the battle zone the combatants became "waiting machines" or little more than "beasts of burden marching under the lash of the ox-driver."[105] The tedium and monoto-

---

[100]Duhamel, *Civilization*, p. 26; Renn, *War*, p. 253.

[101]Manning, *Middle Parts of Fortune*, p. 42. The best historical accounts of day-to-day life in the trenches include John Ellis, *Eye-Deep in Hell* (New York, 1976); Denis Winter, *Death's Men: Soldiers of the Great War* (London, 1978); and Lloyd, *War in the Trenches*.

[102]Louis-Ferdinand Celine, *Journey to the End of Night* (New York, 1960), p. 53.

[103]Wasteland quotations are taken respectively from Jünger, *Storm of Steel*, p. 108; Jones, *In Parenthesis*, p. 39; Remarque, *Western Front*, p. 61; Jünger, *Storm of Steel*, p. 108; Jones, *In Parenthesis*, p. 24; and Brittain, *Testament*, p. 252. See also Sassoon, "The Road," in *War Poems*, p. 51; and Edmund Blunden, *Undertones of War* (London, 1928), pp. 133, 224.

[104]Duhamel, *La possession*, p. 99.

[105]Barbusse, *Under Fire*, p. 17; Léon Cathlin, my translation of the passage cited in

ny of day-to-day existence was recounted by innumerable participants, as was their sense of helplessness in the face of the forces working to destroy them. In *Storm of Steel*, Ernst Jünger captures the feelings of immobility and impotence that gripped all who experienced the intermittent artillery barrages at the front: "You cower in a heap alone in a hole and feel yourself the victim of a pitiless thirst for destruction. With horror you feel that all your intelligence, your capacities, your bodily and spiritual characteristics, have become utterly meaningless and absurd. While you think it, the lump of metal that will crush you to a shapeless nothing may have started on its course. . . . You know that not even a cock will crow when you are hit."[106]

As Eric Leed has shown, psychological studies done during and after the war indicated that this enforced passivity in the face of intense danger was a key source of wartime neurosis. It was found that the incidence of war-related psychic disorders was particularly high among combatants in the trenches and in observation balloons, where they provided relatively immobile targets for the enemy and where there was little opportunity to release anxiety through aggression or "manipulative activities" that would reduce their vulnerability. Fighter pilots, by contrast, who had a high degree of mobility and control over the aircraft in which they dueled for their lives, registered proportionally far fewer cases of mental breakdown. The loss of manipulative capacity and the apparent uselessness of problem-solving skills for the trench soldier also accounted for the low self-esteem found to be widespread among infantrymen, and the pervasiveness of superstition, rumors of miracles, and retreat into myths at the front.[107] Ways of thinking which European writers had come to associate with primitive or savage peoples had great appeal for the trenchbound youth of civilized Europe. Their combat experiences led them to the disheartening conclusion that logic could not account for their predicament, nor could rational calculations better their chances of survival.

At the front even the most elemental bearings were lost. The soldiers' sense of time and space was altered in ways that often brought it close to that attributed to savages or ignorant peasants by proponents of the

---

Barrie Cadwallader, *Crisis of the European Mind: A Study of André Malraux and Drieu la Rochelle* (Cardiff, Wales, 1981), p. 5. The image of soldiers as animals being led to the slaughter is a common one in the combatant literature; for examples, see Owen, "Anthem for Doomed Youth," in *Collected Poems*, pp. 44–45; and Jones, *In Parenthesis*, pp. 2, 31–32, 87.

[106]Jünger, *Storm of Steel*, p. 180.

[107]Leed, *No Man's Land*, esp. pp. 180–86. On the profusion of myths and rumors at the front, see Fussell, *The Great War*, pp. 114–35.

civilizing mission. On the basis of combatants' accounts, Stephen Kern has shown the various ways in which space became disaggregated at the front. The traditional line of battle was blurred by the shell-pocked landscape and "the irregular geometricized forms" that made up the trench fortresses. Camouflage obscured "conventional visual borders," and the necessity for defense in depth broke down the "single vanishing point perspective" that had dominated Western perception since the Renaissance. The trench labyrinth disoriented all who entered, forcing them to confront "the ambiguities and irregular contours of reality."[108]

The "fabricated" sense of time that was linked to the rise of capitalism and measured by some of the Europeans' most ingenious inventions was also exposed as false.[109] Combatants remarked on their indifference to the passage of clock time and the "timeless confusion of battle." Some recounted feeling that time had fallen apart, while others recalled that it had been "indefinitely and most unpleasantly prolonged." Like "primitives" or peasants, the men in the trenches became more sensitive to the natural passage of time, an awareness that as Paul Fussell has shown was expressed in numerous poetic or prose descriptions of sunrises and sunsets.[110] Like the so-called savages of Africa or Melanesia, soldiers at the front became obsessed with mere survival in the present. They clung to memories of a peacetime past, but like civilian life the past tended to recede and become increasingly unreal. Most critically, time no longer had a sense of direction or purpose for those in the trenches. The future was vague, remote, far off.[111]

As a number of intellectuals noted after the war, the Europeans' prewar association of the future with progress and improvement was also badly shaken by the mechanization of slaughter in the trenches. Henry James's poignant expression of the sense of betrayal that Europeans felt in the early months of the war, when they realized that technical advance could lead to massive slaughter as readily as to social betterment, was elaborated upon in the years after the war by such thinkers as William Inge, who declared that the conflict had exposed the "law of inevitable progress" as a mere superstition. The Victorian mold, Inge declared, had been smashed by the war, and "the gains of that age now

[108]Kern, *Culture of Time and Space*, pp. 302–12.

[109]The term and conclusion are Maurice Nadeau's; see *The History of Surrealism* (Harmondsworth, Eng., 1978), p. 50.

[110]Fussell, *The Great War,*, pp. 51–63. For a superb example of the incorporation of the natural passage of time into war narrative, see Barbusse, *Under Fire*, pp. 42–45.

[111]Examples of the combatant's recollections of the meaning of time are taken from Kern, *Culture of Time and Space*, esp. pp. 290–96; Aldington, *Death of a Hero*, pp. 267, 323 (quoted phrases); Leed, *No Man's Land*, p. 104; Terraine, *White Heat*, p. 206; Wohl, *Generation of 1914*, p. 225; Kreisler, *Four Weeks*, pp. 1–2; Manning, *Middle Parts of Fortune*, p. 8; and Cadwallader, *Crisis of the European Mind*, p. 5.

seem to some of us to have been purchased too high, or even to be themselves of doubtful value." Science had produced perhaps the ugliest of civilizations; technological marvels had been responsible for unimaginable destruction. Never again, he concluded, would there be "an opportunity for gloating over this kind of improvement." The belief in progress, the "working faith" of the West for 150 years, had been forever discredited.[112]

Challenges to the Civilizing Mission and the Search for Alternative Measures of Human Worth

Well before the armistice in November 1918, it was clear to virtually all Western thinkers that European civilization had entered a period of profound crisis. But there was little agreement on how to answer the questions raised by the sustained violence and destruction of the war or how to meet the challenges to European global hegemony that were suddenly so widespread and seemingly overwhelming. Much of the debate on these issues centered on the proper roles of science and technology, which many of those engaged in the postwar controversies believed responsible for both the catastrophic effects of the war and Europe's loss of dominance to such rivals as the United States and Japan. A minority, but a highly vocal one, argued that Europe's crisis was brought on by the very nature of scientific and industrial civilization. The ideas of these thinkers are of the greatest relevance to the present study, but it is important to stress that their critiques were by no means the only or even the predominant response to questions linking Europe's dilemmas to science and technology. Most leaders and social commentators who dealt with these issues in the postwar era concluded that the disasters had occurred because science and technology had been misused. Inept political and military leaders had failed to understand the potential and proper application of the machines and weapons that had become decisive in modern warfare. They had blundered into war and then lost control of the instruments by which it was to be waged. The result was the mechanized, indecisive slaughter of the trenches.

The assumption that poor leadership or flawed methods of organization, not industrial civilization itself, were responsible for Europe's crisis was frequently linked to a second supposition: only a renewed de-

[112]*The Letters of Henry James*, ed. Percy Lubboch (London, 1920), vol. 2, pp. 398, 402–3; William Inge, *Outspoken Essays* (London, 1922), vol. 2, pp. 158–69, 171–83. For French variants on this theme, see Lucien Romier, *Explication de notre temps* (Paris, 1925), p. 1; Gustave Le Bon, *Le déséquilibre du monde* (Paris, 1923), pp. 9–11; and Cruickshank, *Variations on Catastrophe*, p. 65.

dication to scientific investigation and technological innovation could reverse Europe's decline and provide the basis for social and economic reconstruction within the individual nations. The leaders and political theorists on both the right and the left who held these views—as well as the engineers, entrepreneurs, managers, and labor leaders who were actually engaged in the rebuilding efforts—tended (at least in the 1920s) to see America as the epitome of the mass-producing and -consuming technocracy of the future. A greatly heightened European interest in the scientific management procedures by which Frederick Taylor and his disciples proposed to increase plant efficiency and in the mass production techniques pioneered by Henry Ford reflected a growing conviction that Europe's salvation could best be won by emulating America's technological successes.[113] Pro-American writers such as Philip Gibbs declared that the fate of industrial Europe depended on the decisions of the wealthy and powerful businessmen and politicians of the United States.[114] Even America's strongest critics were forced to concede its appeal to a crisis-ridden Europe. Georges Duhamel, for example, who after a three-week visit in 1929 wrote a scathing indictment of American culture titled *Scènes de la vie future*, admitted that most Europeans saw America's present as their future, and only a handful were distressed by this prospect.[115] The high regard of large numbers of Europeans for American ways and especially American power and technology had been dramatically demonstrated two years earlier by the adulation bestowed on Charles Lindbergh following his solo flight from New York to Paris. Like the exploits of the pilot heroes of the Great War, Lindbergh's feat served to rekindle faith in the potential inherent in the combination of men and the most advanced machines—if the men were skillful and firmly in control.

[113]The best introduction to these patterns of interaction can be found in Charles S. Maier, "Between Taylorism and Technocracy: European Ideologies and the Vision of Industrial Productivity in the 1920s," *Journal of Contemporary History* 5 (1970), 27–51. Maier's *Recasting Bourgeois Europe: Stabilization in France, Germany, and Italy in the Decade after World War I* (Princeton, N.J., 1975) places these issues in their larger political and social contexts. For a discussion of a number of important postwar authors, including René Loti, Édouard Herriot, and Peter Hamp, who looked to science and industry to rescue Europe, see Albert Schinz, *French Literature and the Great War* (London, 1920), esp. pp. 112–13, 248–53, 281–88.

[114]Philip Gibbs, *The Hope of Europe* (London, 1921), pp. 137–38, 158, 249–50, and *People of Destiny* (London, 1920), pp. vi, 126.

[115]Duhamel's critique was translated into English and published under the provocative title *America the Menace: Scenes from the Life of the Future* (London, 1931). His admission of America's popularity can be found in the preface (pp. xiii–xv) to the English edition. For Duhamel's place in shaping anti-American sentiment in France, see David Strauss, *Menace in the West: The Rise of French Anti-Americanism in Modern Times* (Westport, Conn., 1978), chap. 4.

A very different sort of combination of man and machine was envisioned by the German writer Ernst Jünger, who had spent nearly three years in the trenches. As we have seen, Jünger's earliest recollections of his experience were decidedly ambivalent. His highly autobiographical *In Stahlgewittern* (*Storm of Steel*) both glorified the heroism and chronicled the mutilation and degradation of the soldiers engaged in the struggle. The book also conveyed both a fascination with the power of weapons designed to obliterate humans on a mass scale and a sense of their horrific capacity to brutalize their human targets, mentally as well as physically. But in the many essays and books on the trench experience and the future of Germany that Jünger wrote in the 1920s and 1930s, his focus shifted to the heroism and camaraderie of German males in battle and the power of the instruments of modern warfare. Though politically conservative, Jünger strongly opposed the anti-industrial sentiments that had long been associated with the political right in Germany. He argued that experience at the front had produced a new breed of men who, though their primal passions had been rekindled in the heat of battle, had survived by becoming hard, precise, mechanized— servants or even appendages of the technology that Jünger saw as the key to Germany's revival. His emphasis on the subordination of men to machines may have been the most extreme, but numerous other writers, particularly those on the far right, also viewed a combination of brute force and the most advanced technology as the only way to revitalize their own nations and restore Europe to its former global hegemony.[116]

Gustave Le Bon agreed that Europe's salvation depended on massive technological and scientific innovation, but he doubted that humans had advanced rapidly enough mentally and socially to be true masters of the machines they had devised. His reliance on technology as the agent of Europe's salvation was rather different from that either of Jünger or of those who hoped to rebuild Europe on the American model. Harking back to a notion that had won widespread acceptance before the war, Le Bon argued that Europe's only hope was to develop weapons so destructive that their use would become inconceivable.[117] This "balance of terror" approach was not original, but the Great War had transformed it

[116]A substantial literature has developed on Jünger's thinking. The fullest treatment in English of the ideas relevant here can be found in Jeffrey Herf, *Reactionary Modernism: Technology, Culture, and Politics in Weimar and the Third Reich* (Cambridge, Eng., 1984), chap. 4. See also Leed, *No Man's Land*, pp. 150–62. As his title indicates, Herf treats Jünger's ideas in the context of thought on the German right as a whole and in terms of their impact on the National Socialists.

[117]Gustave Le Bon, *Premières conséquences de la guerre* (Paris, 1916), pp. 321–23, and *Déséquilibre*, pp. 7–10.

from a science fiction fantasy into one of the central realities of a century of unprecedented violence.

Most of the leaders and intellectuals who believed that Europe's recovery from the war hinged upon new scientific discoveries and increased mechanization were confident that a revived Europe would continue to exercise its dominance over the peoples of Africa and Asia. By contrast, a minority of intellectuals and maverick politicians, who traced the calamities that the Europeans had inflicted upon themselves to flaws inherent in the science- and technology-obsessed civilization of the West, tended to a pessimistic prognosis of the colonial powers' chances of maintaining their global dominion. The main concern of many who adopted this position amounted to little more than a revival of prewar fears of competition from industrializing non-Western societies. William Inge, for example, had little to say about what the death of the idea of progress and the crisis of European industrial society, which he so resoundingly proclaimed, meant for Europe's mission to the colonized peoples of Africa and Asia. But in "The White Man and His Rivals," an essay published just after the war, he predicted that the "suicidal conflict" would greatly accelerate the rise of Japan and China as economic *and* military rivals. Citing a range of authorities on Asian issues, Inge declared that there were no physical or intellectual barriers to prevent the "yellow races" from successfully competing with the greatly weakened Europeans, and he repeated the familiar threats of a flood of Chinese and Japanese migrant laborers into the white dominions and of glutted markets throughout the West.[118]

The French poet and essayist Paul Valéry also feared that Europe would be eclipsed by its economic rivals. But his critique of the technological and scientific strains of the civilizing mission contended that the Europeans had only themselves to blame. Competition between the European powers had driven them to export the engines and arms and teach the techniques that had made the West "the sovereign of the world" to the very peoples they hoped to dominate. This transfer not only provided the colonized peoples with the means to challenge Europe's economic dominance and political hegemony; it also forced social and intellectual changes in non-Western societies which broke down the barriers that had once induced the immobility and lethargy that had rendered them more susceptible to Western dominance. Having lost its scientific and technological edge, Valéry reasoned, and weakened by the war, Europe appeared doomed to be reduced again to the second-rank

[118]Inge, *Outspoken Essays*, vol. I, pp. 215–30. See also Gibbs, *Hope of Europe*, p. 251; and André Siegfried, *Europe's Crisis* (New York, n.d.), pp. 111, 116–18.

status that its small size and modest resources had destined it to occupy. It was condemned by its history to be dominated by the ancient civilizations of Asia and the upstart Americans, a "people who have no history."[119]

Georges Duhamel and Maurice Muret concurred with Valéry's view that Europe's suicidal infighting and its squandering of the secrets of material preeminence had contributed much to its decline and the strength of its rivals. But each stressed that the damage the Europeans had done to themselves in disseminating their science and technology was as much psychological as material. Once African and Asian peoples had acquired the technology of the West, that technology and its European creators had been demystified. Finding that they too could run, and increasingly manufacture, the machines and decipher the mathematical formulas of the West, once-subordinated peoples were less in awe of the genius or magic of the Europeans and more and more taken with their own cleverness and competence. Unknowingly reversing Inge's mockery of the sense of superiority the Victorians had derived from their mechanical mastery, Muret observed that when the colonized realized their own ability to operate a telegraph machine and drive an automobile at forty miles per hour, they lost their sense of inferiority to the "white men." The decline in the Westerners' mystique of power and invincibility had been hastened, Muret acknowledged, by the war, which had undermined both the moral and the material authority of the Europeans.[120]

Duhamel also saw the war as decisive in undermining the image of the Europeans as "inscrutable masters," "dazzling and terrible demi-gods." In supporting their colonial rulers in the war, the Africans and Asians had discovered that the Europeans' claim that they possessed attributes which entitled them to dominate the rest of humankind was false. The "men of color" found that the Europeans inhabited a small and divided continent, that their overlords were not gods but "miserable, bleeding animal[s] . . . devoid of even hope and pride." The colonized peoples learned that despite the boast of having mastered the forces of nature, Europeans too submitted to cold and heat, to epidemics, and to "perils without names." Not surprisingly, Duhamel suggested, the colonized felt little pity for their once proud masters; their disdain for Europe and its long-touted accomplishments was quite palpable in their interaction

[119]Paul Valéry, *Regards sur le monde actuel* (Paris, 1931), pp. 39–43, 50–51, and "Europe's Power to Choose" (1925), in *Collected Works*, vol. 10, pp. 324–25. For similar sentiments, see Lucien Romier, *Who Will Be Master: Europe or America?* (London, 1928), pp. 29–34, and *Nation et civilisation* (Paris, 1926), pp. 66–67, 118, 134, 147.

[120]Maurice Muret, *The Twilight of the White Race* (London, 1926), pp. 9, 13, 22, 29–31.

with Europeans who visited the colonies.[121] Both Muret and Duhamel took some comfort in the belief that though the Africans and Asians had borrowed the inventions and scientific learning of the West, they had not really understood them. Like André Siegfried's suggestion that despite Asian competition Europe retained the "creative edge," the idea that the borrowers could not themselves invent or discover still offered some hope, however slight, that Europe's decline might yet be reversed.[122]

The fear that the Europeans' internecine clashes had irreparably broken the spell of the mastery they had long exerted over non-Western peoples was widespread in the decade after World War I. Inge and Muret noted the great interest, even delight, of the colonized in Europe's self-destruction. Even so patriotic a commentator as Henri Massis conceded that the prestige of European civilization had fallen sharply among the Asians. These and other writers also admitted that the Europeans had lost confidence in themselves and the assurance that they alone knew what it meant to be civilized.[123] But many of those concerned with colonial issues went a good deal further: they argued that the Europeans' belief in their innate superiority had been exposed as a delusion. Its corollary, the conviction that the Europeans had a mission to raise up the rest of humankind to their level of civilization, was revealed as little more than a conceit that was no longer tenable. Most authors expressed their doubts in general terms,[124] but some stressed the ways in which the Europeans' betrayal by science and technology had undermined their assumption that they had the duty to shape the course of development for the rest of humankind. Among these, none matched Georges Duhamel and George Orwell in their sustained and skilled assault on the civilizing mission in its scientific and technological manifestations.

Duhamel was haunted by one brief and seemingly insignificant occurrence that he came to see as a moment of revelation. In an instant the

---

[121]Duhamel, *Entretien sur l'esprit européen* (Paris, 1928), pp. 21–22, 29–31.

[122]Muret, *Twilight*, p. 73; Duhamel, *L'esprit européen*, pp. 21–22, 41; and André Siegfried, *Europe's Crisis*, pp. 120–23.

[123]Inge, *Outspoken Essays*, vol. 1, p. 243; Muret, *Twilight*, pp. 9, 21, 30; and Henri Massis, *Defense of the West* (London, 1927), pp. 6, 9, 134.

[124]For examples, see Maurice DelaFosse, *The Negroes of Africa* (London, 1931), pp. xxxii–xxxiii; Girardet, *L'idée coloniale*, pp. 154–56, 162, 167–70; Arthur Mayhew, *The Education of India, 1835–1920* (London, 1926), p. 61; and Mayhew's discussion of the formulation of British educational policies for Africa in the mid-1920s in *Education in the Colonial Empire* (London, 1938), pp. 77–78. As Herbert Sussman has shown, even Rudyard Kipling, the banjo bard of the imperial engineer, came to have doubts about the assumption that scientific and technological advance were necessarily progressive and beneficial; see *Victorians and the Machine* (Cambridge, Mass., 1968), pp. 218–33.

falsehood of the European claim to having fashioned the greatest of all civilizations became unmistakably clear, and he realized that his discovery was shared by African witnesses to the calamities the Europeans had brought upon themselves. In *Civilization, 1914–1917,* the best of the largely autobiographical novels in which Duhamel recounted his experiences as an officer in the French medical corps, he describes the arrival at his unit's operating room of several "little Malagasy" stretcher-bearers carrying badly wounded French cuirassiers. His initial impression of the Africans is typical of the era of European dominance. With their "thin black necks, encircled by the yokes," and their "shriveled fingers," they remind him of "sacred monkeys, trained to carry idols." But after the bearers place the cuirassiers on the operating tables, Duhamel relates:

> At this moment my glance met that of one of the blacks and I had a sensation of sickness. It was a calm, profound gaze like that of a child or a young dog. The savage was turning his head gently from right to left and looking at the extraordinary beings and objects that surrounded him. His dark pupils lingered lightly over all the marvelous details of this workshop for repairing the human machine. And these eyes, which betrayed no thought, were none the less disquieting. For one moment I was stupid enough to think, "How astonished he must be!" But this silly thought left me, and I no longer felt anything but an insurmountable shame.[125]

Though Duhamel instinctively resorts to the imagery of the colonizer—child, dog, savage—the ideology of European dominance is for him, at least, reduced to absurdity. The cannonade in the distance and "the hum of industry" about him in the operating room fix in his mind the "immense harm engendered by the age of the machines." When he recalls the "inexplicable look of the savage," he feels only "pity, anger, and disgust."[126]

In the many works he published after the war, Duhamel tirelessly reiterated and embellished the insights he had gained from this chance encounter with the Malagasy stretcher-bearer.[127] He concluded that the war was the inevitable outcome of the Europeans' centuries-long obsession with scientific and industrial advance. They had been so captivated by mechanical progress and material increase that they had neglected or completely forgotten the needs of the soul and spirit. They had allowed "moral civilization" and ideals to wither, while increasingly subordinat-

---

[125]Duhamel, *Civilization,* pp. 282–83.
[126]Ibid., p. 284.
[127]On at least one occasion after the war, Duhamel explicitly referred to the incident; see *L'esprit européen,* pp. 40–41.

ing themselves to the machines they had created to serve them. They confused the development of industry and science with civilization and became convinced that "good" could be equated with the ability to go a hundred miles per hour. These errors and misunderstandings had inevitably led to Europe's ruin in a war that had devastated its once prosperous lands, thrown its societies into turmoil, and aroused the colonized peoples to resistance. And resist they must, Duhamel insisted—resist not just European colonial dominance but the spread of the "cruel" and "dangerous" civilization that the Europeans (and, after the war, their American progeny) sought to impose on all humankind. The war had revealed the unequaled barbarity of this so-called civilization and the perils of destructiveness and brutality that threatened those who sought to emulate its narrowly materialistic achievements.[128]

In "Shooting an Elephant," George Orwell also uses a single incident, which he weaves into an allegorical tale, to proclaim the approaching end of the white man's dominance and the bankruptcy of the civilizing-mission ideology. Drawing on his experience as a policeman in Burma in the mid-1920s, Orwell describes the killing of an elephant that had gone into heat, trampled an Indian "coolie," and run amuck. In this brief, deceptively matter-of-fact account of a few hours in the life of a minor colonial official in a Burmese backwater, Orwell captures the rising contempt of the colonized for their European overlords—"the sneering yellow faces of the young men"; "the insults hooted after me"—and powerfully illustrates the "dirty work of Empire at close quarters" which had convinced him that "imperialism was an evil thing." But the central message of the story is that for all their lordly pretensions and superior technology, the Europeans are no longer in control of their empires. Because a crowd of spectators has gathered to witness the elephant's execution, the police official is forced to kill a magnificent and valuable animal that he knows is no longer dangerous. In doing so, he realizes that despite the fact that he is the official in charge and the rifle is in his hands ("a beautiful German thing with cross-hair sights"), the Burmese crowd is in command. He must kill the elephant to fulfill his role as "sahib," and that act brings home to him

> the hollowness and futility of the white man's dominion in the East. Here was I, the white man with his gun, standing in front of the unarmed native crowd—seemingly the leading actor of the piece; but in reality I was only an absurd puppet pushed to and fro by the will of those yellow faces behind. I perceived in this mo-

[128]See esp. Duhamel, *La possession*, pp. 140, 241–46, 254–56; *L'esprit européen*, pp. 14–18, 20–21, 30–36; and *Civilization*, p. 288.

ment that when the white man turns tyrant it is his own freedom that he destroys. He becomes a sort of hollow, posing dummy, the conventionalized figure of a sahib. For it is the condition of his rule that he shall spend his life trying to impress the "natives," and so in every crisis he has got to do what the "natives" expect of him. He wears a mask, and his face grows to fit it. I had got to shoot the elephant. I had committed myself to doing it when I sent for the rifle.[129]

The excruciatingly slow death of the elephant, which has to be shot several times before it dies, becomes a metaphor for the futility of the "duty" to civilize, which Orwell and other colonizers once believed gave meaning to the colonial enterprise. Whatever their intentions, and Orwell was far from convinced that these were predominantly benevolent, the colonizers were not in a position to determine the outcome of their policies and actions. Because of their superior technology they still ruled. But they could not ensure that their guns and machines or even their irrigation projects would not victimize the innocent in the same way that the reluctant police officer reduces the becalmed elephant— which is "beating a bunch of grass against his knees, with that preoccupied grandmotherly air that elephants have"—to a creature "suddenly stricken, shrunken, immensely old."

In his earlier *Burmese Days*, a novel built on a biting and extended critique of European colonial society and the sahibs and memsahibs who inhabited it, Orwell sought to expose the falsehood and hypocrisy of the most revered tenets of the imperial ideology. In response to the Indian doctor Veraswami's impassioned recitation of the benefits that British rule has brought to Burma, Flory—Orwell's protagonist and surrogate —dismisses each of the imperial accomplishments that the well-indoctrinated doctor has suggested. Interestingly, most of Flory's rebuttal is devoted to refuting the argument that the colonized, Indian and Burmese alike, have gained greatly from the technology and techniques made available by the imperial connection. Flory spends little time debunking platitudes about good government and justice, contenting himself with hackneyed puns like "Pox Britannica" and allusions to the filthy prisons where he and the gentle doctor have presided over beatings and hangings. He concentrates instead on the poor quality of colonial schools, which he characterizes as "factories for cheap clerks," and the appalling lack of technical progress under colonial rule. Flory avers that the Europeans never taught a "single useful manual trade" to the

[129]George Orwell, "Shooting an Elephant," in *Inside the Whale* (Harmondsworth, Eng., 1968), p. 95.

colonized; on the contrary, they destroyed the indigenous handicraft industries in order to eliminate competition with their own shoddy, machine-made goods. Rather than advancing the technology and technical proficiency of the Indians, he charges, colonization deprived them of their ancient skills in shipbuilding, gunmaking, and textile manufacture. In a final satirical flourish, he expresses misgivings about the environmental impact of the measures the British have taken to civilize the backward "natives." This is the most novel of Orwell's criticisms, but it may well reflect his familiarity with the growing postwar concern of a number of colonial officials, epitomized by Albert Howard, for the long-term ecological consequences of irrigation schemes, new cropping patterns, and deforestation.[130] Flory argues that British rule has "wrecked the whole Burmese national culture" and that "modern progress" would lead to little more than "our dear old swinery of gramophones and billycock hats. Sometimes I think that in two hundred years all this . . . will be gone—forests, villages, monasteries, pagodas all vanished. And instead, pink villas fifty yards apart; all over those hills, as far as you can see, villa after villa, with all the gramophones playing the same tune. And all the forests shaved flat—chewed into woodpulp for the *News of the World*, or sawn up into gramophone cases."[131]

Although Orwell and Duhamel sought to discredit the civilizing mission by underscoring the damage that Europeans had done to the peoples they dominated and by warning of the dangers that European materialism posed for the colonized, neither suggested that solutions to Europe's crisis might be found in the imperiled cultures of Africa and Asia. But many European intellectuals believed that this was possible. Thinkers as disparate as Hermann Hesse and Bronislaw Malinowski argued that the scientifically and technologically "backward" cultures of the non-Western world might well provide viable alternatives to the disoriented and crisis-ridden civilization of the West. Their search for more meaningful world views and life styles was bolstered by the added strength that a relativistic approach to human cultures gained from the

---

[130]Albert Howard expressed these concerns throughout the 1920s and 1930s in numerous articles and addresses; for a representative sample, see *An Agricultural Testament* (London, 1940). Fears relating to environmental degradation were complemented by a more general uneasiness about the tendency of industrialized societies to squander natural resources, which the war had greatly exacerbated, and the potential global scarcities that would result; see Siegfried, *Europe's Crisis*, pp. 14, 22; Valéry, "Remarks on Progress" (1929), in *Collected Works*, vol. 10, p. 165; and Marc Elmer, Introduction to Rabindranath Tagore, *La Machine* (Paris, 1929), pp. 23–24.

[131]George Orwell, *Burmese Days* (New York, 1963), pp. 37–39. On the background to Orwell's service in Burma, see Peter Stansky and William Abrahams, *The Unknown Orwell* (London, 1972), pt. 3.

disillusionment with the West engendered by the war. The human suffering that had resulted from the misuse of the most advanced science and technology compounded the prewar doubts of scientists and philosophers such as Henri Poincaré and Pierre Duhem about the absolute objectivity and potential for "pure observation" of the inductive approach. The forceful contention of such eminent theorists that the principles of scientific inquiry developed in the West were, like all forms of human cognition, culturally constructed left open the possibility that the Europeans had not after all gained a superior understanding of the underlying realities that they assumed made up a single system. The contemporaneous challenges to the Newtonian view of the universe by physicists such as Max Planck and Albert Einstein further strengthened the position of those who argued for relativism and cultural pluralism. Fluidity and uncertainty within the sciences also gave encouragement to thinkers who sought more intuitive approaches to the meaning of human existence.[132]

The search in the cultures of Africa for correctives or alternatives to Western ways amounted to little more than attempts to revive the myth of the noble savage, which had proved such a tempting target for the satirical barbs of Victorian writers. The genre as a whole enjoyed considerable popularity in the bleak postwar years, as is perhaps best illustrated by D. H. Lawrence's stories and especially his novels *The Plumed Serpent* (1926) and *Kangaroo* (1923). In the ancient legends and imagined eroticism of vanquished Amerindian civilizations and among what he considered the rough and unreflective white settlers of Australia, Lawrence sought refuge from the enclosed, overcultivated English countryside and the grimy mining and factory towns that he had depicted in his earliest writings and returned to in works such as "St. Mawr" and *Lady Chatterley's Lover* (1928). In fact, Mellors, a character in the latter through whom Lawrence expounds his own views, is very much a homegrown noble savage—sensual and sexually adept, hostile to the sulfurous mining pits that symbolize the industrial order, and close to nature in his role as gamekeeper. Throughout the novel these qualities in Mellors are set against the reserve and adherence to social convention of

[132]On the upheavals in scientific thinking, see Peter Alexander, "The Philosophy of Science, 1850–1910," in D. J. O. O'Connor, *A Critical History of Western Philosophy* (New York, 1964); Antonio Aliotta, *The Idealistic Reaction against Science* (London, 1914), esp. sec. 2, chap. 1; Biddiss, *Age of the Masses*, pp. 45, 64–75, 83–98; and Kern, *Culture of Time and Space*, pp. 132ff., 151–52, 184–85. For the views of Poincaré and Duhem, see Joseph J. Kockelmans, ed., *Philosophy of Science: The Historical Background* (New York, 1968), chaps. 13 and 15. On the growing importance of relativism, see Leclerc, *Anthropologie et colonialisme*, pp. 64–68; Girardet, *L'idée coloniale*, pp. 160–61; and W. F. Wertheim, *Evolution and Revolution* (Harmondsworth, Eng., 1974), pp. 18–19.

Lord Clifford Chatterley, who has been shipped back from the war "more or less in bits" and permanently paralyzed from the hips down.[133]

In *Civilization, 1914–1917*, Georges Duhamel's protagonist, a sergeant in the French medical corps, fantasizes about escaping the horrors of the trench struggle by fleeing to the mountains to live among the "savage" blacks, who exist, he implies, in a state of nature, free from mechanical outrages. But the sergeant wonders if there are any "real black people" left. Having seen them riding bicycles at Soissons and striving, just like the Europeans, for war decorations, he despairs of finding any that have not been contaminated by European "civilization."[134] In *Batouala*, which won the French Prix Goncourt for literature in 1922, René Maran assured his European readers that there were some noble African savages left but lamented the approaching end of their way of life under the advance of European colonialism. Though Maran was black and had been born in Martinique, from childhood he had attended French schools and had been fully assimilated to French culture. His education opened the way to the French colonial service, where he was employed for decades before taking up a career as a novelist. In *Batouala*, he anticipated many of the themes later favored by the writers of the Négritude movement. The people of Ubangui-Shari, where the novel takes place, are depicted as comely and physically powerful, sensitive to the beauty and rhythms of the luxuriant rainforest that constitutes their world, and unabashedly devoted to the pleasures of the flesh. Though the protagonist Batouala admits to an "admiring terror" of the whites' technology—including their bicycles and false teeth—he hardly sees them as "demigods." He mocks their pale and sensitive skin, their meager sexual endowment, their smelly bodies, and especially their feet inexplicably encased in "black, white, or banana-colored skins." From the African viewpoint the coming of the Europeans means only hard work and the abuse of overseers. Their paradisiacal existence disintegrates in the face of the demands of the foreign overlords. In an angry harangue Batouala dismisses the colonizers' promises to use the labor and revenue extracted from his people to build "villages, roads, bridges and machines which move by fire on iron rails." Of these, he charges, the Africans have seen "nothing, nothing" except useless bridges across rivers that are easily forded. The whites are liars; they have come to Ubangui-Shari not as friends or civilizers but "just to

[133]D. H. Lawrence, *Lady Chatterley's Lover* (London, 1972), p. 1. For samples of Lawrence's anti-industrial invective, see pp. 3–4, 106–9, 115–16, 149–55, 345–47, 366–67.
[134]Duhamel, *Civilization*, pp. 271–72.

suppress" and exploit the Africans, whom they dislike and treat as slaves or beasts of burden.[135]

Nonfiction writers also strove to resurrect a Rousseauian image of "primitive" peoples free of the civilized constraints imposed by the dehumanizing industrial order. In a 1930 essay on anthropology and colonial rule (responding to an earlier article by a British colonial official), Bronislaw Malinowski argued that he and many of his anthropological co-workers increasingly regarded primitive societies as a refuge from the "mechanical prison" of Europe and North America. Charging that "science is the worst nuisance and greatest calamity of our days," he launched into a pithy diatribe against men turned into robots, the pace of life in Western societies, and the "aimless drive of modern mechanization." For the anthropologist, Malinowski asserts, the cultures of Africa or the South Pacific provided a "romantic escape" from the passive, standardized, and shoddy mass-goods culture that science and industry had imposed on Europe.[136]

Though R. Austin Freeman did not himself take refuge in "exotic" lands, he agreed with Malinowski's assessment of the dehumanizing effects of mechanization In many ways Freeman's 1921 study *Social Decay and Regeneration* was little more than a restatement of the criticisms of industrial society that had been raised more eloquently by Ruskin, Morris, and others in the nineteenth century. Much of Freeman's account was devoted to familiar discussions of the dehumanizing effects of the factory system, the poor quality of machine-made goods, and the filth and ugliness of factory towns. But like Lewis Mumford, a young American author with similar concerns, Freeman stressed the ill effects of industrialization on the environment, both from pollution and by depletion of the earth's resources. With the "cataclysm of the war" very much in mind, he also warned of the industrial machine's capacity for the "wholesale physical destruction of man and his works and the extinction of human culture." Though he does not say so explicitly, Freeman implies that without mechanization these calamities would not be possible.

He also argues, here explicitly, that on a day-to-day basis preindustrial Africans are vastly better off than the pale British workers who have been stripped of skills and self-esteem by the machines they serve. The idealized "African negro" whom Freeman compares to the "British sub-man" is the epitome of the mythic noble savage. He is "usually

---

[135]René Maran, *Batouala* (London, 1972), pp. 29–31, 47–48, 50, 74–76.

[136]Bronislaw Malinowski, "The Rationalization of Anthropology and Administration," *Africa* 3/4 (1930), 405–6. Much of the rest of the article is devoted to the dangers of "dehumanizing" "pre-logical" peoples through the "rationalization" of anthropology.

sprightly and humorous. He is generally well-informed as to the flora and fauna of his region, and nearly always knows the principal constellations. He has some traditional knowledge of religion, myths and folklore, and some acquaintance with music. He is handy and self-helpful; he can usually build a house, thatch a roof, obtain and prepare food, make a fire without matches, spin yarn, and can often weave cotton cloth and make and mend simple implements. Physically he is robust, active, hardy and energetic."[137] Interestingly, though traces of the "happy sambo" image persist and Freeman remains convinced of the racial superiority of the ordinary Englishman, his noble savage is an accomplished individual with knowledge of the natural environment and numerous technical skills—the skills of the preindustrial craftsmen eulogized by Morris and Ruskin. As Freeman renders it, the African embodiment of the European past suggests a return not to the brutality of the Stone Age but to the idyllic stage of "savagery" that Rousseau had preferred over the stifling artificiality of "civilized" Europe.[138]

Of those intellectuals who sought alternatives to the "wounded" civilization of Europe in the postwar era,[139] the great majority were drawn to Asia, particularly India. The late nineteenth-century interest in Hinduism and Buddhism became something of a fad among the disenchanted youth of western Europe. The works of Hermann Keyserling and Hermann Hesse were prominently displayed in the bookstalls of German railway stations; large and enthusiastic audiences throughout Europe nodded approvingly when Rabindranath Tagore told them that their materialistic civilization was floundering and could be saved only if they turned to the spiritualism of "the East."[140] Henri Massis thought the danger of what he termed "Asiaticism" so great that he wrote an extensive polemic against the teachings of the German philsopher and Swiss novelist, as well as their French allies such as Romain Rolland. In

---

[137]R. Austin Freeman, *Social Decay and Regeneration* (London, 1921), pp. 251–52.

[138]As Arthur Lovejoy has shown ("The Supposed Primitivism of Rousseau's *Discourse on Inequality*," *Modern Philology* 21 [1923], 165–86), Rousseau distinquished four stages of human development from the most primitive to the one that prevailed after plants were first domesticated. Contrary to the impression given in most discussions of Rousseau's defense of primitivism, he exalted the third or patriarchal stage, when government was centered on the family and tools for hunting and fishing had been devised.

[139]Massis, *Defense of the West*, p. 67.

[140] For observations on the popularity of Hesse and Keyserling, see Paul Morand's *Bouddha Vivant*, translated as *The Living Buddha* (London, 1927), p. 11. On Tagore's tours of the West, see his *Diary of a Westward Voyage* (Bombay, 1962), pp. 67–86; and Kripalani Krishna, *Tagore: A Life* (New Delhi, 1971), chaps. 12 and 14. And for a sense of Tagore's message, see his *Nationalism* (London, 1917), esp. pp. 29–33, 43–45, 91–92. Morand, Massis, and other French writers could not resist the inference that the failure of the Germans to win world mastery by arms had much to do with their heightened interest in otherworldly philosophies.

his *Défense de l'Occident*, Massis warned that the threat was no longer merely a concern for intellectuals; it had become an issue that politicians ought to take seriously because its victory would have far-reaching consequences for European societies. He conceded that as a result of the war, "mechanism was in full flight" and the West threatened with destruction by the very means that had once promised prosperity and well-being; he also admitted that the sciences were in turmoil. But he warned that those who advocated journeys to the East to find ways to revive European civilization could only bring about its utter ruin and contribute to the spread of Bolshevism. The spiritual revival that would save the West, he insisted, must come from its own Catholic traditions and medieval precedents.[141]

The actual results of a number of the more prominent journeys to the East in this period suggests that perhaps Massis was unduly alarmed. Renaud, the protagonist of Paul Morand's 1927 novel *Bouddha Vivant*, flees Europe to find meaning in the teachings of the Asian sages. But rather than wisdom, most of the Asians he encounters exhibit the shortcomings that ardent colonizers had been denouncing for decades: they are ignorant and afraid of machines, indifferent to precision, lethargic.[142] Renaud comes to despise China, finds that most of India has been spoiled by Westernization, and eventually returns to Europe—his quest unfulfilled. Jali, an Asian prince who joins Renaud in his wanderings, finally abandons his own pilgrimage and returns home to take the throne of his deceased father. Jali is thoroughly disillusioned and embittered by what he has seen of the world—both West and East.

André Malraux also traveled to China in search of alternatives to the "ordered barbarism" and fundamentally absurd life of the industrialized West. He looked to an ill-defined Chinese "wisdom" and reflectiveness to fill the vacuum created by the Europeans' failure to discover the meaning of existence through the sciences, a failure that he believed had deprived Western civilization of any spiritual purpose.[143] Like Morand, he found China as much in crisis as Europe, collapsing under the impact of contacts with the West and wracked by poverty, famine, and civil war. He despaired of finding solutions for Europe's dilemmas in the

[141]Massis, *Defense of the West*, esp. pp. 1–28, 41–44, 51–54, 101–6, 151–65, 184–86, 191–203. For a detailed discussion of Massis's exchanges with the advocates of Asiaticism and his defense of the West, see Cadwallader, *Crisis of the European Mind*, chap. 1 and pp. 106–25.

[142]Morand, *Living Buddha*, pp. 34, 36, 91, 96.

[143]Malraux, *Tentation*, pp. 20–26, 31–33, 67–68, 81, 100–102, 109–17; and "D'une jeunesse européenne," in *Écrits* (Paris, 1927), pp. 133–36, 145–46.

ancient wisdom and values of the Taoists or Confucians who had failed to save Chinese civilization itself.[144]

Unlike Malraux and Morand, Hermann Keyserling was confident that remedies for the crisis of the West could be found in the philosophies of China and especially India, a position that he had already adopted before the outbreak of the war. But it is unlikely that his rambling postwar discourses on subjects ranging from the "culture of the future" to the "true" problem of progress provided intelligible alternatives to any but the devotees who gathered about him at his villa at Darmstadt.[145]

The more enduring of the thinkers who proposed Asian alternatives to Western empiricism and materialism tended to allow the Asians to speak for themselves. Romain Rolland, who was often the target of Massis's polemical assaults, helped organize European speaking tours for such luminaries as Tagore; wrote introductions to works expounding the Indian world view, such as Ananda Coomaraswamy's studies of Indian art and philosophy; and taught by example through his biographies of Gandhi, Vivekananda, and Ramakrishna, whom he depicted as living embodiments of "Eastern" wisdom and values.[146]

Though much less prolific and famous than Rolland, Marc Elmer also sought to promote the works of Asian writers, most notably Tagore's allegorical one-act play on the dehumanizing effects of mechanization, *La Machine*. In a lengthy introduction to the play, Elmer denounced machines as the despoilers of nature and humans. He pieced together citations from an impressive array of writers of many nationalities who blamed industrialization for innumerable vices, from ugly and polluted cities and the ruin of the family to imperialism and the Great War. Elmer charged that the war had shattered the "mystique of progress," which had been mistakenly associated with greater productivity and material

[144]Malraux's disenchantment is evident in his early novels *Les conquerants* (1929) and *La condition humaine* (1933). For more detailed discussions of his search for personal meaning, see Joseph Hoffman, *L'humanisme de Malraux* (Paris, 1963), esp. chaps. 2 and 3; Cadwallader, *Crisis of the European Mind*, esp. chaps. 6 and 7; and H. Stuart Hughes, *The Obstructed Path* (New York, 1966), pp. 137–48.

[145]For a sample of Keyserling's prewar views, see *East and West and Their Search for the Common Truth* (Shanghai, 1912). Though less famous than his travel memoirs, *The World in the Making* (New York, 1927) is perhaps more representative of his efforts to suggest ways out of Europe's postwar dilemmas.

[146]Again, Cadwallader's *Crisis of the European Mind* has the best discussion in English of Rolland's writings and ongoing debate with Massis and other writers; see esp. chaps. 1 and 4. For a sense of Rolland's approach and the fervor of his conversion, see his introduction to the English edition of Ananda Coomaraswamy's *Dance of Shiva* (London, 1924).

accumulation. It had revealed that machines were in control, not men, and that machines and the sciences that improved them had advanced blindly and inexorably, with little regard for the real needs of the humans they were supposed to serve. Elmer urged Western readers of Tagore's play to turn to the ideas and values of the "Orient" (though he did not attempt to identify them), which would permit them to bridle their runaway machines and balance their material obsessions with pursuits that enlarge and enrich the life of the spirit.[147]

Hermann Hesse and E. M. Forster sought through fables and novels to suggest that there was another, higher reality beyond that which the Victorians were so confident their sciences had permitted them to discover. Hesse contrasted the arid social and intellectual wasteland of a Europe "devastated by war" with the vibrant and engaged but (at least in Hesse's rendering) rather confused search for self-realization that he depicted as the central preoccupation of "Eastern" peoples and pilgrims from the West.[148]

Though the nebulous quality of the higher truths that E. M. Forster sets against the pettiness of the everyday lives of the colonizers in *A Passage to India* (1924) has provided openings for numerous doctoral dissertations, the nature of his alternatives to European thinking is as obscure as those explored in Hesse's fables. But the meaning of the echo that Mrs. Moore hears in the Marabar cave is less important than Forster's underlying and at the time unsettling contention that the colonizers' platitudes about their duty to civilize the "natives" are nonsense in view of the fact that for millennia the natives in question had been probing psychic and spiritual depths that most Europeans did not even imagine existed. In his unreflective complacency, self-assurance, and lack of subtlety, Mrs. Moore's son Ronny Heaslop, who is a judge in the Indian Civil Service, epitomizes the sort of European who had for centuries dominated both society in the West and the overseas empire. Events in the novel belie Ronny's conviction that only his sort of Briton is able to rule and administer justice fairly to the "slack," disorganized, and dishonest Indians. They suggest that the colonizers are dwarfed and that their civilizing mission will ultimately be frustrated by more profound and powerful forces: "The triumphant machine of civilization may suddenly hitch and be immobilized into a car of stone, and at such moments the destiny of the English seems to resemble their predeces-

---

[147]Elmer, Introduction to Tagore, *La Machine*, pp. 7–26.

[148]Hermann Hesse's most explicit treatment of the disenchantment with a materialistic Europe shattered by the war can be found in *Die Morgenlandfahrt* (1932), translated as *The Journey to the East* (New York, 1961), esp. pp. 5–7, 12–13, 18–19. For his idealized vision of the "East," see *Siddhartha*, originally published in 1921.

sors', who also entered the country with intent to refashion it, but were in the end worked into its pattern and covered with dust."[149]

Perhaps because he was a philosopher rather than a novelist or poet, René Guénon explored the Indian alternatives to the scientific and industrial civilization of the West more thoroughly and cogently than any other postwar writer. Guénon's early studies in mathematics gave his spirited critiques of Western empiricism and inductive reasoning an authority that the vague or sweeping strictures of Malraux and Elmer simply could not match.[150] His extensive and serious study of Hindu philosophy[151] made it possible for him to explain to Western readers the Indian teachings and meditational techniques that he believed could help the Europeans work their way out of the *Kali Yuga* (Dark Age) in which he believed they had been entrapped for centuries. Guénon's vehement rejection of what he regarded as the pseudo-Hinduism of the Theosophists and his quarrels with the Freemasons gave his exposition of the teachings of the Indian sages an added aura of authenticity.

Guénon explicitly challenged the validity of scientific and technological advances as the supreme measures of individual human worth or of civilized achievement. He insisted that "there are other ways of showing intelligence than by making machines."[152] Guénon premised his challenge on the supposition that there was a three-tiered hierarchy of human cognition; the highest of these was metaphysics, not the sort of scientific inquiry that for centuries had been dominant in the West. He contended that only metaphysics was based on genuine intellectualism and that metaphysics alone allowed one to explore the highest truths, the ultimate reality. He considered science a lesser form of reasoning which, because its objects were material and its methods empirical, shared much with the lowest form of human cognition, based primarily on sensory perceptions. Guénon believed that the approach to scientific thought that had developed in the West since the Renaissance was narrower and more applied than that followed by such thinkers as Aristotle and the sages of Asian civilizations. Science had been transformed into the handmaiden of technology, and both pressed into the service of

[149]E. M. Forster, *Passage to India* (New York, 1952), p. 211. At another point (pp. 251–52) Forster comments that the grace and "restfulness" of an Indian gesture "reveals a civilization which the West can disturb but will never acquire."

[150]Biographical information is taken from Paul Serant, *René Guénon* (Paris, 1953), esp. pp. 7–26.

[151]See esp. René Guénon, *Man and His Becoming According to the Vedanta* (New York, 1958), which Ananda Coomaraswamy judged the best work on the Vedanta school in any language.

[152]René Guénon, *East and West* (London, 1941), p. 11 (preface to the English edition).

European inventors and engineers bent on manipulating the natural world. The understandings achieved through experiment were necessarily limited and provisional, as evidenced, Guénon surmised, by a profusion of conflicting and transitory theories. He argued that this sort of thinking posed little problem for Indian philosophers, who regarded it as an inferior form of intellectual endeavor and preferred to concentrate on the "immutable truths" that could be probed only at the highest metaphysical level. But in the West, the sciences had come to dominate thinking, despite the fact that they dealt with only the most superficial aspects of existence.[153] This dominance had resulted in an obsession with material improvement and increase, which led in turn to greatly intensified cravings for creature comforts and to discontent when these could not be fulfilled. It had also brought into being a civilization that was driven at an ever more frenetic pace by the machines and empirical knowledge that its best minds had produced. The Great War and the crisis of the modern, or scientific-industrial, order were the inevitable result. But Guénon foresaw even greater calamities ahead, unless the West fundamentally reordered its priorities and restored the primacy of metaphysical thought.[154]

In questioning the superiority of scientific-industrial civilization, which he equated with the modern, Guénon reversed many of the attributes that the advocates of the civilizing mission had seen as vital sources of European global dominance. He argued that change was a symptom of instability; that progress was an illusion. What the Victorians had disdained as stagnation, Guénon celebrated as the "Eastern" striving for equilibrium and beyond that for the immutable—the highest of all states. He saw the Westerners' need for incessant activity as a sign of their shallowness and immaturity; he connected their restlessness to an inability to concentrate or synthesize. Guénon felt that the civilizing-mission ideology itself was little more than "moralist hypocrisy, serving as a mask for designs of conquest or for economic ambitions." It is "an extraordinary epoch," he said,

> in which so many men can be made to believe that a people is being given happiness by being reduced to subjection, by being robbed of all that is most precious to it, that is to say of its own civilization, by being forced to adopt manners and institutions that were made for a different race, and by being constrained to the most

---

[153]Ibid., pp. 23–26, 36–39, 43–44, 51, 57–62, 68; and *The Crisis of the Modern World* (London, 1942), pp. 24–26, 55, 66–67, 125–26.

[154]Guénon, *East and West*, pp. 106–7; *Crisis*, pp. 11–25, 46–47, 56–57, 129, 131, 153–54.

distasteful kinds of work, in order to make it acquire things for which it has not the slightest use. For that is what is taking place; the modern West cannot tolerate that men should prefer to work less and be content to live on little; as it is only quantity that counts, and as everything that escapes the senses is held to be non-existent, it is taken for granted that anyone who is not in a state of agitation and who does not produce much in a material way must be an "idler."[155]

Guénon also denounced the "savage competition" and social leveling that had taken hold of modern societies. He yearned for the restoration of social stratification, which he deemed attuned to the natural order of things. He envisioned the creation of an intellectual elite that could lead Europe from the abyss of materialism and violence into which it had fallen.[156]

Having dismissed the notion that the scientific-industrial order of the West was the model to be emulated by all humankind, Guénon argued that only the "East," which had preserved the essentials of "traditional" society, could show Europeans the error of their ways and help them relearn how to think at the highest level of cognition—relearn, because Guénon believed that until the late medieval period the values and pursuits of the Europeans had been compatible with those of Indian and other premodern civilizations. Like Massis, Guénon admired Catholicism, the religion of his youth, but he rejected Massis's conviction that a Catholic revival would suffice to save Europe. He believed that if the hold of science and technology over the West was to be broken and the final catastrophe averted, Europeans must master the ancient teachings of the Indians, who had come the closest to understanding the fundamental truths common to all humans and to creation as a whole. Like Massis, Guénon conceded that these teachings and India's traditional civilization were threatened by the encroachments of the West. But unlike Massis, he refused to concede that modernity had made serious gains beyond small circles of Western-educated Indians, whom Guénon dismissed (in passages reminiscent of Valentine Chirol and Jules Harmand) as rootless rabble-rousers. Because it represented a higher stage of intellectual development, the real India was impervious to threats from the West. If the Europeans were willing to borrow its wisdom, they might yet save themselves from the "materialist barbarism" that

[155]Guénon, *East and West*, pp. 45, 82–86; *Crisis*, pp. 50–55, 133–34 (quoted portion), 135–36.

[156]Guénon, *East and West*, pp. 66, 191, 205. See also Serant, *Guénon*, pp. 33–41, for a discussion on the antiegalitarian, antidemocratic aspects of Guénon's thinking.

had taken hold of their civilization. If they refused, Europe would perish, but India and the "East" would endure.[157]

The challenges posed to Western ideologies of dominance by Orwell, Guénon, and other intellectuals did not, of course, persuade European politicians and generals to abandon overseas empires in the decades after the Great War. In fact, the British and French empires increased substantially in size as the victorious allies divided up the former possessions of the defeated Germans and Turks. And even though Egypt, India, and Indochina were shaken in the postwar years by growing social unrest and nationalist agitation, the Europeans who governed them appear to have been only marginally affected by the literature of disillusionment produced by the generation of 1914. Orwell's misgivings about the colonizers' mission may have been shared by other officials, but Forster's Ronny Heaslop was more representative than John Flory of the postwar colonial functionary in his absorption in the tasks of day-to-day administration and his conviction that the "natives" could not possibly rule as honestly and efficiently as their British overlords.[158]

The defenders of empire stressed that war had indisputably demonstrated the dependence of Britain and France on their colonies for the maintenance of their position as great powers. They argued that soldiers, workers, and war matériel drawn from the colonies had been essential to the success of the allied war effort, and that the markets and resources of the overseas empire were vital to postwar reconstruction. Some even sought to rekindle zeal for the civilizing mission. Albert Sarraut and Ernest Lavisse, for example, extolled the mutually beneficial impact of the colonial connection. They pointed to the educational and medical advances that had occurred in Africa and Indochina as a result of French rule; they increasingly linked colonization with the concept of "modernization," which came into widespread use for the first time in the interwar era. In school texts and at popular colonial expositions such as those at Wembley in 1924 and Vincennes in 1931, law and order continued to receive their due. But technical progress in the form of port facilities, roads and railway lines, and the "scientific management" of labor loomed larger for the defenders of colonialism in an era when

[157]Guénon, *East and West*, pp. 110, 175, 202, 205, 251–54; and *Crisis*, pp. 22–23, 28, 42–46, 133–34, 141–42, 152.

[158]At least if one can judge from the colonial records I have consulted for this period relating to Burma, Bihar, the Panjab, and the Netherlands Indies. There is evidence, however, that recruiting British youths for the Indian Civil Service was difficult in the interwar years, partly because of the disenchantment, engendered by the war, with patriotism and with the imperial mission. See Hugh Tinker, "Structure of the British Imperial Heritage," in Ralph A. Braibanti, ed., *Asian Bureaucratic Systems Emergent from the British Imperial Tradition* (Durham, N.C, 1966), pp. 61–63.

charges of imperialist exploitation were pervading the writings of African and Asian nationalists and the European left.[159]

Postwar efforts to defend Europe's colonial mission and the fact that the antiindustrial stance remained the position of a small minority of European writers caution against overstating the extent to which the trench debacle undermined the European acceptance of machines as the measure of men. The postwar faith in the engineer as the savior of Europe and the transformer of backward colonies demonstrates the great strength and staying power of scientific and technological gauges of achievement and capacity. But the awesome forces of destruction that the Europeans had unleashed on their own civilization during the Great War did raise (for all but the most myopic) troubling questions about the assumptions on which their sense of racial superiority and commitment to the civilizing mission had been based. Their confidence that they alone were rational, in control, and civilized was shaken. Their belief that invention and scientific discovery were inherently progressive was cruelly exposed as a chimera. Their conviction that their unprecedented material achievements entitled them to dominate the globe and chart the course of development of subjugated peoples was increasingly challenged both by the Europeans themselves and by the Africans and Asians they ruled. The reversal of Europe's fortunes and the prevailing mood of uncertainty are perhaps most tellingly captured in an image that recurs in postwar writings: that of a mighty continent, reveling in its global dominion, suddenly reduced to a "projecting peninsula" (Inge) or "little promontory" (Valéry) of the great Asian landmass.

[159]For a contemporary author who explored these themes in great detail, see Etienne Richet, *Le probleme coloniale* (Paris, 1918). On the French generally, see Girardet, *L'idée coloniale*, pp. 117–32; and Manuela Semidei, "De l'Empire à la décolonisation à travers les manuels scolaires français," *Revue Française de Science Politique* 14/1 (1966), pp. 58–71. For the British, see Thomas G. August, *The Selling of the Empire: British and French Imperialist Propaganda, 1890–1940* (Westport, Conn., 1985), esp. pp. 126–40.

Modernization Theory and the
Revival of the Technological Standard

IN THE decades after the First World War the bitter debates between leading European intellectuals over the perils of mechanization and Americanization and the fate of industrial society contrasted sharply with the upsurge of enthusiasm across the Atlantic for inventors and innovation, for the mass consumer products of industrial technology, and for the same visions of progress and unlimited improvement that the war had brought so many European thinkers to dismiss as cruel delusions. While European leaders struggled to shore up empires under siege and find ways to rebuild their shattered societies, the Americans grew increasingly assertive in the exercise of their newly won political and economic influence. Just as the decline of Europe's global hegemony opened the way for the emergence of the United States as the premier world power, the Europeans' doubts about their civilizing mission strengthened the Americans' growing conviction that *they* knew best how to reform "backward" societies that were racked by poverty, natural calamities, and social unrest.

Though the term "modernization" was rarely used until after World War II, American educators, missionaries, and engineers of the 1920s and 1930s advocated political, economic, and cultural transformations in China, the Philippines, and Latin America which were as fundamental and wide-ranging as those proposed by development specialists in the 1950s and 1960s. In the interwar period, as in the decades when modernization was in vogue, industrial, democratic America was assumed to be the ideal that less fortunate societies ought to emulate. America's path to political stability and prosperity through the rational management of its resources, through the application of science and technology to mass production, and through efforts to adapt the principles of scientific investigation to the study of human behavior was increasingly held up as the route that "underdeveloped" and unstable societies were destined to travel as they "entered the modern age." Though the ideology of modernization would not begin to be fully articulated until the 1950s, its basic tenets had begun to be formulated long before. Through it the scientific and technological measures of men[1] and cultures were reworked and revived. In modernization theory they found their broadest application and their most elaborate expression.

Critics of modernization theory, who argue that it is primarily a Cold War response, an attempt by American social scientists and policymakers to counter the appeal of Communism to the peoples of the underdeveloped world,[2] distort the origins and significance of the tradition-to-modernity paradigm. As the term "Non-Communist Manifesto"— the subtitle of Walter W. Rostow's influential work *The Stages of Economic Growth*—suggests, there were in fact links between the Cold War struggle and the great emphasis on development and modernization in the works of American social scientists in the 1950s and 1960s. But virtually all the assumptions and perspectives that informed the writings of those who proposed the many and diverse theories of modernization that appeared in this period long predated the global rivalry between the United States and the Soviet Union. Though recast in development jargon in the post–World War II era, most of the ideas associated with modernization theory had been formulated decades, sometimes centuries, earlier. These ideas were deeply rooted in both America's own historical experience and the currents of European thought that have been the focus of this study.[3]

[1]Like the thinking of Europe's colonial civilizers, that of American social scientists on modernization has been overwhelmingly oriented to problems relating to male attitudes, economic activities, and political participation. Science and technology continue to be seen as spheres meant to be dominated by men. Hence, in discussing various approaches to modernization, I have retained the references to "modern man" and "men" which have been indicative of this bias, both in development theory and all too often in the resulting programs.

[2]This connection is most explicitly made in Dean Tipps, "Modernization Theory and the Comparative Study of Societies: A Critical Perspective," *Comparative Studies in Society and History* 15/2 (1973), esp. 200, 208–11. For an early expression of modernization as an antidote to Communism, see Morris Watnick, "The Appeal of Communism to the Peoples of Underdeveloped Areas," *Economic Development and Cultural Change* 1/1 (1952–53), 36.

[3]A full history of the origins and impact of modernization ideology has yet to be written. Though my European focus clearly renders that considerable task beyond the scope of the present study, I seek to trace in a thematic way some of the connections between earlier expressions of the scientific and technological measures of human achievement and potential and the underlying assumptions of modernization theory in its various guises.

From the time of the earliest European settlement in North America, chroniclers, missionaries, and colonizers made much of the technological gap between the peoples of the Old World and those of the New. As had been the case in Africa, the European perception that the Amerindians lacked religion and followed strange customs had more to do with the relegation of the New World "natives" to the status of savages than with what they lacked in material culture. But the myth that the Indians were hunters rather than agriculturists, which persisted despite considerable evidence of their skill as farmers, and the vision of America as a land of abundant resources that the Indians had not begun to tap because their technology was too primitive, buttressed the arguments of those who sought to justify the Indians' subjugation and displacement. The settlers' association of civility with human domination over nature and their view of the new continent as a sparsely populated wilderness led thinkers on both sides of the Atlantic to the conclusion that its Indian inhabitants were savages, much in need of assistance from the "industrious men" and "engins" that only Europe could provide.[4]

Though Jefferson's "agrarian idealism" predominated in the decades when America was transformed from a patchwork of settlement colonies into a nation, technological advance was increasingly seen as essential to the growth and well-being of the fledgling republic. As John Kasson has argued, Jefferson, himself an inventor of considerable ingenuity, came to see the controlled introduction into America of the industrial technology being developed in England as "a welcome ally in the republican enterprise."[5] Many of Jefferson's contemporaries assigned technology an even more essential role. Merchants and politicians, ministers and educators hailed the machine as the answer to the new nation's shortage of labor, as the means by which the unity of the rapidly expanding republic would be preserved, and as a vital source of prosperity and progress. Inventors were compared to magicians and men to "republican machines." Machines were lauded "not simply as functional objects but as signs and symbols of the future of America."[6] Technological development was increasingly equated with the rise from barbarism to civilization, and machines were viewed as key agents for

---

[4]James Rosier as quoted in Karen Kupperman, *Settling with the Indians: The Meeting of English and Indian Cultures in America, 1580–1640* (Totowa, N.J., 1980), pp. 80 (quoted portion), 81–90. On these themes, see also Francis Jennings, *The Invasion of America* (Chapel Hill, N.C., 1975), pp. 15–31, 60–61, 73–75. Kupperman points out (pp. 86, 104–6) that a number of observers thought the Indians' technology admirably suited to the American environment.

[5]John Kasson, *Civilizing the Machine: Technology and Republican Values in America, 1776–1900* (New York, 1982), pp. 22–25 (quoted portion p. 25).

[6]Ibid., pp. 29–30, 32 (quoted phrase), 35, 38, 41 (quoted portion), 46–47.

the spread of this civilization in the New World wilderness. Prominent politicians, writers, and artists of the day caricatured the Indians as slothful, technologically poor, and unprogressive vestiges of savage societies that must either adopt the white man's ways or perish. The shortcomings of the Indians were set against the virtues of the expansive European pioneers whose hard work, discipline, thrift, foresight, and technological ingenuity were transforming the undeveloped wilderness into a land of unprecedented prosperity.[7]

As America industrialized, the "special affinity between the machine and the New Republic" was increasingly emphasized.[8] Though some prominent American thinkers, including Nathaniel Hawthorne and Henry Adams, shared their European counterparts' hostility to the industrial order as dehumanizing and environmentally degrading, and a few anti-industrialists followed Ruskin and Morris in attempting to resurrect artisan production,[9] the great majority of Americans exulted in the power and productivity of the new technology: "they grasped and panted and cried for it."[10] As in Europe and the colonial world, the railway became the premier symbol of the advance of industrialism in the United States. John Stilgoe has shown how extensively railroads had transformed the American landscape by the last decades of the nineteenth century, pervading virtually all aspects of American life from commerce and advertising to education and recreation.[11] In this same period, industry more generally eclipsed agriculture as the dominant sector in the American economy and the main influence on American social life. By the late 1880s the combination of science and technology found in American industry was matched only by that of Germany, and America was outstripping its European competitors in the production of iron and steel. Many Americans regarded machines as objects of aesthet-

---

[7]For attitudes toward the Amerindians in this period, see Roy Harvey Pearce, *The Savages of America* (Baltimore, Md., 1953), esp. pp. 66–71, 82–91, 165–66; and Robert E. Berkhofer, Jr., *The White Man's Indian* (New York, 1978), pp. 91–96. A good introduction to the values that had come to be identified with the advance of the expansive republic is provided in Thomas L. Haskell, "Capitalism and the Origins of the Humanitarian Sensibility," *American Historical Review* 90/2–3 (1985), 339–61, 547–66. As Berkhofer points out (p. 90), in the American literary and artistic tradition the noble savage was portrayed as "safely dead and historically past."

[8]Leo Marx, *The Machine in the Garden* (London, 1964), p. 203.

[9]Ibid., chap. 5, contains superb discussions of early intellectual hostility to the coming of the machine age. For the late nineteenth and early twentieth centuries, see Jackson Lears, *No Place of Grace: Antimodernism and the Transformation of American Culture, 1880–1920* (New York, 1981); chap. 2 includes a fine analysis of the origins and weaknesses of the craft movement.

[10]Perry Miller, quoted in Marx, *Machine*, p. 208.

[11]John Stilgoe, *Metropolitan Corridor: Railroads and the American Scene* (New Haven, Conn., 1983).

ic pleasure; others proclaimed them divinely ordained instruments for building the nation and strengthening its moral resolve.[12]

If it was natural for the Europeans who had excelled all other peoples in the mastery of the material world to view scientific and technological accomplishments as key measures of human worth, it was inevitable that Americans would do so. Though they might concede European superiority in the fine arts, philosophy, and the other pursuits of "high culture," Americans came to regard invention and technological innovation as endeavors in which they could surpass all other peoples, including those of western Europe.[13] As we have seen, American thinkers had played significant roles in promoting scientific and technological standards for judging human worth, from the contributions of Morton and Nott to "scientific" racism to the emphasis placed by anthropologists such as Morgan on the role of technology in the evolution of society. In the last decades of the nineteenth century, when Americans became increasingly involved overseas, assumptions of their scientific and technological superiority became integral components of their own version of the civilizing mission. In response to the backwardness and social turmoil they encountered in China and other countries that they hoped would become major trading partners for the United States, medical missionaries and military advisers provided the earliest expressions of this reforming vision.[14]

But the fullest elaboration of America's civilizing-mission ideology was prompted by the conquest of the Philippines at the turn of the century. The decision to retain the islands as a colonial possession forced politicians and colonial officials to develop policies and programs that would justify the subjugation of the recalcitrant Filipinos. Anticipating an emphasis later found in the writings of modernization theorists, American colonizers stressed the political dimensions of reform and reconstruction in the new colony. Their vision of good government, however, extended far beyond the "peace, order, and justice" formula of British and French defenders of imperialism. From the outset, the U.S. role was defined as one of tutelage rather than paternalistic domination. American officials viewed economic reforms and education as

---

[12]Kasson, *Civilizing the Machine*, chap. 4; Marx, *Machine*, pp. 190–226. Robert Rydell has shown how late nineteenth-century international expositions held in American cities became showplaces for American scientific and technological achievements; see *All the World's a Fair* (Chicago, 1984), esp. pp. 5–6, 14–15, 80–82, 121–23, 160–62. A useful though controversial survey of the integration of scientific research into American industry is provided in David Noble, *America by Design: Science, Technology, and the Rise of Corporate Capitalism* (Oxford, 1977), esp. chaps. 1–3.

[13]Marx, *Machine*, p. 205.

[14]See Chapters 4 and 5, above.

ways of creating a prosperous Filipino middle class from which moderate political leaders, committed to representative democracy and continuing economic ties to the United States, could be drawn. Though the timetable was vague, the Americans' well-publicized intent was to prepare the Filipinos for self-rule.[15]

The preparation of the Filipinos for independence was never seen in purely political terms. It was clear to most American policymakers that legislatures and elections would have little meaning until the backward and impoverished society that was the legacy of centuries of Spanish rule was thoroughly reconstructed. Perhaps more than in any other colony, the role of the engineer as civilizer was touted by politicians in Washington and officials in the Philippines. No American personified this ideal more than the energetic W. Cameron Forbes, an impatient businessman turned colonial administrator. Forbes, whose main task was to oversee the rebuilding and expansion of the road and railway networks of the islands, was "very much a man of the age of steel and machine . . . who believed unquestioningly that things modern were things progressive" and whose journals were full of "celebrations of the new, the speedy, the mechanical, and, perhaps above all else, the efficient." As he saw it, the Filipinos could not get enough of the new technology that colonization made available to the islands. No sooner had he met them, Forbes remarked, than the Filipinos asked immediately for railroads: "They are crying for them, and from all sides I am pressed with questions as to their probability, how soon can they have them, etc."[16] The centrality of the technological and scientific components of America's colonial mission in the Philippines was summed up concisely by J. Ralston Hayden in the mid-1920s in an early historical assessment of American rule: "The old Spanish legal codes were largely rewritten and modernized, a modern government was organized and successfully operated, a great system of popular education was created, a census taken, a modern currency system was established, a program of public works including the construction of roads, bridges, port improvements, irrigation works, artesian wells, school houses, markets and other public buildings were laid down and carried out, an admirable public health service was inaugurated."[17]

[15]Of the substantial literature on political developments in the Philippines in the early decades of American rule, the best studies include Peter W. Stanley, *A Nation in the Making: The Philippines and the United States, 1899–1921* (Cambridge, Mass., 1974); Bonifacio S. Salamanca, *The Filipino Reaction to American Rule, 1901–1913* (n.p., 1968); and Oscar M. Alfonso, *Theodore Roosevelt and the Philippines, 1897–1909* (Quezon City, 1970).

[16]Quoted in Stanley, *Nation in the Making*, pp. 96–106 (quoted portions pp. 99, 104).

[17]Quoted in Michael Onorato, *A Brief Review of American Interest in Philippine Development and Other Essays* (Manila, 1972), p. 3.

Involvement in World War I further enhanced the already high esteem in which Americans held inventors and machines. Despite some misgivings about the war's indecisive outcome, it strengthened their conviction that they were a people destined by virtue of their scientific and technological prowess to shape the development of less fortunate societies. There were, of course, individuals such as Ernest Hemingway and Ezra Pound who fled from the growth and consumer-crazed society that America had become by the 1920s, but the great majority of Americans reveled in the spectacles and creature comforts of "the machine age." The Americans' experience in the Great War had been very different from that of the European adversaries. In the two and a half years before the United States entered the war, American production, profits, and employment had soared in response to the insatiable demand for food and war matériel on the part of its future allies. American merchants competed with Japanese traders for the overseas markets that hard-pressed European combatants had been forced to abandon.[18]

America's late entry into the war also meant that its soldiers were spared the long years of futile slaughter in the trenches. By the time substantial numbers of doughboys were in combat in the spring of 1918, German offensives had broken the trench stalemate and restored a war of motion and maneuver. In the counteroffensives that forced the Germans to sue for peace, the allies made extensive use of the new military technology, particularly tanks and airplanes, which had developed during the war years. The lessons of the final offensives were not lost on George Patton and George Marshall and other future commanders, or on Billy Mitchell and other champions of the new weapons in the interwar years. In their view, American industry had done much to keep the allies in the war; European and American scientific research and inventiveness had devised the weapons that carried them to victory.[19]

[18]On the economic impact of the war on the United States, see David M. Kennedy, *Over Here: The First World War and American Society* (Oxford, 1980), chaps. 2 and 6. For the precombat phase, see Frederick Paxson, *American Democracy and the World War* (New York, 1966), vol. 1, chaps. 9, 12, 13. On the economy in wartime, the fullest accounts remain Grosvenor B. Clarkson, *Industrial America in the World War* (Boston, 1923); and Benedict Crowell and Robert F. Wilson, *The Armies of Industry* (New Haven, Conn., 1921), 2 vols.

[19]For the most detailed survey of American military participation in the war, see Edward M. Coffman, *The War to End All Wars* (Madison, Wis., 1968). For the role of technology in the allied victory and America's military future, see William Mitchell, *Memoirs of World War I* (New York, 1960), pp. 291–92, and *Winged Defense* (New York, 1925); George C. Marshall, *Memoirs of My Services in the World War, 1917–1918* (Boston, 1976), chaps. 10 and 11; M. Blumenson, ed., *The Patton Papers, 1885–1940* (Boston, 1972), pp. 446–59; and H. Essame, *Patton: A Study in Command* (New York, 1974), pp. 10–18 and chap. 2.

As a result of their late entry into combat and the end of the trench stalemate, American losses were lower than those of the other powers: just over 50,000 dead—fewer than the number of American servicemen who died of disease during the war, and fewer than the number of those killed in the Vietnam conflict. These totals suggest that America's "lost generation" was literary rather than literal, in contrast to the losses of the European powers, whose young men were slaughtered and maimed by the millions. Hemingway and John Dos Passos nothwithstanding, postwar American disillusionment was a product of the failure of the peace process more than a reaction against modern mechanized warfare. Thorstein Veblen's scathing 1922 critique of the folly of intervention best summed up American frustration over involvement in a conflict that had ended so indecisively as to make future wars appear inevitable. One of Veblen's main concerns was the great boost the war had given to Billy Sunday's brand of religion and the threat that the religious revival posed for the "material sciences," whose methodology he viewed as the "most characteristic and most constructive factor engaged in modern civilization."[20]

In the decades after World War I, applied science and technology pervaded American life to a degree that greatly exceeded that experienced by any other society. Between 1917 and 1940 the number of American households that were electrified increased from less than 25 to over 90 percent. The automobile, produced on Henry Ford's moving assembly line for the first time in 1913, became an item of mass consumption. The family car and commercial airlines gave Americans unprecedented mobility; the radio, the great expansion of telephone networks, motion pictures, and mass advertising radically transformed American communications. A great proliferation of laborsaving appliances from vacuum cleaners and electric stoves to toasters and washing machines brought the American home irretrievably into the machine age. The impact of machine design was evident in American architecture and the fine arts. Streamlining and Art Deco graced the most fashionable homes and offices, and automakers vied to perfect the new styles. Henry Ford was widely regarded as the prophet of a new age of "heroic optimism," in which science and invention were hailed as the key to American prosperity and the best solution for social ills. Ameri-

---

[20]Veblen's essay "Dementia Praecox" originally appeared in *Freeman* 5 (1922). I have used the reprint in Leon Ardzrooni, ed., *Essays on Our Changing Order* (New York, 1934), pp. 423–36 (quoted portion pp. 430–31). Even Hemingway's disillusionment had more to do with the bungling of Italian commanders and the Italian general staff than with mechanization; see, e.g., the analysis of his wartime experiences in Michael S. Reynolds, *Hemingway's First War: The Making of "A Farewell to Arms"* (Princeton, N.J., 1976), esp. chaps. 4 and 5.

can politicians and intellectuals celebrated factories as the modern equivalents of medieval cathedrals and praised industry for its simplicity and truth. Alfred North Whitehead judged successful those organisms that "modified their environments." Charles Beard proclaimed science and "power-driven machinery" the hallmarks of Western civilization. Even when the Great Depression brought major challenges to the capitalist foundations of the American industrial order, few questioned the primacy of the machine in American or any progressive society. The 1930s were dominated by massive construction projects—dams, highways, tunnels, and bridges—designed to generate energy for and to extend the range of the technology that Americans continued to view as the key to their rise to global power and as an essential means of eradicating poverty both at home and abroad.[21]

In the era between the two world wars, the long-standing assumption that technological innovation was essential to progressive social development came to be viewed in terms of a necessary association between mechanization and modernity. As Richard Wilson has argued, in American thinking the "machine in all of its manifestations—as an object, a process, and ultimately a symbol—became the fundamental fact of modernism."[22] Modernization theory represented an extension of this association—which was grounded in the American and European historical experience—to the peoples and cultures of the non-Western world. Though the lexicon of American educators and policymakers in overseas areas remained rudimentary and their conceptualizations crude by the social science standards of the 1960s, many of the presuppositions that later informed paradigms of tradition and modernity were evident in their curricula and proposals for reform.

L. G. Morgan, for example, viewed the teaching of "modern science" as indispensable to China's efforts to bring its antiquated society into the "modern world." Morgan, who taught the sciences to Chinese students for many years, believed that beyond advances in practical knowledge and technique, training in the Western sciences would instill a much-

---

[21]The comparisons to cathedrals and truth are quoted in Richard G. Wilson, D. H. Pilgrim, and D. Tashjian, *The Machine Age in America, 1918–1941* (New York, 1986), pp. 24, 30; see also Alfred North Whitehead, *Science and the Modern World* (New York, 1925), pp. 295–96; and Charles Beard, *Whither Mankind* (New York, 1929), pp. 14–15. The Wilson volume and the exhibit at the Brooklyn Museum which it places in historical perspective leave little doubt that the postwar decades were indeed "the machine age." This summary relies heavily on the fine essays by Wilson and the other contributors to the volume, as well as William Leuchtenburg's useful survey *The Perils of Prosperity, 1914–32* (Chicago, 1958); Walter Polakov, *The Power Age* (New York, 1933); and Siegfried Giedion's detailed exploration of the transformation of the American home in *Mechanization Takes Command* (New York, 1969), pt. 6.

[22]Wilson, *Machine Age*, p. 23.

needed sense of discipline and precision in the youth of China. He averred that it would enhance their critical faculties, render them more "cool and logical," and enable them to overcome the passivity that he identified as a major impediment to Chinese innovation and growth. Morgan suggested that a modern mind-set, if propagated in China by the study of Western scientific learning, would enable the Chinese people to root out the sources of their backwardness and poverty, from bad roads to corrupt government. He was confident that at least some Chinese were capable of mastering even the most advanced scientific theories of the West. He urged the Chinese to emulate the Japanese, who had built a strong and prosperous nation within a generation because they had adopted the scientific approach of the West.[23]

Morgan's fellow educator George Twiss provided a more detailed analysis of the obstacles in "old" China that would have to be surmounted if the living standards of the Chinese people were to be improved and the country's great resources "developed." His discussion of such impediments singled out many of the "barriers to development" stressed by later modernization theorists: ancient customs and beliefs, poor communications, a low level of control over the natural environment. His solutions were standard fare among modernizers of the post–World War II era: Western education, "modern" science, and industrialization.[24]

After World War II the modernization paradigm supplanted the beleaguered civilizing mission as the preeminent ideology of Western dominance. American social scientists were the main exponents of the new ideology, which was much more systematically and coherently articulated than its predecessor. Competing theories of the dynamics and stages of the transition from "tradition" to "modernity" were debated by academics, and their jargon-laden discourse played a major role in policy formulation with respect to the "underdeveloped," "developing," or "emerging" nations of the "Third World." New hierarchies of the levels of social development—the first, second, third, and (somewhat later) fourth worlds; postmodern, modern, traditional, primitive; mature, developing, underdeveloped—replaced the civilization/barbarian/savage scale that had long served as the standard. Like the nineteenth-century improvers, the modernizers rejected the long-standing convictions that innate deficiencies were responsible for the lowly position that most non-Western peoples occupied in these hierarchies and that these shortcomings would make improving their rank-

---

[23]L. G. Morgan, *The Teaching of Science to the Chinese* (Hong Kong, 1933), esp. pp. xii–xiii, 51–52, 55–62, 69.

[24]George Twiss, *Science and Education in China* (Shanghai, 1925), pp. 12, 39–41, 48–60.

ing difficult if not impossible. The postwar convergence of revulsion against Nazi atrocities perpetrated in the name of racial purification, and African and Asian nationalist challenges to the claims of racial superiority by their colonial overlords had much to do with the decidedly antiracist premises of the new ideology. Like the improvers, the modernizers assumed that all peoples and societies not only could but would "develop" along the scientific-industrial lines pioneered by the West. As Ali Mazrui has argued, a social evolutionist teleology informed this view.[25] But the modernizers drastically reduced the time frame in which the process of social advance was to occur. In contrast to the centuries envisioned by the improvers (or the millennia calculated by those inclined to racist explanations for human disparities), the transition to modernity was plotted in decades. In the more advanced "emerging" nations, it could conceivably occur within one person's lifetime.

The nonracist assumptions of the modernization theorists were linked to a second major premise that distinguished their models of development from the programs of both the racists and improvers of the nineteenth century. Though the modernizers regarded American and European capital and technical assistance as vital to Third World development, they envisioned Africans and Asians—not Westerners—as the main agents of the transformation of underdeveloped societies. The very nature of the colonial relationship had dictated that the European colonizers consider their leadership essential for the task of civilizing savage and barbaric societies. But the process of decolonization, which coincided with the decades when modernization theory peaked in acceptance, rendered this vision of Western paternalism obsolescent. Concerns about the composition and methods of the new elites who were to oversee the transition from tradition to modernity, as well as anxiety about the appeal of Communism in the generally impoverished former colonies, explain the much greater emphasis on political aspects of development in modernization theory. Proponents of this approach have seen social mobility, expanding political participation, and the democratization of political institutions as key measures of the degree of modernity attained by non-Western societies.[26] Even the most radical of nineteenth-century reformers had given little attention to these issues. The political context in which the improvers worked was a given:

[25] Ali Mazrui, "From Darwin to Current Theories of Modernization," *World Politics* 21/1 (1968), 69–83.

[26] This is clearly evident in many of the classics of the modernization genre: e.g., David E. Apter, *The Politics of Modernization* (Chicago, 1965); Gabriel A. Almond and James S. Coleman, *The Politics of Developing Areas* (Princeton, N.J., 1960); and Samuel P. Huntington, *Political Order in Changing Societies* (New Haven, Conn., 1968).

though limited reforms might be suggested, the colonial administration was normally considered beyond the purview of their civilizing efforts. Thus, contrary to Walter Rostow's conclusions in assessing the impact of imperialism,[27] modernization as it has been understood since World War II would have been inconceivable in the colonial context.

The ultimate outcome of the transformation of traditional into modern societies was also unthinkable in the colonial era. Even in its most benevolent manifestations, the underlying aim of the civilizing mission was to reshape colonial economies in ways that would make them more compatible with the metropolitan economies of each imperial system. Modernization paradigms have almost invariably been based on the assumption that industrialization is essential for full development, for genuine modernity—even though their creators have rarely thought through the implications of this outcome for the economies of the developed nations. Their ideal is a world of industrially competitive nations; the reality of the postcolonial world has been the continuing economic dependence of virtually all the primary-product-producing former colonies on the industrialized West (and, in recent decades, Japan).

Though modernization theory differs in important ways from its civilizing-mission predecessors, in many of its more influential formulations it stresses the scientific and technological measures of human achievement that have long been vital components of ideologies of Western dominance. Given the varying emphases of different advocates of the modernization paradigm and their predilection for scientific-sounding stage sequences, corollaries, and axioms, the centrality of scientific-technological gauges has often been obscured. But the continuing importance of those gauges is underscored by the very characteristics that have been used to distinguish the traditional from the modern. Modernity is associated with rationality, empiricism, efficiency, and change; tradition connotes fatalism, veneration for custom and the sacred, indiscipline, and stagnation. Joseph Kahl, who has provided one of the more detailed listings of these attributes, argues that traditional men are passive and fatalistic largely because they lack the sophisticated technology required to "shape the world to their own desires." Modern men, by contrast, make use of sophisticated technology to remake their environment and change their social systems in ways intended to advance both their own careers and the development of their societies as a whole.[28] Two of the key criteria by which Alex Inkeles has distin-

---

[27]Walter W. Rostow, *The Stages of Economic Growth: A Non-Communist Manifesto* (Cambridge, Eng., 1964), p. 27.
[28]Joseph Kahl, *The Measurement of Modernism* (Austin, Tex., 1968), esp. pp. 4–6, 18–20. For an early expression of these criteria, see Bert F. Hoselitz, "Non-Economic

guished "Modern Men" in six "developing countries" are their belief in the efficacy of science and medicine (and a corresponding rejection of fatalism and passivity), and their insistence that people be "on time" and "plan their affairs in advance." In an earlier essay, Inkeles also stressed the modern man's ability to "dominate his environment in order to advance his own purposes and goals," and the high level of technology normally associated with modern cultures.[29]

Most of Inkeles's remaining measures of modernity relate to human qualities that Daniel Lerner established as definitive of the "modern personality" in his pioneering work *The Passing of Traditional Society*. Lerner viewed advanced communications as the key to making societies modern. Newspapers, radios, and other forms of media would expose the increasingly urbanized and literate populations of underdeveloped areas to a flood of information about the world beyond family and community. By making them aware of other perspectives and modes of behavior, the media revolution would enhance their capacity to respond effectively to new ideas and ways of doing things.[30] Though Lerner's theories were based on extensive field surveys and census "audits," in his stress on communications as the key to modernity he shared much with Marx and nineteenth-century improvers, who predicted that the telegraph and railway would promote social and economic advance by breaking down such ancient obstacles as caste barriers and village myopia.

The stress on the pivotal roles of applied science and technology in the modernizing process was repeatedly affirmed by the foremost champions of the paradigm in its mid-1960s heyday. Marion Levy, who produced one of the most elaborate discussions of the effects of the transition from tradition to modernity, began with the assumption that the degree to which a society has been modernized could be measured by the extent to which it made use of inanimate power and employed tools "to multiply the effect of effort."[31] C. E. Black regarded as essential preconditions for the "ascent" to modernity the "revolution in sci-

Barriers to Economic Development," *Economic Development and Cultural Change* 1/1 (1952–53), 14–15.

[29]Alex Inkeles, "Making Men Modern: On the Causes and Consequences of Individual Change in Six Developing Countries," *American Journal of Sociology* 75/2 (1969), esp. 210–11; and "The Modernization of Man," in Myron Weiner, ed., *Modernization: The Dynamics of Growth* (New York, 1966), pp. 141–44.

[30]Daniel Lerner, *The Passing of Traditional Society* (New York, 1958).

[31]Marion Levy, *Modernization and the Structure of Societies* (Princeton, N.J., 1966), pp. 11–12. After reviewing varying definitions of the essence of modernity, Wilbert Moore analyzed the modernization process as if it were synonymous with industrialization; see *Social Change* (Englewood Cliffs, N.J., 1963), chap. 5.

ence" in early modern Europe and the later application of scientific discoveries to "the practical affairs of man in the form of technology." Though the "Consolidation of Modernizing Leadership" was the first of the sequential transformations Black sought to plot for all existing nations, in his formulation the core changes in the modernizing process occurred during the phase of "Economic and Social Transformation" which in most cases (with Japan the most obvious exception) corresponded to an era of rapid industrial growth.[32] Walter Rostow, whose views had perhaps the greatest impact on the actual implementation of policy in this era, also stressed as preconditions for modernity the importance of the rise of scientific thinking and man's growing confidence that he could successfully manipulate his natural environment. The "take off" and "drive to maturity," which Rostow argued must follow if meaningful modernization were to occur, were measured in purely economic terms: in the ability of societies to save for investment in industrial growth a designated portion of what they produced.[33]

In the 1970s and 1980s, modernization theory has come under heavy criticism for everything from basic methodological flaws to its failure to account accurately for the actual experience of the peoples of the "developing world."[34] Though case evidence and specific points of attack have varied widely, most major critiques of the ideology of modernization have challenged its ethnocentric vision of a single route to development and the artificiality of its dichotomous vision of traditional and modern social systems.[35] As Dean Tipps has pointed out: "Regardless of how well-intentioned or critical of American policy abroad a modernization theorist might be, the limited cultural horizons of the theory tend to involve him in a subtle form of 'cultural imperialism,' an imperialism of values which superimposes American or, more broadly, Western cultural choices upon other societies, as in the tendency to subordinate all

[32]C. E. Black, *The Dynamics of Modernization* (New York, 1967), pp. 7, 10–11, 76–77, 89–94.

[33]Walter W. Rostow, *Economic Growth*, chaps. 2–5.

[34]For a sample of criticisms from a variety of perspectives, see Tipps, "Modernization Theory," pp. 199–226; Reinhard Bendix, "Tradition and Modernity Reconsidered," *Comparative Studies in Society and History* 9/3 (1967), 292–346; Albert O. Hirshman, "Obstacles to Development: A Classification and a Quasi-Vanishing Act," *Economic Development and Cultural Change* 13/4 (1965), 385–93; Henry Bernstein, "Modernization Theory and the Sociological Study of Development," *Journal of Development Studies* 7/2 (1971), 141–60; and Donald Levine, "The Flexibility of Traditional Culture," *Journal of Social Issues* 24/4 (1968), 29–41.

[35]It is noteworthy that these aspects of the critique have been stressed by area specialists who have conducted extensive research in "developing" societies. For important examples, see Lloyd and Susanne Rudolph, *The Modernity of Tradition* (Chicago, 1967); and Milton Singer, "Beyond Tradition and Modernity in Madras," *Comparative Studies in Society and History* 13/2 (1971), 160–95.

other considerations (save political stability perhaps) to the technical requirements of economic development."[36]

Ironically, African and Asian peoples have been faced with a similar sort of cultural imperialism in the alternative socialist programs espoused by the Soviets and their allies—programs that the modernizers regard as the antithesis of the path to development they advocate. Proponents of the two approaches may clash over the role of the state and the market in the development process, but they share the conviction that traditional or "feudal" beliefs, customs, and institutions are little more than impediments to the inevitable transformation of backward non-Western economies and societies. Perhaps even more than modernization paradigms, the "social engineering" methods of the Soviet Union are centered on heavy industrialization and the application of science to everything from production to social organization.[37] African and Asian leaders may deplore the draconian means by which this approach was pursued in its Stalinist phase, but they have been impressed by its success in rapidly transforming the backward Soviet society, ravaged by war and revolutionary struggles, into one of the world's leading industrial and military powers. The influence of the social-engineering approach is evident in the five-year plans, giving heavy industrialization top priority, which were so often adopted by the leaders of newly independent states.[38]

The ethnocentrism of ideologies built on the assumption that development in Africa and Asia must proceed along the scientific-industrial lines pioneered in western Europe presents a dilemma in the truest sense of the term—a choice between undesirable alternatives—to those engaged in the struggle against injustice and poverty in the Third World. If the industrialized West and the Soviet bloc refuse to share their technology and scientific understandings with non-Western nations in order to

[36]Tipps, "Modernization Theory," p. 210.

[37]The literature on the Comintern and the publications generated by the organization itself are primarily devoted to questions involving the struggle to spread the revolution. But Kermit E. McKenzie, *The Comintern and World Revolution, 1928–1943* (London, 1964), esp. chap. 5, includes some coverage of the type of economic planning and transformations envisioned for areas won to socialism and the emphasis on technicians and engineers in the Soviet Union itself. Robert Lewis, *Science and Industrialization in the USSR* (New York, 1979), discusses the integral roles of applied science and state coordination of research and production in Soviet industrialization.

[38]Perhaps Jawaharlal Nehru best exemplified both the admiration and the unease that the Soviet example aroused in the nationalist leaders of Africa and Asia; see his *Toward Freedom* (New York, 1941), pp. 229–31, 348–51; and B. R. Nanda, *The Nehrus* (Chicago, 1974), pp. 257–59. For the most striking example of the influence on a developing economy of Soviet centralized planning and stress on heavy industry, see Maurice Meisner, *Mao's China: A History of the People's Republic* (New York, 1977), esp. pp. 114–21.

avoid neocolonial domination or cultural imperialism, that refusal would deny essential tools and knowledge to societies just beginning the quest for development—and beginning with much greater demographic and economic liabilities than those that first industrialized. But the full industrial transformation of the Third World along either Western or Soviet lines could prove disastrous, given the limited and diminishing resources now available and the new burdens such a change would impose on an already endangered global ecology.[39] To escape from this predicament by encouraging Third World peoples to forgo industrial development, on the grounds that the earth cannot stand further pollution and resource depletion, is obviously self-serving. In any case, this argument has little chance of winning much African or Asian support, particularly since neither Western nor Soviet leaders have ever seriously considered stabilizing, much less reducing, the level of mechanization in their own societies.

There is unlikely to be a single grand solution to the dilemma posed by the perils of Third World industrial development on the one hand and, on the other, the consequences of limiting the transfer of technology and scientific learning to non-Western societies. But the makings of a middle way between these untenable options may be found in the increasing emphasis given by development specialists to projects that in scale and the nature of their technological input are attuned to the community needs and ecological constraints of the local areas where they are introduced. This search for "appropriate technologies" offers the possibility of approaches to development that are independent of those followed by either the West or the Soviet Union.[40] Alternative strategies have often proved better suited to the actual needs and socioeconomic conditions in developing areas than programs oriented to large-scale, capital-intensive projects. They may also be more compatible with the

[39]The most extreme exponents of this view have been authors associated with the "Club of Rome" in the mid-1970s; for examples, see Donella H. Meadows, Dennis Meadows, Jørgen Randers, and W. Behrens III, *The Limits to Growth* (New York, 1972); and M. Mesarovic and E. Pestel, *Mankind at the Turning Point* (New York, 1974). The Club of Rome's positions have sparked a good deal of controversy and equally extreme responses, including the optimistic prognoses of the possibilities for unlimited growth associated with Herman Kahn and the Hudson Institute.

[40]The essays by Paul Bourrières and Partha Dasgupta in Austin Robinson, ed., *Appropriate Technologies for Third World Development* (New York, 1979), and A. K. N. Reddy's "Alternate Technology: A View from India," *Social Studies of Science* 5/3 (1975), 331–42, provide a good introduction to this approach; Ken Darrow and Rick Pam, *Appropriate Technology Sourcebook* (n.p., 1977), gives a good sense of the kinds of technology and techniques employed; and David Dickson, *Alternative Technology and the Politics of Technological Change* (Glasgow, 1974), describes some of the obstacles faced by its early advocates.

different estimates of African and Asian peoples as to what is necessary for human well-being and fulfillment, estimates shaped by historical experiences and cultural emphases that vary greatly from those of the West. Consequently, alternative strategies of development may prove conducive to the affirmation of measures of achievement, potential, and human worth which are very different from but every bit as valid as the scientific and technological standards that have been such a powerful source of Western dominance in recent centuries.

INDEX

Abban, Jouffroy d', 135

Abbott, Jacob, 231

Abolition, 108–10, 276. *See also* Slavery; Slave trade

About, Edmond, 152, 214, 218

Académie des Sciences, 75, 139–40

Accuracy (precision): European perceptions of, 63–64, 84–85, 91, 259, 262–65; and scientific instruments, 31–32, 60, 82–83, 141

Adanson, Michael, 75, 115

Africa, 2, 11, 15–16, 18(fig.), 33, 35, 70, 94, 126, 143, 165, 204; colonial policy in, 203, 339, 383; exploration of, 37–38; housing in, 36; mining in, 39, 112–13; societies of, 34, 36, 40, 353–54, 390–92; towns in, 260–61; transportation in, 22, 37, 223. *See also* Africans; Benin, Ethiopia, Maghrib; Senegal; South Africa

Africans, 36, 38, 41, 108–10, 274, 338, 353–54; accuracy of, 263–64; anthropological views of, 355–56; as "backward," 124, 126, 302–3; belief systems of, 40–41; as "childlike," 305–7; culture of, 346–47; development of, 118–19, 127; education of, 286–87, 291, 324–25, 334–36; European attitudes toward, 110–11, 113–14, 125–26, 163–64; and European technology, 113–15, 130(fig.), 158–60; exploitation of, 353–56; and mechanical devices, 160–64; and nature, 216–17; perceived intelligence of, 116–18, 297–300, 304–5, 316–17; religions of, 40–41; as "savages," 67–68, 195; science of, 53, 115–16; scientific education of, 287, 289, 336; and scientific instruments, 1–3; social development of, 110, 315–16; spatial perception of, 259, 261–62; stereotypes of, 154–57; as superior, 392–93; technological development of, 39–40, 153–54, 158–60; textile production by, 38–39, 155–56; time consciousness of, 61, 242–46; tools of, 22, 39–40, 111–12; views on skin color of, 65–67; weapons of, 39–40; work habits of, 252–55

Agriculture, 27, 39, 45, 77, 92–93, 404

Akbar, 42

Alchemy, 32

Alembert, Jean d', 92; *Elements of Philosophy,* 72

Algarotti, Francesco: *Newtonianism for the Ladies,* 72

Algeria, 201, 325–26

Amerindians, 2, 4, 126, 201, 253, 338, 404

Ancestral Rites Controversy, 80, 82

Anglo-Saxons, 150, 190–93, 273

Anson, George, 71, 124, 125; on China, 90–92, 184; *A Voyage around the World,* 90

Anthropology, 10, 209, 330, 392; on African culture, 335, 355–56; and evolutionary theory, 312–13

Arabs, 34, 157, 176, 196, 326

Archaeology, 156, 157, 273

Architecture, 6, 260; Chinese, 45, 181; Indian, 99, 174; Zimbabwe, 156–58. *See also* Cities; Civil engineering

Aristocrats and colonization, 194, 197–98, 207–9; and technology, 28–29, 209

Arnold, Edwin, 226

Arnold, William: *Oakfield; or Fellowship in the East,* 172–73

"Arrested development," 302–4

Library of Congress Cataloging-in-Publication Data

Adas, Michael, 1943-
  Machines as the measure of men: science, technology, and ideologies of Western
dominance/Michael Adas.
    p.   cm.—(Cornell studies in comparative history) Includes bibliographical ref-
erences and index.
  ISBN 0-8014-2303-1 (alk. paper)   1. Technology—History. 2. Technology—Philosophy. I. Title. II. Series.
  T15.A33 1989   609—dc19   89-845